J.M. Pajares A.S. Peña P. Malfertheiner (Eds.)

Helicobacter pylori and Gastroduodenal Pathology

With 71 Figures

Springer-Verlag
Berlin Heidelberg New York
London Paris Tokyo
Hong Kong Barcelona
Budapest

Pajares, José M., M.D., Ph.D.
Professor of Medicine
Head of Department of Gastroenterology
Hospital de la Princesa
Autonoma University of Madrid
Diego de Leon, 62
28006 Madrid, Spain

Peña, A. Salvador, M.D., Ph.D.
Professor of Gastrointestinal Immunology
Free University of Amsterdam
De Boelelaam 1117
1081 HV Amsterdam, The Netherlands

Malfertheiner, Peter, Prof. Dr. med.
Abt. Innere Medizin II, Universitäts Klinik Ulm
Robert-Koch-Str. 8
W-7900 Ulm, Germany

ISBN-13:978-3-642-77488-1 e-ISBN-13:978-3-642-77486-7
DOI: 10.1007/978-3-642-77486-7

Library of Congress Cataloging-in-Publication Data. Helicobacter pylori and gastroduodenal pathology/J.M. Pajares, A.S. Peña, P. Malfertheiner (eds). p. cm. Includes bibliographical references and index. 1. Helicobacter pylori infections—Congresses. I. Pajares, J.M. (José M.), 1934-. II. Peña, A.S. III.—Malfertheiner, P. (Peter), 1950-. QR201.H44H45 1992 616.9'31—dc20 92-30270

Softcover reprint of the hardcover 1st edition 1993

Typesetting: Macmillan India Ltd., Bangalore-25

19/3130/SPS-543210-Printed on acid-free paper

Preface

In 1987, a group of researchers interested in the field of *Campylobacter pylori* and gastroduodenal pathology organized a meeting in Copenhagen to present and to discuss the current research and findings on the subject.

This informal meeting was the start of the European Campylobacter Study Group (ECSG). When 3 years later,the name *Campylobacter* was changed to *Helicobacter* the name of the group changed to EHSG. From its very start this working party has been a multidisciplinary group with the aim of stimulating research on the role of *Helicobacter* in animal and human pathology. One of the group's commitments is an annual International Workshop: the first official workshop was held in 1988 in Bordeaux, France, and the next two followed in 1989 and 1990 in Ulm, Germany, and Toledo, Spain.

The EHSG has published the proceedings of these workshops, collecting the new data which are generated each year. This third proceedings book has been named "*Helicobacter pylori* and Gastroduodenal Pathology".

We have introduced new sections in this volume dealing with malignancies and disease in children. Other sections have maintained their titles, but they contain new data and also give information on original methods or approaches in basic research and practical clinical applications. It is not unusual for proceedings volumes to present data and hypotheses which have not yet been validated at the time of publication. The editors support this idea, and feel that the publication of preliminary data will help to stimulate the interchange of ideas between the scientists involved.

The workshops and the proceeding are the result of the scientific contributions, the suggestions, and continuous cooperation of the European Group. The financial support of pharmaceutical and commercial companies is gratefully acknowledged. We also wish to express our most sincere appreciation and thanks to the authors, who have contributed their manuscripts to create this book.

July 1992 — The Editors

Contents

I. Microbiology

II. Epidemiology

IV. Pathology

V. Immunology

List of Contributors

Addresses are given at the beginning of the respective contributions

I. Microbiology

Helicobacter pylori Detection by Polymerase Chain Reaction of the Gene Encoding 16S Ribosomal RNA in Fresh and Paraffin-Embedded Material

S.A. Ho[1], F.A. Lewis[1], D.A. Secker[2], D.S. Tompkins[2], G.R. Taylor[3], J.I. Wyatt[4], M.F. Dixon[1], and P. Quirke[1]

Introduction

The emergence of *Helicobacter pylori* as an important aetiological agent in gastroduodenal disease has led to the introduction of many methods for the detection of this organism. Current methods for *H. pylori* detection can be divided into two main categories, endoscopic and non-endoscopic. Endoscopy provides biopsy specimens for subsequent culture [3], histology [5], in situ hybridisation [20] or rapid urease testing [10]. The non-endoscopic techniques include serological tests [12] and the ^{14}C or ^{13}C breath tests [4, 23]. All of these techniques are unsatisfactory because they are insensitive to low bacterial densities and cannot identify "non-culturable" coccoid forms of *H. pylori.* Research has been hampered through the lack of a sensitive assay and the invasive nature of the best methods, and thus many key questions remain unresolved. The development of an extremely sensitive technique which can be applied to a wide variety of materials (including saliva or faeces), may allow the sources and routes of transmission of *H. pylori* to be elucidated.

The polymerase chain reaction (PCR) is a revolutionary technique for amplifying minute quantities of nucleic acid. The nature of amplification is extremely sensitive and precise, allowing millions of copies of a highly specific area of DNA or RNA to be replicated from amongst large chains of nucleic acid such as the human or bacterial genome. Amplification only occurs when the target nucleic acid sequence is present. Thus PCR has proved to be invaluable for identifying the absence or presence of specific gene sequences and has formed the basis of many diagnostic tests.

The gene encoding 16S ribosomal RNA (rRNA) was selected as a specific target for PCR. This was because it fulfilled several important criteria. The 16S rRNA gene is not transferred between bacteria. The availability of data demonstrating areas of greatest dishomology on 16S rRNA sequences between *H. pylori* and closely related bacteria eased the task of designing specific oligonucleotide primers. Most importantly, *H. pylori* was recently established as a new and separate genus based in part on 16S rRNA studies which showed that

[1] Department of Pathology, University of Leeds, LS2 9JT, UK
[2] Department of Microbiology, Bradford Royal Infirmary, Bradford, BD9 6RJ, UK
[3] Yorkshire Regional DNA Laboratory, Clarendon Wing, Leeds General Infirmary, Leeds, LS1 3EX, UK
[4] St. James's Hospital, Leeds, LS9 7TF, UK

J.M. Pajares et al. (Eds.)
H. pylori and Gastroduodenal Pathology

significant sequence differences existed between *H. pylori* and its closest relatives [8, 15, 18, 19]. In addition, "universal" sequences are present to which PCR primers could be targeted to act as a positive control. This would eliminate false negatives arising from inhibition of amplification, which is a problem encountered particularly with paraffin-embedded tissue blocks [9]. Specificity can be obtained using 16S rRNA. A synthetic oligonucleotide probe complimentary to this gene has recently been described for specific detection of *H. pylori* [11].

The urease gene was not utilised because, until recently, sequence information has been limited, especially concerning other urease producing bacteria. Specific primer design may have been difficult since DNA probe hybridisation analysis has revealed that urease genes exhibit conserved sequences among phylogenetically distant Gram-negative bacteria [1]. Particular difficulty may be encountered in the amplification of closely related urease producers such as *Helicobacter mustelae*, *Helicobacter felis* or other gastric organisms such as *Gastrospirillum hominis.*

This paper describes the development of a PCR for the highly sensitive and specific detection of *H. pylori* utilising the gene encoding 16S rRNA.

Materials and Methods

Design of Oligonucleotide PCR Primers

Universal Primer Design

Two sets of "universal" primers were synthesised from highly conserved areas on the 16S rRNA sequence. These primers possess a broad specificity for a wide range of eubacteria. The fragment lengths amplified were 124 bp (U1 : U2) and 992 bp (U1 : U3) (Table 1).

Design of H. pylori-Specific Primers

Published sequences for *H. pylori* and its closest relatives (Campylobacters and *Wolinella succinogenes*) were compared, and regions where their 16S rRNA sequences are highly dishomologous were identified [8, 15, 18, 19]. Particular attention was paid to areas of the 16S rRNA sequence where *H. pylori* is highly

Table 1. Primer sequences

Primer	Sequences (5′-3′)	Region of 16S rRNA compared with *E. coli*
U1	CGG TTA CCT TGT TAC GAC TT	1491–1510
U2	CCT TGT ACA CAC CGC CCG TC	1386–1405
U3	CAG CAG CCG CGG TAA TAC	518–535
Hp1	CTG GAG AGA CTA AGC CCT CC	834–853
Hp2	ATT ACT GAC GCT GAT TGT GC	744–763
Hp3	AGG ATG AAG GTT TAA GGA TT	407–426

dishomologous to *W. succinogenes*. This was because *W. succinogenes* was the closest relative of *H. pylori* from which 16S rRNA sequences were available at that time [19]. Region 834–853 of the 16S rRNA sequence was found to be suitable for designing an *H. pylori*-specific primer. This primer was called Hp1 (Table 1).

In the region 834–853, nine bases are found in *H. pylori* which do not occur in the other closely related organisms: there are 13 base differences between *H. pylori* and *Escherichia coli* and ten base dishomologies between *H. pylori* and *W succinogenes*. In addition, there is one deletion present in *E. coli*, *W. succinogenes* and the Campylobacters which is not present with *H. pylori*. Hp1 produces an amplification PCR product of 335 bp when used in conjunction with U3.

A second *H. pylori*-specific primer, called Hp2 (Table 1), was designed in order to eliminate weak amplification products produced by Hp1 : U3. Areas of greatest mismatch of *H. pylori* with *Campylobacter laridis*, *Campylobacter jejuni* and *Campylobacter coli* were identified. Region 744–763 was found to be highly dishomologous to these organisms. Sequence information concerning *H. mustelae* and *H. felis* was not available at this time. A PCR product of 109 bp was amplified by Hp1 : Hp2.

A third primer (Hp3) was designed for nested PCR [14, 17] (Table 1). Primer Hp3 was targeted to a further area of dishomology (407–426) on the 16S rRNA sequence.

Bacteria Tested

DNA was extracted from the following bacteria: a human clinical isolate of *H. pylori*, *H. mustelae* (F1 and F8 isolated by Dr. D.S. Tompkins), "*C. cinaedi*" (Seattle 48), *C. fetus* subsp. *fetus* (NCTC 10348), *C. laridis* (NCTC 11352), *C. jejuni* (NCTC 11168), *C. coli* (NCTC 11353), *C. sputorum* subsp. *sputorum* (NCTC 11528), *E. coli* (JM 101) and *W. succinogenes* (NCTC 11488). DNA was qualitatively assessed by 0.7% agarose gel electrophoresis and quantitated using the TKO-100 dedicated mini-fluorometer (HSI) and the fluorochrome Höescht 33258 (Polysciences).

DNA Extraction

Bacterial colonies were suspended in 567 μl Tris-ethanol buffer in an Eppendorf tube with a flamed inoculating wire loop. To this was added 30 μl 10% Sodium dodecylsulphate (SDS) (Sigma) and 3 μl (2 mg/100 μl) proteinase K (Bethesda Research Laboratory). After thorough mixing, the tube was incubated at 37°C for 1 h in a water bath. Then 100 μl 5 M NaCl and 80 μl hexadecyltrimethylammonium bromide (CTAB) were added and the tube incubated for a further 10 min at 65°C. The contents were then extracted with an equal volume of chloroform/isoamyl alcohol [24], spun at 9000 g in a microcentrifuge for 3 min and the aqueous layer transferred to a fresh Eppendorf tube. Two further extractions were performed using phenol/chloroform/isoamyl alcohol followed by chloroform/isoamyl alcohol. Isopropanol was added (0.6 vol) to precipitate the DNA at $-20^\circ C$. The DNA was collected by microcentrifugation at high speed (16000 g), washed with 70% (vol/vol) ethanol and briefly desiccated before being dissolved in 100 μl of molecular biological grade water (BDH). Quantitative analysis of the extracted DNA was performed using the TKO-100 dedicated mini-fluorometer (HSI) and the fluorochrome Höescht 33258 (Polysciences), and a 0.7% agarose electrophoresis gel stained with ethidium bromide was used for qualitative analysis.

Preparation of Coccoids

A suspension of the non-culturable coccoid form of *H. pylori* was kindly supplied by Dr. H. Oosterhom (Gist-Brocades, Holland) and DNA extracted by the method described.

DNA Extraction from Paraffin-Embedded Material

A proteinase K, SDS extraction protocol was used, modified by an increased proteinase K incubation time of 5 days as described previously by Jackson et al. [6]. To avoid the problems of DNA contamination, the microtome blade was cleaned with xylene before and after cutting, and a negative control was also "extracted" alongside the samples to exclude contamination. DNA was assessed as described before and then amplified by PCR.

Preparation of Faecal Suspensions Seeded with *H. pylori*

A paediatric clinical specimen was selected since it was less likely to harbour *H. pylori* than an adult sample. The faecal sample was emulsified in saline (1 : 1) in order to obtain a consistency which could be measured using 1.0-ml plastic tips. Using 0.7 ml aliquots of the faecal sample in sterile plastic 1.0-ml cryovials (Nalgene Cryoware) ten fold serial dilutions of known quantities of *H. pylori* in 0.3 ml Brucella broth containing 5% fetal calf serum were added. Also, 0.3 ml plain Brucella broth was added to an extra tube to act as a negative control. DNA was then extracted with the protocol utilising CTAB before analysis by PCR.

PCR Reaction Conditions

PCR was performed in 0.5 ml Eppendorf tubes in 50-μl reaction volumes; 1 μl of each oligonucleotide primer was aliquoted (50 pmol/μl for each primer) in an Eppendorf tube and dried in a vacuum desiccator before the addition of: 5 μl 10XPCR buffer (500 mM KCl, 100 mM TRIS-Cl, 15 mM $MgCl_2$, 0.1% (w/v) gelatin pH 8.3), 8 μl dNTP mixture (final concentration of 1.25 mM dATP, dCTP, dGTP and dTTP : BRL or Pharmacia), 2.5 U Taq DNA polymerase enzyme (Promega) and distilled water (BDH) and finally 1 μl of the extracted DNA sample. Liquid paraffin was then added (40 μl) to seal the reaction mixture, and the tube was then placed in a programmable thermal cycler (Grant: Autogene). The temperature profile was: 30 s at 95°C, 30 s at 55°C or 60° and 30 s at 72°C. The final cycle used an extended period at 72°C (5 min), and the mixture was subsequently refrigerated at 4°C before analysis. The number of cycles was 20, 30 or 40. For nested PCR, 25 cycles were used for each set of primers. PCR amplification products were analysed by 2% agarose gel electrophoresis stained with ethidium bromide.

Assessment of Sensitivity

H. pylori DNA were amplified in ten fold serial dilutions with Hp1 : Hp2 from a concentration of 1 ng/μl–1 fg/μl. The thermal profile employed 40 cycles and an annealing temperature of 60°C. Similarly for nested PCR, Hp1 : Hp3 were used for the first reaction followed by Hp1 : Hp2 to amplify 1 μl reaction product from the initial PCR.

Results

Universal Primers. Amplification fragments of 124 bp and 992 bp were successfully produced by each of the paired universal primers (U1 : U2 and U1 : U3). The universal primers U1 : U2 also amplified the correct fragment size of 124 bp from DNA of all the bacteria included in the screening group. This confirmed the gene encoding 16S rRNA was capable of PCR amplification in all of the samples under study, and negative results caused by the failure of the PCR reaction could be excluded.

*Primers Hp*1 : *U*3. Using a thermal cycle of 30 cycles and an annealing temperature of 55°C only *H. pylori* DNA amplified strongly. Weak amplification was observed in DNA samples from *H. mustelae* (F1 and F8), *C. laridis*, *C. jejuni* and *C. coli*. These weak cross-reactive bands were completely eliminated by reducing the number of cycles to 20 or by increasing the annealing temperature to 60°C.

*Primers Hp*1 : *Hp*2. *H. pylori* DNA was successfully amplified, producing a product size of 109 bp utilising 40 cycles and an annealing temperature of 60°C with the two specific primers. No other bacterial samples were amplified (Fig. 1). This indicated that Hp1 : Hp2 were highly specific for *H. pylori* DNA. Further PCRs incorporating primers Hp1 : Hp2 utilised 40 cycles and an annealing temperature of 60°C to ensure stringent specificity.

Nested PCR. Hp1 : Hp3 were used for 25 cycles followed by 25 cycles of Hp1 : Hp2. This revealed stronger amplification of the 109-bp final product.

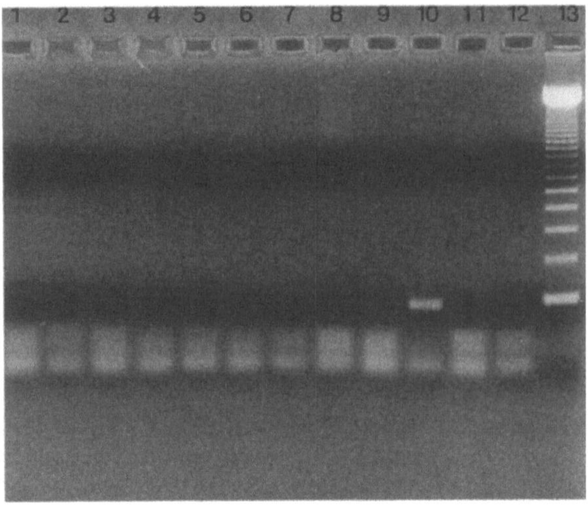

Fig. 1. A 2% agarose electrophoresis gel stained with ethidium bromide demonstrating the specificity of H. pylori detection obtained by PCR using primers Hp1 : Hp2. *Lane* 1, H. mustelae (F1); 2, "C. cinaedi"; 3, C. fetus subsp. fetus; 4, C. laridis; 5, C. jejuni; 6, H. mustelae (F8); 7, C. coli; 8, C. sputorum subsp. sputorum; 9, E. coli; 10, H. pylori; 11, W. succinogenes; 12, negative control; 13, 123-bp DNA ladder

Sensitivity. Hp1 : Hp2 could amplify from as little as 0.1 pg of starting bacterial DNA. Nested PCR enhanced the sensitivity by ten fold, and 0.01 pg of DNA was detectable.

Detection of Coccoids. DNA extracts from non-culturable coccoid forms of *H. pylori* were successfully amplified with Hp1 : Hp2. No amplification was observed in the negative control from the DNA extraction protocol indicating the absence of PCR reagent contamination.

Paraffin-Embedded Material. Ten gastric biopsies which were histologically verified as *H. pylori* positive were obtained for DNA extraction. They produced the correct fragment size of 109 bp. Other human tissues from outside the human gastrointestinal tract were used for negative controls and these did not produce any positive reaction, showing that the Hp primers did not amplify human DNA. To exclude false negatives caused by Taq inhibition, factor VIII primers [7] were used to confirm that the DNA from the negative tissue controls were amenable to amplification by PCR.

Exclusion of Cross-reactivity. DNA extracted from human faeces alone were not amplified, whilst those seeded with *H. pylori* from 1.89×10^7 to 7.40×10^4 organisms per millilitre yielded a 109-bp fragment. A similar experiment was performed where *H. pylori* DNA was added to 1 μg human genomic DNA in ten fold serial dilutions. Amplification was observed only in those samples which contained added *H. pylori* DNA, down to a concentration of 1 pg. This represents a $1:10^6$ concentration ratio of *H. pylori* DNA to human genomic

DNA. These experiments prove that *H. pylori* detection is extremely sensitive even when a large background of "stray" DNA is present.

Discussion

Ribosomal RNA sequence data is highly suitable for demonstrating bacterial phylogenetic diversity because it is a common but distinctive cellular component [2, 15, 18, 19, 22]. Several PCR tests have recently been reported for a range of bacteria, some of which utilise the 16S rRNA gene as their target [13, 21]. The PCR assay we have developed for *H. pylori* DNA amplification is highly sensitive and specific utilising the gene encoding 16S rRNA. Our experiments show that a high degree of specificity is obtainable using only one specific primer (incorporating a 10-base mismatch in addition to a 1-base deletion) in conjunction with a universal primer if the annealing temperature is relatively high at 60°C or the number of cycles employed is reduced. A considerable amount of 16S rRNA data are readily available which should enable this gene to be used as an extremely versatile target for PCRs, and specific amplification can be achieved even when there is little sequence dishomology. A second and a third Hp primer was designed because high sensitivity and specificity were major requirements.

The level of sensitivity of *H. pylori* detection achieved by our PCR assay easily surpassed those of existing methods. Further increased sensitivity by up to 1000-fold may be possible by the inclusion of a preliminary reverse transcription step in the protocol since each bacterium contains approximately 15 000 copies of 16S rRNA. Such improvements have been reported in a study investigating Johne's disease [21].

Using Hp1 : U3, weak cross-reactions occurred with *H. mustelae*, *C. laridis*, *C. jejuni* and *C. coli*. Elevation of the annealing temperature from 55°C to 60°C abolished all these cross-reactions except from *H. mustelae*. In the area of the 16S rRNA gene to which primer Hp1 flanks, the sequence between *H. pylori* and *H. mustelae* contains five base mismatches and one deletion, whereas between *H. pylori* and *H. felis* there are six base mismatches [16]. However, bacterial DNA from *H. felis* was not available to us for amplification. The design of Hp2 eliminated cross-reactions to *H. mustelae*. Hp2 contains one base mismatch and one deletion to *H. mustelae* and two base mismatches to *H. felis*; thus the latter is unlikely to be recognized by this PCR if *H. mustelae* fails to amplify. In addition, if nested PCR is used, the problem of concatamerisation is overcome [14, 17]. No cross-reactions occurred when human genomic or faecal DNA extracts were subjected to PCR, which reaffirmed the specificity of the Hp primers.

Existing techniques for *H. pylori* identification have been unable to elucidate the sources and mode of transmission of *H. pylori*. The high sensitivity and versatility of PCR will be invaluable in searching for oral bacterial niches and detecting small numbers of bacteria which may persist in the stomach after therapy. Current methods are unable to detect low bacterial densities and may then give rise to false-negative results. PCR can be used for confirmation of complete eradication of *H. pylori* which is of considerable importance in patient management and perhaps in finding new, more effective means for treating *H. pylori* infection. Finally, if PCR can detect *H. pylori* in minute amounts of excreted material, not only will this shed some light on the possible modes of transmission of infection, but PCR may then become a useful non-invasive test for confirming current *H. pylori* infection.

Acknowledgements. We express our gratitude to Gist-Brocades for their generous funding and also to Dr.H. Oosterhom for providing the coccoid forms of *H. pylori*. We thank Mrs. J. Fearnley for her help in preparing this manuscript.

References

1. Blanchard A, Barile MF (1989) Cloning of Ureaplasma urealyticum DNA sequences showing genetic homology with urease genes from gram-negative bacteria. Res Microbiol 140:281–290
2. Estrada ICE, Lamb FI, Colston MJ, Cox RA (1988) Partial nucleotide sequence of 16S ribosomal RNA isolated from armadillo-grown Mycobacterium leprae. J Gen Microbiol 134:1449–1453
3. Goodwin CS, Blincow ED, Warren JR, Waters TE, Sanderson CR, Easton L (1985) Evaluation of cultural techniques for isolating Campylobacter pyloridis from endoscopic biopsies of gastric mucosa. J Clin Pathol 38:1127–1131
4. Graham DY, Klein PD, Evans DJ, Evans DG, Alpert LC, Opekun AR, Boutton TW (1987) Campylobacter pylori detected noninvasively by the ^{13}C-urea breath test. Lancet I:1174–1177
5. Gray SF, Wyatt JI, Rathbone BJ (1986) Simplified techniques for identifying Campylobacter pyloridis. J Clin Pathol 39:1279–1280
6. Jackson DP, Lewis FA, Taylor GR, Boylston AW, Quirke P (1990) Tissue extraction of DNA and RNA and analysis by the polymerase chain reaction. J Clin Pathol 43:499–504
7. Kogan SC, Doherty M, Gitschier J (1987) An improved method for prenatal diagnosis of genetic diseases by analysis of amplified DNA sequences. N Engl J Med 317:985–990
8. Lau PP, DeBrunner-Vossbrinck B, Dunn B, Miotto K, MacDonell MT, Rollins DM, Pillidge CJ, Hespell RB, Colwell RR, Sogin ML, Fox GE (1987) Phylogenetic diversity and position of the genus *Campylobacter*. System Appl Microbiol 9:231–238
9. Lo Y-MD, Mehal WZ, Fleming KA (1989) *In vitro* amplification of hepatitis B virus sequences from liver tumour DNA and from paraffin wax embedded tissues using the polymerase chain reaction. J Clin Pathol 42:840–846
10. Marshall BJ, Warren JR, Francis GJ, Langton SR, Goodwin CS, Blincow ED (1987) Rapid urease test in the management of Campylobacter pyloridis-associated gastritis. Am J Gastroenterol 82:200–210
11. Morotomi M, Hoshina S, Green P, Neu HC, LoGerfo P, Watanabe I, Mutai M, Weinstein IB (1989) Oligonucleotide probe for detection and identification of Campylobacter pylori. J Clin Microbiol 27:2652–2655
12. Newell DG, Stacey AR (1989) The serology of Campylobacter pylori infections. In: Rathbone BJ, Heatley RV (eds) Campylobacter pylori and gastroduodenal disease. Blackwell, London, pp 74–82
13. Olive DM (1989) Detection of enterotoxigenic Escherichia coli after polymerase chain reaction amplification with a thermostable DNA polymerase. J Clin Microbiol 27:261–265
14. Pääbo S, Highuchi RG, Wilson AC (1989) Ancient DNA and the polymerase chain reaction. The emerging field of molecular archaeology. J Biol Chem 264:9709–9712
15. Paster BJ, Dewhirst FE (1988) Phylogeny of Campylobacters, Wolinellas, Bacteroides gracilis, and Bacteroides ureolyticus by 16S ribosomal ribonucleic acid sequencing. Int J Syst Bacteriol 38:56–62
16. Paster BJ, Lee A, Fox JG, Dewhirst FE, Tordoff LA, Fraser GJ, O'Rourke JL, Taylor NS, Ferrero R (1991) Phylogeny of Helicobacter felis sp. nov., Helicobacter mustelae, and related bacteria. Int J Syst Bacteriol 41:31–38

17. Porter-Jordan K, Rosenberg EI, Keiser JF, Gross JD, Ross AM, Nasim S, Garrett CT (1990) Nested polymerase chain reaction for the detection of cytomegalovirus overcomes false positives caused by contamination with fragmented DNA. J Med Virol 30:85–91
18. Romaniuk PJ, Zoltowska B, Trust TJ, Lane DJ, Olsen GJ, Pace NR, Stahl DA (1987) Campylobacter pylori, the spiral bacterium associated with human gastritis is not a true Campylobacter sp. J Bacteriol 169:2137–2141
19. Thompson LM III, Smibert RM, Johnson JL, Krieg NR (1988) Phylogenetic study of the genus Campylobacter. Int J Syst Bacteriol 38:190–200
20. Van den Berg FM, Zijlmans H, Langenberg Wies, Rauws E, Schipper M (1989) Detection of Campylobacter pylori in stomach tissue by DNA in situ hybridisation. J Clin Pathol 42:995–1000
21. Vary PH, Andersen PR, Green E, Hermon-Taylor J, McFadden JJ (1990) Use of highly specific DNA probes and the polymerase chain reaction to detect Mycobacterium paratuberculosis in Johne's disease. J Clin Microbiol 28:933–937
22. Ward DM, Weller R, Bateson MM (1990) 16S rRNA sequences reveal numerous uncultured microorganisms in a natural community. Nature 345:63–65
23. Weil J, Bell GD (1989) Detection of Campylobacter pylori by the ^{14}C-breath test. In: Rathbone BJ, Heatley RV (eds) Campylobacter pylori and gastroduodenal disease. Blackwell, London, pp 83–93
24. Wilson K (1987) Preparation of genomic DNA from bacteria. In: Ausubel FM, Brent R, Kingston RE, Moore DD, Smith JA, Seidman JG, Struhl K (eds) Current protocols in molecular biology 1987–88. Greene, Los Angeles

Polymerase Chain Reaction for *Helicobacter pylori*

B. Gobert[1], J.D. Korwin[2], M.C. Conroy[3], M.C. Bene[1], and G.C. Faure[1]

Introduction

The recent development of gene amplification techniques has considerably expanded the field of investigations at the DNA level in biology. The understanding and epidemiology of infectious diseases will be greatly improved in the coming years by the now possible identification of microorganism-specific sequences in media or conditions not suitable for classical culture or microscopic identification.

Helicobacter pylori, the microorganism identified as associated with chronic gastritis [1], is slow growing but can be easily visualized in microscopy in the mucus of infected stomach biopsies. The urease test, based on the strong enzymatic activity of *H. pylori* is useful for a rapid assessment of infection but requires live organisms [2]. Serologic studies, especially when they discriminate anti-*H. pylori* isotypes, provide an indirect but non-invasive means of follow up, showing the appearance or decrease of a humoral systemic immune response [3–5]. These variations are correlated with infection and healing [5].

Great difficulties have, however, been encountered in two circumstances in the study of *H. pylori* infection: (a) assessing post-therapeutic eradication; and (b) identifying the mode of contamination. Gene amplification could provide answers to both these problems.

Here we present a preliminary study designed to test the applicability of the polymerase chain reaction (PCR) to biologic samples for the identification of *H. pylori*-specific DNA. Our data were compared with classical methods of *H. pylori* identification and suggest that PCR could indeed be useful in prospective epidemiologic and post-therapeutic studies.

Patients and Methods

Patients

Biopsy samples were obtained in 11 patients whose clinical condition justified gastric endoscopy (Table 1). These specimens were collected at the same time as

[1] Laboratoire d'Immunologie, CHU de Nancy, BP 184, 54500 Vandoeuvre Les Nancy, France
[2] Médecine H, CHU de Nancy, Hôpital Central, 54000 Nancy, France
[3] Laboratoire de Bactériologie, CHU de Nancy, Hôpital Central, 54000 Nancy, France

J.M. Pajares et al. (Eds.)
H. pylori and Gastroduodenal Pathology

Table 1. Indications for endoscopy in the 11 biopsied patients

Patient No.	Indication	Histology	PCR
1	Peptic ulcer	Positive	Positive
2	Epigastralgia	Negative	Positive
3	Epigastralgia, ulcer	Positive	Positive
4	Dysphagia, gastritis	Positive	Positive
5	Peptic ulcer	Positive	Positive
6	Gastritis, hiatus hernia	Positive	Negative
7	Treated ulcer (control)	Negative	Negative
8	Treated ulcer (control)	Negative	Negative
9	Treated ulcer (control)	Negative	Negative
10	Gastritis, phlebitis	Negative	Negative
11	Epigastralgia	Negative	Negative

biopsies used for standard histology, Giemsa staining, *H. pylori* culture, and the urease test. They were snap frozen in liquid nitrogen, processed for immunohistology on frozen-cut sections, and the remainder, always maintained below $-30°C$ and mostly at $-80°C$, were used for PCR.

Gastric juice was obtained from these 11 patients at the time of endoscopy and saliva from nine of them. In addition 39 saliva samples and 34 samples of gastric juice were obtained from other patients whose biopsy was not tested in PCR. A total of 44 paired saliva and gastric juice samples were analyzed in PCR.

For all patients, the current analyses of the gold standard were performed. Anti-*H. pylori* immunoglobulin G (IgG), IgA, and IgM were assayed in serum samples from all of them using an enzyme-linked immunosorbent assay (ELISA) with whole crude extract as antigen [5]. Six saliva samples were obtained from healthy controls without gastric complaints.

Methods

Polymerase Chain Reaction

Biopsy samples were treated immediately upon thawing, after removal from the $-80°$ freezer. Thawed samples were placed in sterile Eppendorf tubes and mixed with 9 μl bi-distilled sterile water (BDS-H_2O) containing 0.15% of proteinase K (Sigma, St Louis, USA). After incubation for 30 min at 65°C, the samples were vortexed and boiled at 100°C for 5 min.

Freshly collected or thawed gastric juice or saliva samples were boiled for 5 min in BDS-H_2O and were not submitted to enzymatic digestion.

One microliter of each preparation was sampled and placed in a sterile Eppendorf microtube. Eleven microliters of *H. pylori* primers [6, 7] in BDS-H_2O buffer was added to the sample, followed by 1.2 μl Taq polymerase (Perkin Elmer Cetus, Norwalk, USA), 57 μl nucleic acids and 11 μl DMSO. After thorough vortexing, the mixture was covered with one drop of mineral oil, and the tube was capped and placed in the Perkin Elmer automated amplifier (Cetus). The programmed amplification procedure included 35 cycles with 40 s denaturation at 92°C, 30 s annealing at 57°C, and 40 s synthesis at 72°C. The program included a stepwise 2 s increment of annealing and synthesis times.

DNA from wild *H. pylori* colonies, cultured from biopsies of infected patients and treated in similar conditions, was always used as a positive control.

BDS-H_2O alone was always processed as a sample to provide a negative control and check the absence of contamination.

PCR Product Identification

Polyacrylamide Gel Electrophoresis and Silver Staining. Upon completion of the amplification cycles, the samples were vortexed, and 1 μl was removed for molecular weight identification of the amplified products. This procedure was carried out on Phast system (Pharmacia, Uppsala, Sweden) polyacrylamide mini-gels. This polyacrylamide gel electrophoresis (PAGE) allows the reproducible separation of minute amounts of proteins or nucleotides. Eight to twenty-four samples can be run simultaneously. In this study, a mixture of phage lambda restriction fragments (Boehringer, Mannheim, Germany) was run as a base-pair length control. Electrophoresis was performed on 10–15 gradient gels, with native Veronal buffer strips, for 20 min, and a cumulative voltage of 60 Vh. Immediately after the end of the electrophoretic separation, the gels were stained in the automated coloration chamber of the Phast system, using a standard silver-staining procedure (Pharmacia). Results were obtained approximately 1 h after the end of the amplification cycles.

Restriction. The primers used in this study allow the specific amplification of *H. pylori* urease-C gene. The sequence of this gene contains a restriction site for *Cfo*I (*Hha*I) that will yield two fragments of approximately 150 bp. Aliquots of amplification products (10 μl) were incubated with 1 unit of this restriction enzyme (Boehringer) for 1 h at 37 °C. The resulting products were submitted to PAGE and silver staining in the same conditions as described above.

Dot Blot. A 45mer *H. pylori*-specific probe, provided by A. Labigne [7], was used for a nonisotopic dot blot. Amplification products were dotted on Nytran membrane (Schleicher Schuell, Dassel, FRG) using the BioRad Dot-Blot vacuum apparatus (BioRad, Ivry, France). The membrane was air dried and the dotted DNA fixed by exposure to ultraviolet light. Nonspecific sites of the blotted membrane were blocked by salmon sperm DNA (Boehringer) overnight at 65 °C. Hybridization was carried out with the biotinylated probe in the same conditions and revealed with peroxidase-labeled streptavidin. Resulting spots were visualized by incubating the membrane in hydrogen peroxide and diamino-3, 3′-benzidine tetrachlorhydrate (Sigma).

Results

All the amplification products obtained had a length of approximately 290 bp identified using the migration reference curve provided by the oligonucleotide standards run with each gel (Fig. 1). This band was always obtained after

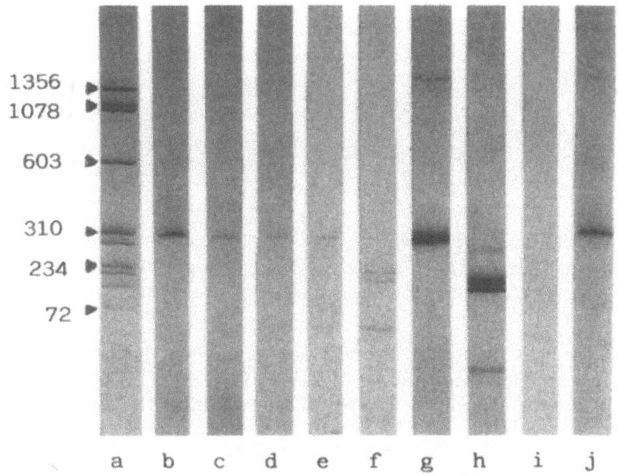

Fig. 1. Identification of PCR products by PAGE and silver staining. Lane a, base pair length markers; lane b, positive biopsy sample; lane c, positive gastric juice sample; lane d, positive saliva sample; lane e, positive saliva sample; lane f, saliva sample from lane e after enzymatic restriction of the amplification product with Cfo I; lane g, wild colony of *H. pylori*; lane h, wild colony from lane g after enzymatic restriction of the amplification product with Cfo I; lane i, amplification of bi-distilled water (negative control); lane j, wild colony of *H. pylori* (positive control)

amplification of *H. pylori* DNA and is consistent with the 294-bp sequence of the Ure-C gene [8]. Replacement of DNA or biologic samples by BDS-H_{20} in the process yielded no amplification product. This method further allowed the appreciation of the amount of *H. pylori* DNA amplified, since bands of varying intensities were observed, related to the amount of DNA initially available for amplification. Thus control lanes where amplification of *H. pylori* colonies were run always yielded the strongest bands, while some saliva samples produced fainter bands.

Digestion of amplification products with the *Cfo*I restriction enzyme yielded two shorter fragments of 170 and 125 bp, respectively, consistent with the *H. pylori*-specific sequence expected to be amplified (Fig. 1). Finally, as exemplified in Fig. 2, all the samples which failed to produce a detectable band in PAGE were negative in dot blot, while all those scored as positive by PAGE and enzymatic cleavage yielded clear dots after specific hybridization.

As a whole, *H. pylori*-specific DNA was amplified from six biopsies, 33 saliva and 27 gastric fluid samples. None of the six control saliva yielded positive amplification. These data were correlated with each other in paired series and with other elements allowing the appreciation of current or previous *H. pylori* infection. Correlations, evaluated as positive–positive or negative–negative were observed with the frequencies indicated in Table 2.

The best correlation was obtained for paired samples of gastric juice and biopsy material, with only one discordant result. The latter (negative gastric juice, positive biopsy) could be interpreted by a strong adherence of *H. pylori* to the overlying or intraglandular mucus. Conversely, the 82% correlation obtained when comparing biopsy PCR and histology suggests that focal areas of

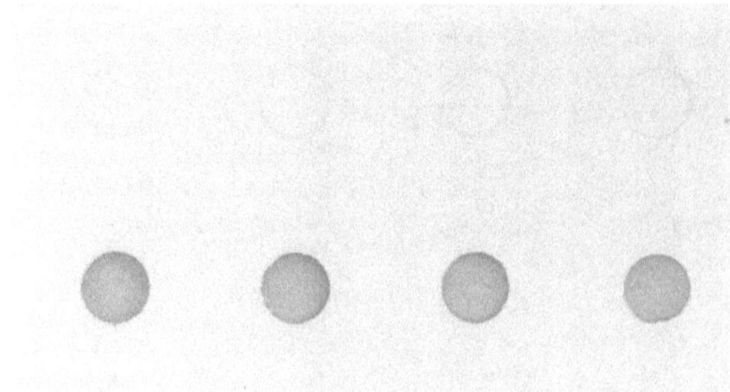

Fig. 2. Identification of PCR products by dotblot hybridization. *Upper row*, bi-distilled water (negative control) and two biopsy samples detected as negative by PAGE; *lower row*, amplification product of a colony of *H. pylori* (positive control) and two biopsy samples yielding a 294-bp band in PAGE

Table 2. Consistent correlations (positive/positive or negative/negative) between the various PCR tests and between PCR and other criteria of *H. pylori* infection

	Polymerase chain reaction on:		
	Biopsy	Saliva	Gastric Juice
PCR biopsy	—	44	91
PCR saliva	—	—	77
Histology	82	61	66
One of gold standard	82	82	67
Serology	56	74	60

H. pylori infection may lead to misleading interpretations, in both methods, related to the size of the sample analyzed.

Only nine paired samples of gastric mucosa and saliva were studied, and the poor correlation obtained needs to be tested on a larger series, since the 44 pairs of gastric juice and saliva indicate a higher level of consistency. Discordant data in this series were more often (eight times) related to positive saliva versus negative fluid, suggesting that the presence of *H. pylori* in the saliva might occur in healthy carriers and contribute to contamination. This is further suggested both by the 61% correlation obtained between saliva samples and histology, and by the 74% observed with serology, suggesting that protective antibodies may be raised following a chance contact with this microorganism. Although more data are again needed, one of the features leading to *H. pylori*-related gastric diseases could be found in inadequate levels of such antibodies, should the very similar 56% and 60% correlation between biopsy PCR or gastric juice PCR and serology be confirmed.

Discussion

This study reports on the efficiency of gene amplification to identify *H. pylori*-specific DNA in biologic fluids. The specificity of the amplification product obtained was checked by the length of the sequence amplified, by its sensitivity to enzymatic restriction, and by its ability to hybridize a specific probe.

Amplification of the urease-C gene with the primers used in this method has been demonstrated of high specificity to discriminate *H. pylori* from related microorganisms [6]. The PCR technology used is a standardized procedure, but the methodology chosen for identifying amplification products has been less widely used [9, 10]. It has the advantages of requiring minute amounts of the samples tested, of being nonisotopic, and of yielding results within a few hours. It was validated in this study by the two other nonisotopic tests used to confirm the specificity of amplified products. Electrophoresis identification of nonisotopic amplification products is increasingly used in vegetal, animal, and human studies for the investigation of exogenous DNA sequences.

From a fundamental point of view, the correlative data reported here are consistent with the few similar studies performed for other microorganisms. In their work on HIV infection, for instance, Horsburgh et al. [11] report a 95% consistency in paired serum and mononuclear cell samples, which could be compared to our smaller series of biopsies and gastric juices. When more samples have been studied, PCR might appear as a less cumbersome alternative to in situ hybridization for the identification of residual bacteria within the mucosa after therapy [12].

Finally, regarding the ecology of *H. pylori*, our results support other studies suggesting that oral-oral or oral-fecal transmission could be involved in the pathogenesis of *H. pylori* infection. More samples will be studied with this method, and comparison of future results with those obtained using other probes will undoubtedly be highly informative.

Acknowledgements. The authors are grateful to Dr. A Labigne for her gift of oligonucleotides and for her help in the preparation of this manuscript.

References

1. Marshall B (1983) Unidentified curved bacilli on gastric epithelium in active chronic gastritis. Lancet i: 1273–1275
2. McNulty CAM, Dent JC, Uff JS, Gear MWL, Wilikinson SP (1989) Detection of Campylobacter pylori by the biopsy urease test: an assessment in 1445 patients. Gut 30: 1058–1062
3. Rathbone BJ, Wyatt JI, Worsley BW, Shires SE, Trejdosiewicz LK, Heatley RV, Losowsky MS (1986) Systemic and local antibody responses to gastric Campylobacter pylori in non-ulcer dyspepsia. Gut 27: 642: 347
4. Barthel JS, Everett ED (1990) Diagnosis of Campylobacter pylori infections: "the gold standard" and the alternatives. Rev Inf Dis 12: S107–S114
5. Gobert B, Béné MC, DeKorwin JD, Faure G (1989) Isotype evolution in the follow-up study of patients with Campylobacter pylori associated gastritis. Gastroenterol Clin Biol 13: 880–883

6. Courcoux P, Freuland C, Piemont Y, Fauchére JL, Labigne A (1990) Polymerase chain reaction and direct DNA sequencing as a method for distinguishing between different strains of Helicobacter pylori. Rev Esp Enferm Dig 78 [Suppl 1]: 29A
7. Brisou P, Courcoux P, Labigne A (1990) Detection of Helicobacter pylori by polymerase chain reaction. Rev Esp Enferm Dig 78 [Suppl 1]: 30A
8. Labigne A, Cussac V, Courcoux P (1991) Shuttle cloning and nucleotide sequences of Helicobacter pylori genes responsible for urease activity. J Bacteriol 173: 1920–1931
9. Lampel KA, Jagow JA, Trucksess M, Hill WE (1990) Polymerase chain reaction for detection of invasive Shigella flexneri in food. Appl Environ Microbiol 56: 1536–1540
10. Jestin A, Foulon T, Pertuiset B, Blanchard, Labourdet M (1990) Rapid detection of pseudorabies virus genomic sequences in biological samples from infected pigs using polymerase chain reaction DNA amplification. Vet Microbiol 23: 317–328
11. Horsburgh CR, Ou CY, Jason J, Holmberg SD, Lifson AR, Moore JL, Ward JW, Seage GR, Mayer KH, Evatt BL, Schochetman G, Jaffe HW (1990) Concordance of polymerase chain reaction with human immunodeficiency virus antibody detection. J Infect Dis 162: 542–545
12. Van den Berg FM, Ziljmans H, Langenberg W, Rauws E, Schipper M (1989) Detection of Campylobacter pylori in stomach tissue by DNA in situ hybridization. J Clin Pathol 42: 995–1000

Helicobacter pylori strains from Duodenal Ulcer Patients Differ at the Genomic Level from Those Patients with simple Gastritis

H.H. Yoshimura and D.Y. Graham

Introduction

The frequent presence of *Helicobacter pylori* in healthy individuals without peptic ulcer disease suggests that as yet unidentified specific microbe-host interactions are important in determining whether an infected individual does or does not develop ulcer disease. Development of a peptic ulcer following *H. pylori* infection may require the genetic predisposition of the infected individual, infection with an "ulcer-virulent" *H. pylori* strain, or both.

We investigated whether differences in the infecting organism result in different clinical outcomes of infections with *H. pylori.* We compared the DNA base sequences of *H. pylori* isolates by DNA–DNA hybridization to determine whether hybridization levels allowed classification of the *H. pylori* isolates into groups corresponding to the types of diseases (e.g., duodenal ulcer vs. asymptomatic gastritis) from which they were recovered.

Materials and Methods

Target DNAs were prepared from *H. pylori* strains cultured from gastric biopsy specimens of 20 patients, nine with duodenal ulcers, and 11 from asymptomatic volunteers endoscopically proven not to have peptic ulcer disease. DNA–DNA hybridization was performed with whole genomic probes made from one isolate from each of the disease categories.

Bacteria were scraped from the culture plates and transferred to approximately 25 ml 0.007 *M* phosphate-buffered saline, pH 7.3. The organisms were pelleted by centrifugation for 12 min at 8000 rpm in a Sorvall SS-34 rotor, then washed twice with phosphate-buffered saline and once with TEN (50 m*M* Tris, pH 8.0, 100 m*M* ethylenediaminetetra-acetic acid (EDTA), 150 m*M* NaCl). The bacteria were resuspended in 3–5 ml TEN containing 30 mg/ml lysozyme. After incubation for 30 min at 37 °C, Sodium dodecyl sulfate (SDS), pronase, and proteinase K were added to final concentrations of 1% (w/v), 3 mg/ml, and 0.5 mg/ml, respectively. The mixture was incubated overnight at 37 °C. Another

Department of Medicine and Division of Molecular Virology, Baylor College of Medicine, and Veterans Affairs Medical Centre, Houston, TX, USA

J.M. Pajares et al. (Eds.)
H. pylori and Gastroduodenal Pathology

increment of pronase equal to that used earlier was added, and incubation was continued for an additional 3 h. DNA was purified from the cell lysate as described previously [2], through a process involving phenol extractions, ethanol precipitations, ribonuclease digestions, and spooling.

Whole genomic probe DNAs were labeled with tritium by nick translation as described previously [2] with the exception that a higher concentration (10^{-2} μg/ml) of DNase I was used. Specific activities of the probes ranged from 2.8×10^6 cpm/μg DNA to 5.9×10^6 cpm/μg DNA. DNA–DNA hybridization was carried out as previously described [2] with the following modifications. The reaction mixtures contained 25 ng probe DNA and 25 μg target DNA in 760 μl of 0.01 *M* hydroxyethylpiperazine ethanesulfonic acid (HEPES)-buffered solution (pH 7.0) containing 0.6 *M* NaCl and 0.002 *M* EDTA. The DNAs were denatured by heating for 10 min at 105°C, followed by renaturation at 66°C. Samples were removed after 0, 2, 4, and 7 days of incubation; 20-μl aliquots were assayed for hybridized probe DNA by S1 nuclease digestion, trichloroacetate (TCA) precipitation, and liquid scintillation counting as described previously [2].

Results

Melting point (T_m) determinations were carried out in hybridization buffer (0.6 *M* NaCl) in order to determine the optimal temperature for hybridization. T_m of 91°C–93°C were obtained. Thus, the optimal temperature for hybridization, $T_m - 25°C$ [3], was between 66°C and 68°C; we used 66°C.

Preliminary hybridization kinetics experiments showed that hybridization was essentially complete after incubation for 30 h and that hybrids remained stable for at least 7 days at 66°C. Hybridization levels for each reaction mixture were determined at three time points (2, 4, and 7 days) to ensure that maximal hybridization was attained in each mixture.

Two hybridization groups were defined using the DU probe 88–44 (Fig. 1). The mean levels of hybridization of DNA from *H. pylori* isolates from duodenal

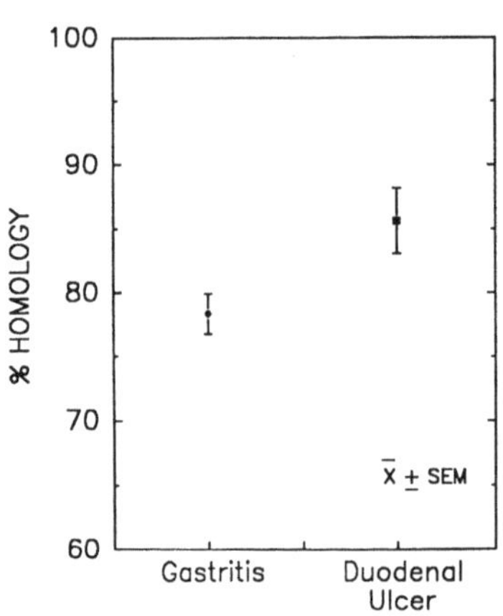

Fig. 1. The results (% homology) of DNA–DNA hybridization in solution between probe 88-44 obtained from a patient with duodenal ulcer and 19 individual target DNAs (11 asymptomatic gastritis and eight other duodenal ulcer patients) is shown. The difference in DNA homology between *H. pylori* obtained from duodenal ulcer patients and those with asymptomatic gastritis was significant ($p = 0.025$)

ulcer patients and asymptomatic gastritis were 85.5% ± 7% and 78.3% ± 5%, respectively (mean ± SD, $p = 0.025$).

Discussion

Our studies confirm that all *H. pylori* isolates are closely related to one another and are very different from *Campylobacter jejuni*. These findings support the recent reclassification of the organisms previously known as *Campylobacter pylori* in a new genus [1].

Some of the hybridization levels we observed were low enough (67%) that one could question, according to generally accepted guidelines of what constitutes a species in hybridization experiments [1], whether the *H. pylori* isolates involved actually belonged to the same species. It remains to be determined whether the various isolates currently designated *H. pylori* actually belong to more than one species within the genus *Helicobacter*.

Our results indicate that, using whole genomic probes from certain *H. pylori* isolates, statistically significant differences may be observed in hybridization with target DNAs from *H. pylori* isolates obtained from patients with different clinical conditions. It is possible that the outcome of infection (e.g., ulcer or no ulcer) may be due to virulence factors encoded by genomic DNA. If such differences exist, it should be possible to produce probes that would identify the ulcer virulence gene(s) and clearly distinguish between ulcerogenic and non-ulcerogenic strains of *H. pylori*.

Acknowledgements. This work was supported by the Department of Veterans Affairs, by grant DK 39919 from the National Institute of Diabetes and Digestive and Kidney Diseases, and by the generous support of Hilda Schwartz.

References

1. Goodwin CS, Armstrong JA, Chilvers T, Peters M, Collins MD, Sly L, McConnell W, Harper WES (1989) Transfer of Campylobacter pylori and Campylobacter mustelae to Helicobacter gen. nov. as Helicobacter pylori comb. nov. and Helicobacter mustalae comb. nov., respectively. Int J Syst Bacteriol 39 : 397–405
2. Graham DY, Yoshimura HH, Estes MK (1983) DNA hybridization studies of the association of Pseudomonas maltophilia with inflammatory bowel diseases. J Lab Clin Med 101 : 940–945
3. Wetmur JG, Davidson N (1968) Kinetics of renaturation of DNA. J Mol Biol 31 : 349-370

Perspectives on the Microbiology of *Helicobacter pylori*

F. Mégraud

Considerable data have been obtained from the newly discovered bacterium named *Helicobacter pylori*. If we compare it to other bacterial pathogens, the optimists will say that much progress has been made in less than 10 years. This is true, and papers presented in this section of the book can give an idea of our state of knowledge. However, the pessimists will argue that we still do not know much and that the real science is just on the way. They are also right since new approaches are now being used. As a tool, molecular biology is taking off in the field. This powerful technique allows us to work at the gene level and, therefore, to gain the maximal insight. All of our past experience must be reconsidered and re-evaluated.

What Can We Expect to Gain by Using Molecular Biology in the Future?

For Diagnosis. Molecular biology can be used to improve the sensitivity of detecting the organism. Currently there is no quick and sensitive method to detect *H. pylori* in biopsy specimens. The urease test and Gram stain probably need 10^4 or more organisms in the biopsy specimen to elicit a positive response. Polymerase chain reaction (PCR) should circumvent this problem, being a quick method which is adaptable to automats.

When one considers other specimens such as stools, another challenge must be faced. The thinking is that *H. pylori* should be present but may be not as cultivable forms. In fact, no culture has even revealed itself positive. Therefore it will be necessary to develop molecular biology techniques in order to detect these noncultivable, but presumably viable, forms. The use of magnetic beads labeled with specific antibodies could also be a useful method.

For Typing. The traditional methods such as serogrouping have not yet been decisive in the proposal of a schema. A concentrated effort should be given to such a schema in the future. Genotyping techniques seem to be the only ones to be highly discriminatory. However, the restriction endonuclease assay is difficult to interpret and other techniques should be proposed. Ribotyping, which is a derivative of the former method using 16S–23S ribosome-labeled RNA to probe

Laboratoire de Bactériologie, Hôpital Pellegrin Enfants, 33076 Bordeaux, France

J.M. Pajares et al. (Eds.)
H. pylori and Gastroduodenal Pathology

the DNA, is a possibility that should be explored. However, the actual preparation procedure of the DNA is time consuming and is therefore, a limiting factor. The possibility of studying products of PCR amplification avoids this problem and may one day give us an answer to the fundamental questions that we are now asking concerning the natural history of the infection and its transmission. The complex technique of pulse field electrophoresis has been used for the typing of other bacteria. There is no doubt that this technique will also be applied to *H. pylori*.

For Our Understanding of Pathogenesis. This is an area where genetic manipulation of the organism will generate answers. How can one prove that a particular property, let us say production of urease or a cytotoxic factor, has a role in the pathogenesis and is not relevant to the area of phenomenology? The same strain possessing a character or not must be used in experiments because, if different strains are used, many properties can be different. It is possible to use spontaneous mutants of the parent strain without the given character; however, it can be argued that it is not the best way because other properties may have changed. The best way will be to construct genetically isogenic mutants for the character under study and to restore the gene for control. Such technology will certainly be available soon for *H. pylori*. It will allow the exploration of different pathogenic mechanisms such as motility, urease, and adherence. But to explore a relationship, it is necessary to have the counterpart, and so the need for a good animal model is crucial. The physiology of this animal should be as close as possible to that of humans. A small animal model with a known genetic background and easy maintenance for a long period of time should theoretically be perfect but may be unrealistic. A key point would be to identify ulcerogenic strains. Current data do not favor this hypothesis. *H. pylori* is not a clonal pathogen but looks more like a highly diverse organism engaged in a commensal relationship. However, much more must be done in this direction.

Other more traditional areas will also be to focus research in the future, for example, on the susceptibility to antimicrobial agents. Metronidazole is a key drug in *H. pylori* eradication, but we do not know the mechanism of resistance. Proton pump inhibitors have a selective bactericidal action on this bacterium. How can this be explained?

We are entering into another exciting realm of research. This fascinating organism still has a lot of mysteries to deliver. We can postulate that microbiological literature will be more and more abundant. This is a fantastic adventure for those participating in the field.

II. Epidemiology

A Descriptive Study of the Characteristics of People with *Helicobacter pylori* Infection

L. Basso[1], J. Clune[2], S. Beattie[1], S. Lawlor[3], and C. O'Morain[1]

Introduction

In 1893 the Italian pathologist G. Bizzozzero first described the presence of "curved bacilli" in the stomach of dogs [1]. During the following years other authors reported the detection of curved bacilli in the stomach of mammals such as rats and cats [2], in the stomach of patients with ulcerated neoplasms [3] and in the stomach of subjects with gastroduodenal peptic ulcers [4]. These findings were confirmed by later studies [5, 6]. However, none of these curved bacilli was ever isolated, and the theory of their role in human diseases progressively declined. However, in 1983 Warren and Marshall [7] succeeded in isolating spiraliform bacteria from human stomachs. These were first called *Campylobacter pyloridis*, then *Campylobacter pylori* and, eventually, *Helicobacter pylori* (HP). These two Australian researchers also suspected that antral gastritis and duodenal ulcers were of infectious origin and have to their credit the definition of the culture conditions necessary for the growth of this microorganism, so that it has been thereafter possible to perform several clinical studies on HP. Thus, we currently know that some 70%–92% of people with chronic gastritis are carriers of HP, and that it is found in about 86% and in 65% of subjects with duodenal and gastric ulcer [8], respectively.

Since 1983 HP has been detected in all human populations investigated. However, there is an astonishing lack of information about the true paths of transmission of this bacterium and about the general features of those people having a higher risk of developing the infection. The aim of the present study was to ascertain the prevalence of HP infection within a selected community of healthy adult males and to describe some of the characteristics of those people who were found to be infected by HP.

Materials and Methods

A total of 133 males serving in the Irish Armed Forces (age: mean = 28.0 years, range = 19–50 years) were studied from 1 April to 31 May, 1990 to assess the

[1] Department of Gastroenterology and Endoscopy, Meath Hospital, Dublin, Ireland
[2] Department of Defence, Dublin, Ireland
[3] St. James's Hospital, Dublin, Ireland

J.M. Pajares et al. (Eds.)
H. pylori and Gastroduodenal Pathology

prevalence of HP infection among them. Each subject was first submitted a standard questionnaire and, following this, a 5-ml venous blood sample was taken.

A standard enzyme-linked immunosorbent assay (ELISA) of specific immunoglobulin G IgG was thereafter performed on the serum to ascertain the presence of HP infection, and a dilution titre of 1:3200 or greater was considered as a positive indicator.

Results

We found HP infection in 51/133 (38.3%) of the subjects studied. Mean age, height, and weight in the HP-positive and in the HP-negative populations are reported in Table 1. Prevalence of HP infection within each age group is reported in Table 2. Prevalence of HP infection was also assessed in relation to rank, education, family or medical history of gastritis and/or of peptic ulcer

Table 1. Mean age, height and weight in the positive and in the negative groups

	Positives ($n = 51$)	Negatives ($n = 82$)
Mean age (years)	29.1	27.4
Mean height (cm)	175	177
Mean weight (kg)	78.8	75.6

Table 2. Prevalence of HP infection within each different age group

	HP infection		
Age (years)	Total (n)	Positives (n)	Positives (%)
19–24	53	17	32.1
25–30	40	15	37.5
31–35	17	8	47.1
36–50	23	11	47.8

Table 3. Prevalence of HP infection in relation to rank

	HP infection		
Rank	Total (n)	Positives (n)	Positives (%)
Troops/sergeants	116	46	39.6
Officers	14	4	28.6

Table 4. Prevalence of HP infection in relation to level of education

Education	HP infection		
	Total	Positives	
	(*n*)	(*n*)	(%)
Primary	13	6	46.1
Secondary	104	40	38.5
Tertiary	13	4	30.8

Table 5. Prevalence of HP infection in relation to family/medical history of disease

Family/Medical History	HP infection		
	Total	Positives	
	(*n*)	(*n*)	(%)
Family history of peptic ulcer	15	8	53.3
Medical history of gastritis	6	4	66.7
Medical history of peptic ulcer	4	3	75.0

Table 6. Prevalence of HP infection in relation to social habits

Social Habit	HP infection		
	Total	Positives	
	(*n*)	(*n*)	(%)
Beer	110	40	36.4
Spirits and/or wine	16	5	31.2
Tea	127	49	38.6
Coffee	66	24	36.4
Cigarettes	94	34	36.2

Table 7. Prevalence of HP infection in relation to history of contact with domestic animals

Animals	HP infection		
	Total	Positives	
	(*n*)	(*n*)	(%)
Dogs	62	34	54.8
Cats	19	10	52.6
Cattle	21	12	57.1
Horses	8	5	62.5
Sheep	9	6	66.7
Pigs	4	3	75.0
Birds	5	3	60.0
None	45	15	33.3

(Tables 3–5). We also enquired about social habits and about history of contact, at home and/or at work, with domestic animals (Tables 6, 7).

Discussion

Serological diagnostic tests are highly reliable, sensitive and specific. Standard ELISAs of specific IgG are sensitive indicators of the presence of HP infection. These also differentiate infections due to HP from those caused by *Campylobacter jejuni*. A titre of 1 : 3200 or greater is usually considered a positive indicator of HP infection.

Several factors may combine to influence the true prevalence of a disease and/or condition. We therefore examined certain subsets to see if there would be any major identifiable patterns. Certainly the extremely limited number of subjects enrolled in our research does not allow us to reach any conclusive statement, which would be far beyond the purposes of our merely descriptive study. Furthermore, the role of age as a confounding variable cannot be excluded from the analysis of our data, and, again, the limited size of our sample did not allow any statistical significance at the 95% level to be achieved. Nevertheless, some of our findings are interesting. As in virtually all other major series [9–11], age proved to be an important factor in increasing the likelihood of developing HP infection. However, it has been found that the relationship between HP infection and age is inversely related to the level of sanitation of a certain environment, since a higher rate of HP infection is found at young ages in Third World countries [12, 13]. Thus, it can be postulated that each population has a sort of "critical" threshold of individual exposure to HP: the more frequent and the more intense the contacts with the bacteria, the higher the likelihood of developing the infection earlier in life.

HP infection causes antral gastritis, and this can turn into gastric atrophy. The latter is commonly considered a precancerous condition. A study [14] was recently carried out on 1882 males in the People's Republic of China to establish any geographical association between HP infection and gastric cancer. This research confirmed a moderate but significant geographical association between gastric cancer mortality and the prevalence of HP infection in 46 rural Chinese counties. Thus, the reported sequence (HP infection–chronic gastritis–gastric atrophy–gastric neoplasm) could, at least partially, explain the fact that gastric cancer during the last 30 years has been declining in developed countries and increasing in Third World environments. In our research we analysed some simple socioeconomic indicators such as rank and level of education, and they both showed a similar relationship, with the occurrence of HP infection being higher in lower socioeconomic classes (i.e. education and/or rank). A previous study had showed conflicting results [15].

Mean height and weight both suggested a higher body mass index in people with HP infection. However, this finding could be merely related and dependent on differences in eating and drinking habits in people with HP infection. Family and/or medical history of gastrointestinal disorders such as gastritis and peptic ulcer confirmed the now well-established clinical relationships between gastroduodenal pathology and HP infection. It is interesting to note that in our questionnaires the nonspecific term of gastritis was used to identify gastric and upper digestive tract symptoms of unknown origin and, therefore, also non-ulcer dyspepsia (NUD).

We also analysed some social habits, some of which are commonly reported to have a relationship with the occurrence of gastritis and of peptic ulcer. These included: alcohol intake from beer, wine, and spirits; consumption of tea or coffee; and cigarette smoking. Alcohol, tea, coffee and smoking all have a direct action on the gastric wall, implying some major metabolic changes [16–18]. However, none of these factors showed any real trend to suggest a true association with HP infection.

An interesting link was detected between HP infection and a history of contact with domestic animals. Previous research [19] had also suggested that HP infection could, indeed, be a zoonosis. However, it is certainly difficult to ascertain whether the animal world represents a sort of "compulsory" reservoir for HP or, rather, whether it is simply an independent pathway for the biological course of the bacteria. In conclusion, no assumption has been made that certain environmental and socioeconomic conditions play a true aetiological role in the development of HP infection. Nevertheless, they should be regarded as contributory to or even necessary for the development of this condition.

References

1. Bizzozzero G (1893) Ueber die schlauchförmigen Drüsen des Magendarmkanals und die Beziehungen ihres Epithels zu dem Oberflächenepithel der Schleimhaut. Arch Mikr Anat 42:82–152
2. Salomon H (1896) Ueber des Spirillum des Säugetiermagens und sein Verhalten zu den Belegzellen. Zentralbl Bakteriol Hyg., 19:433–442
3. Kreinitz W (1906) Ueber des Aufreten von Spirochaten verschiedener Form in Magenihalt bei Carcinoma ventriculi. Dtsch Med Wochenschr 32:872
4. Rosenow EC, Sanford AH (1915) The bacteriology of ulcer of the stomach and duodenum in man. J Infect Dis 17:219–216
5. Doenges JL (1938) Spirochetes in the gastric glands of macacus rhesus and humans without definite history of related disease. Proc Soc Exp Med Biol 38:536–538
6. Freedburg AS, Barron LE (1940) The presence of spirochetes in human gastric mucosa. Am J Dig Dis 7:443–445
7. Warren JR, Marshall BJ (1983) Unidentified curved bacilli on gastric epithelium in active chronic gastritis. Lancet i: 1273–1275
8. McKinlay AW, Upadhyay R, Gemmell CG, Russell RI (1990) Helicobacter pylori: bridging the credibility gap. Gut 31:940–945
9. Dooley CP, Cohen H, Fitzgibbons PL, Bauer M, Appleman MD, Perez-Perez GI, Blaser MJ (1989) Prevalence of Helicobacter pylori infection and histologic gastritis in asymptomatic persons. N Engl J Med 321:1562–1566
10. Kosunen TU, Hoeoek J, Rautelin HI, Myllylae G (1989) Age-dependent increase of Campylobacter pylori antibodies in blood donors. Scand J Gastroenterol 24(1):110–114
11. Graham DY, Klein PD, Opekun AR, Boutton TW (1988) Effect of age on the frequency of active Campylobacter pylori infection diagnosed by the 13C urea breath test in normal subjects and patients with peptic ulcer disease. J Infect Dis 157(4):777–780
12. Graham DY, Klein PD, Opekun AR, Boutton TW, Evans DJ Jr, Evans DG, Alpert LC, Michaletz PA, Yoshimura HH, Adam E (1988) Epidemiology of Campylobacter pylori infection: ethnic considerations. Scand J Gastroenterol [Suppl] 142:9–13

13. Mégraud F, Brassens-Rabbe MP, Denis F, Belbouri A, Hoa DQ (1989) Seroepidemiology of Campylobacter pylori in various populations. J Clin Microbiol 27(8): 1870–1873
14. Forman D, Sitas F, Newell DG, Stacey AR, Boreham J, Peto R, Campbell TC, Li J, Chen J (1990) Geographic association of Helicobacter pylori antibody prevalence and gastric cancer mortality in rural China. Int J Cancer 46: 608–611
15. Gastrointestinal Physiology Working Group (1990) Helicobacter pylori and gastritis in Peruvian patients: relationship to socioeconomic level, age, and sex. Am J Gastroenterol 85(7): 819–823
16. Jørgensen T (1989) Gall stones in a Danish population. Relation to weight, physical activity, smoking, coffee consumption, and diabetes mellitus. Gut 30: 528–534
17. Criqui MH, Wallace RB, Heiss G, Mishkel M, Schonfeld G, Jones GTL (1980) Cigarette smoking and plasma high-density lipoprotein cholesterol. The lipid research clinics program prevalence study. Circulation 62 [suppl IV]: 70–76
18. Stamford BA, Matter S, Fell RD, Sady S, Cresanta MK, Papanek P (1984) Cigarette smoking, physical activity, and alcohol consumption: relationship to blood lipids and lipoproteins in premenopausal females. Metabolism 33: 585–590
19. Vaira D, D'Anastasio C, Holton J, Dowsett JF, Londei M, Bertoni F, Beltrandi E, Grauenfels P, Salmon PR, Gandolfi L (1988) Campylobacter pylori in abbattoir workers: is it a zoonosis? Lancet 2 (8613): 725–726

High Prevalence of *Helicobacter pylori*-Induced Gastritis in China: Results of a Screening Study in 194 Adolescents

R. Raedsch[1], M. Crespi[2], R. Waldherr[1], J. Chang-Claude[3], Y.G. Rei[4], N. Muñoz[4], and P. Correa[4]

Introduction

Helicobacter pylori (HP) infection has been reported in populations all over the world. Whereas the direct source of infection is still unknown, there is increasing evidence that HP is spread via a contaminated water or food supply [1]. The prevalence of HP infection in the general population is correlated to the socioeconomic status and standard of hygiene in specific subpopulations. In some countries up to 90% of healthy adults may be infected [2, 3]. In Western countries the prevalence of HP infection increases with age and reaches a plateau at about 50%.

HP has been identified as the most important factor of chronic gastritis [4]. This fact has led to new classifications of gastritis (Sydney System) [5]. HP-induced gastritis presents most commonly as chronic gastritis of the antrum with activity, but it may also cause chronic pangastritis. Glandular atrophy and intestinal metaplasia are often found and may increase in extent and degree with advancing age or time course of HP infection. There is some evidence that chronic gastritis with atrophy may represent a precursor of gastric cancer [6]. The role of HP in gastric cancer is still undefined, but HP gastritis and HP infection in the early years of life may represent risk factors of gastric cancer.

In the present study we investigated the prevalence of gastritis and HP infection in adolescents of Huixian, Henan Province. Huixian is a rural area in the People's Republic of China which is known as a high-risk area for esophageal cancer. We had the unique opportunity of investigating HP infection and gastritis in young adolescents chosen at random from the general population, irrespective of clinical symptoms.

Patients and Methods

In the course of a screening program for precursor lesions of esophageal cancer, upper gastrointestinal endoscopy was performed in a total of 538 adolescents

[1] Department of Pathology, University of Heidelberg, 6900 Heidelberg, FRG
[2] National Cancer Institute, Rome, Italy
[3] German Cancer Research Center, 6900 Heidelberg, FRG
[4] Department of Medicine, University of Heidelberg, 6900 Heidelberg, FRG

J.M. Pajares et al. (Eds.)
H. pylori and Gastroduodenal Pathology

aged 15–25. One third of the subjects were selected from households with a family member presenting with esophageal cancer within the past 6 years, and two thirds came from control households. The data on precursor lesions of esophageal cancer in these subjects have been published elsewhere [7, 8]. Besides screening for the precursor lesions of cancer in the esophagus, in a subgroup of 194 subjects additional biopsies were taken from the corpus and antrum of the stomach to investigate the presence of gastritis and HP infection.

Histopathological grading of gastritis and determination of HP were performed with hematoxylin and eosin and silver staining. Gastritis was classified according to the criteria of Correa [9] as: superficial, diffuse, atrophic, and active or inactive gastritis. Reading of the pathology slides was performed independently by R.W. and P.C.

Results

In these adolescents, selected irrespective of clinical symptoms, a very high prevalence of gastritis was found. In 182 of the 194 subjects (93.8%) gastritis was present. Figure 1 demonstrates the histopathological grading of chronic gastritis. By histology 71 (36.6%) of cases presented with superficial gastritis (14 active, 57 inactive), 101 (52%) with diffuse gastritis (96 active, five inactive), and in ten (5.1%) adolescents chronic gastritis plus focal atrophy (nine active, one inactive) was demonstrated. In only 12 of the 194 adolescents (6.2%) was normal mucosa found in the antrum and corpus of the stomach.

HP was present in 166 (85.6%) of these young study subjects. All 119 cases showing histologically chronic active gastritis were HP positive (Fig. 2). Of the 63 cases with chronic inactive gastritis, HP infection was found in 37 of 57 (64.9%) of subjects with superficial, three of five (60%) with diffuse, and one adolescent with atrophic inactive gastritis. HP was also demonstrated in six of the 12 (50%) adolescents with histologically normal mucosa.

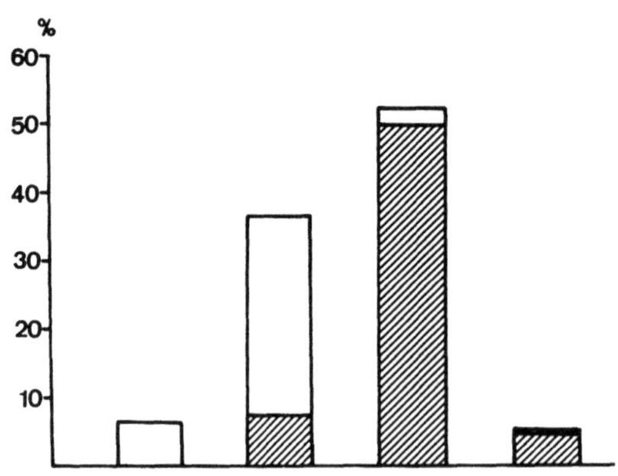

Fig. 1. Prevalence of chronic gastritis in 194 adolescents. In about 94% of these cases gastritis was present. Grading of gastritis showed activity and even atrophy in a considerable number of subjects. *Open columns*, inactive gastritis; *hatched columns*, active gastritis

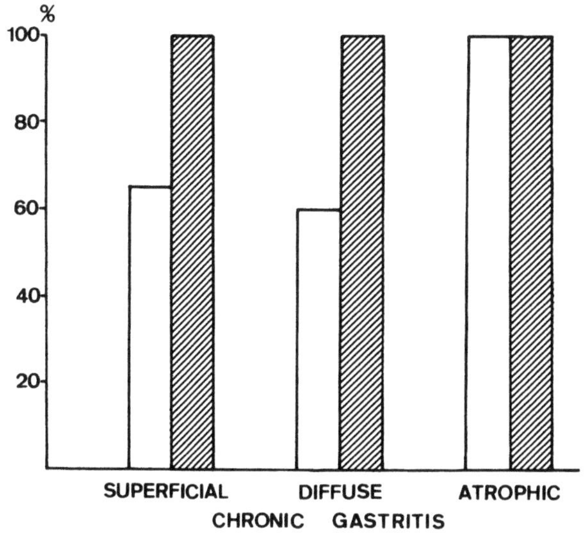

Fig. 2. Prevalence of HP as judged by histology. HP was present in all subjects with active gastritis. *Open columns*, inactive gastritis; *hatched columns*, active gastritis

Conclusions

These data demonstrate a very high prevalence of gastritis and HP infection in young adolescents in China. The prevalence of gastritis and HP infection in adolescents in China is much higher than in comparable age groups in Western countries. These findings suggest an HP infection early in life which led to atrophic gastritis in a considerable number of these adolescents. A possible significance of HP infection at a young age for the development of gastric cancer has to be elucidated.

References

1. Klein PD, Graham DY, Pekun AR, Sekely S (1989) Helicobacter pylori is a waterborne disease in Peruvian children. Gastroenterology 5: A69
2. Megraud F, Brassens-Rebbe M-P, Denis F, Belbouri A, Hoa DQ (1989) Seroepidemiology of C.pylori infection in various populations. J Clin Microbiol 27: 1870–1873
3. Graham DY (1991) Helicobacter pylori in human populations: The present and predictions of the future based on the epidemiology of polio. In: Menge H et al. (eds) Helicobacter pylori 1990, Springer, Berlin Heidelberg, New York pp 97–102
4. Stolte M, Heilmann KL (1989) New classification of gastritis. Leber Magen Darm 19: 220–226
5. Misiewicz JJ, Tytgat GNJ, Goodwin CS, Price AB, Sipponen P, Strickland RG, Cheli R (1990) The Sydney system: a new classification of gastritis. Working party reports, World Congress of Gastroenterology. Blackwell, Melbourne, pp 1–10
6. Talley NJ, DiMagno E, Zinsmeister AR, Perez-Perez G, Blaser M (1990) Helicobacter pylori and gastric cancer: a case-control study. Rev Esp Enferm Dig 78: S17

7. Wahrendorf J, Chang-Claude J, Liang Q, Rei YG, Munoz N, Crespi M, Raedsch R, Thurnham D (1989) Precursor lesions of esophageal cancer in young people in a high-risk population in China. Lancet ii: 1239–1241
8. Chang-Claude J, Wahrendorf J, Liang Q, Rei YG, Munoz N, Crespi M, Raedsch R, Thurnham D, Correa P (1990) An epidemiological study of precursor lesions of esophageal cancer among young persons in a high-risk population in Huixian, China. Cancer Res 50: 2268–2274
9. Correa P (1988) Chronic gastritis: a clinico-pathological classification. Am J Gastroenterol 83: 504–509

Seroepidemiology of *Helicobacter pylori* Infection in Vegans and Meat Eaters

M.J. Webberley, J.M. Webberley, P. Lowe, V. Melikian, and D. Newell

Introduction

The acquisition and transmission of *Helicobacter pylori* infection is poorly understood. Considerable evidence supports the person-to-person spread of the organism within close communities [1, 2], but the initial source of *H. pylori* remains unknown. Volunteer studies have shown that the organism can be acquired by ingestion, implying that diet may be important [3, 4]. Although a natural animal host (other than human) has not yet been found, some workers have demonstrated an association between occupational exposure to animals or meat and *H. pylori* colonisation [5, 7]. The presence of meat in the diet may therefore be a risk factor for *H. pylori* infection. We investigated this hypothesis by measuring anti-*H. pylori*-specific antibodies in Asian life-long vegans, Asian meateaters and Caucasian meateaters.

Subjects and Methods

A total of 67 Asian life-long vegans (age range 18–74 years), 105 Asian meateaters (20–69 years) and 67 Caucasian meateaters (22–70 years) were recruited from routine medical outpatient clinics. All subjects were asymptomatic for gastrointestinal disease and were of similar socioeconomic background. Clotted venous blood (10 ml) was acquired from each patient, and an enzyme-linked immunosorbent assay (ELISA) system [6], using an acid-extractable cell surface *H. pylori* antigen, was used to measure serum *H. pylori*-specific antibodies.

Results

A group of 54 Asian vegans were age and sex matched with Asian meateaters. There was no statistical difference in anti-*H. pylori* antibody levels between these

Department of Gastroenterology, Dudley Road Hospital, Dudley Road, Birmingham, B18, UK
CAMR, Porton Down, Salisbury, UK

J.M. Pajares et al. (Eds.)
H. pylori and Gastroduodenal Pathology

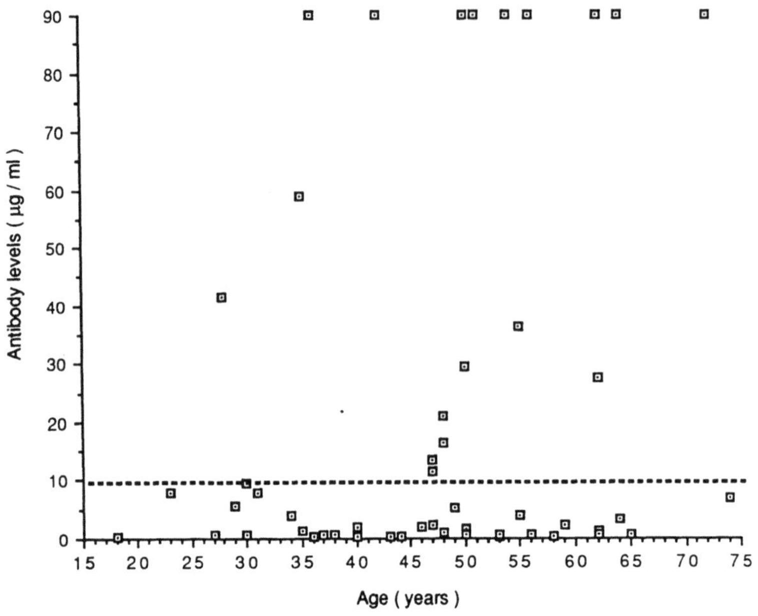

Fig. 1. Anti-*H.pylori* antibody levels in Asian vegans

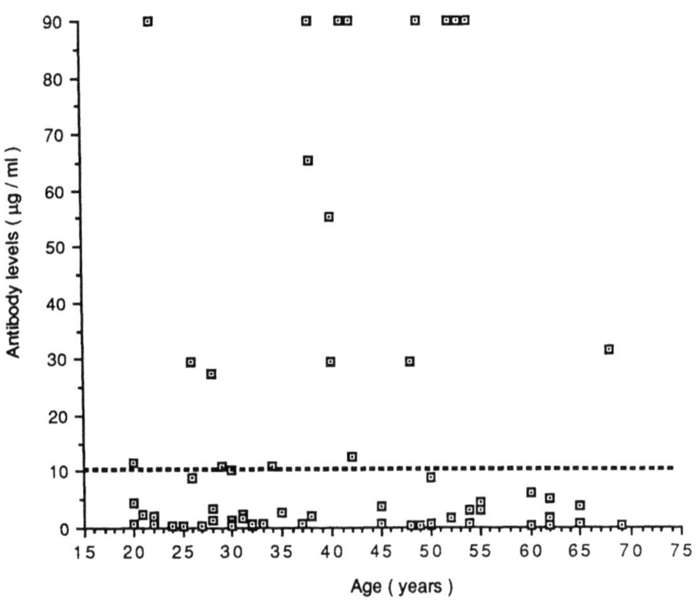

Fig. 2. Anti-*H.pylori* antibody levels in Asian meat eaters

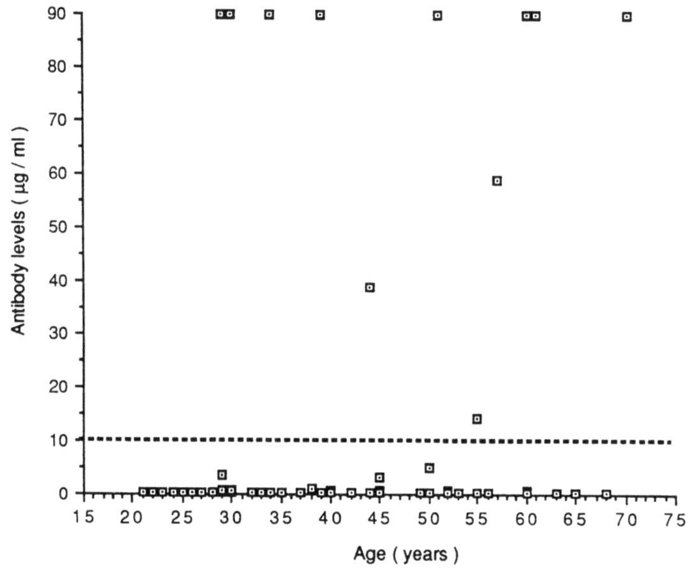

Fig. 3. Anti-*H.pylori* antibody levels in Caucasian meat eaters

two groups. A group of 65 Asian meateaters were age and sex matched with Caucasian meateaters. Significantly higher antibody levels were detected in the Asian group ($p < 0.01$), where 32.3% were found to be seropositive compared with 17% in the Caucasian group.

The distribution of results was significantly different between the two groups. In the Caucasians, the results were clear-cut, either strongly positive or negative, using an established cut-off level of 10 μg/ml. However, amongst both Asian meateaters and vegans, a wide spread of levels was observed (Figs. 1–3).

The cross-reactivity between the flagellar antigens of *H. pylori* and anti-*C. ampylobacter jejuni* antibodies has been shown to influence serological data in populations where there is high asymptomatic carriage of *C. jejuni*. In order to exclude this as a potential cause for the observed difference in spread of antibody levels between the two groups, anti-*C. jejuni* antibodies were measured in all subjects. No correlation was observed between anti-*H. pylori* and anti-*C. jejuni* antibody levels, with no significant differences in anti-*C. jejuni* antibody levels being seen between any of the Asian or Caucasian groups.

Discussion

These results suggest that ingestion of meat is not a risk factor for acquisition of *H. pylori*. The study does highlight differences in distribution of antibody levels as well as seropositivity between Asians and Caucasians. Other studies have shown ethnic differences in *H. pylori* seropositivity but have not demonstrated variations in the spread of antibody levels [2]. Observed intrafamilial clustering of *H. pylori* infection may be important in populations where there are large families, but there is no evidence that this is a significant factor in our study.

More work is needed to investigate different ethnic groups within the same society. The spread of antibody levels within our Asian group may indicate that different criteria for seropositivity may be needed for various ethnic groups. This is currently under investigation.

References

1. Drumm B, Pérez-Pérez GI, Blaser MJ, Sherman PM (1990) Intrafamilial clustering of Helicobacter pylori infection. N Engl J Med 322 (6): 359–363
2. Lee A, Hazell S L (1988) Campylobacter pylori in health and disease. An ecological perspective. Microb Ecol Health Dis 1: 1–16
3. Marshall BJ, Armstrong JA, McGechie DB, Clancy RJ (1985) Attempt to fulfill Koch's postilates for pyloric Campylobacter. Med J Aust 142: 436–439
4. Morris A, Nicholson G (1987) Ingestion of Campylobacter pyloridis causes gastritis and raised fasting pH. Am J Gastroenterol 82: 192–199
5. Morris A, Nicholson G, Lloyd G, Haines D, Rogas A, Taylor D (1986) Seroepidemiology of Campylobacter pyloridis. N Z Med J 89: 657–659
6. Newell DG, Johnston BJ, Ali MH, Reed PI (1988) An enzyme-linked immunosorbent assay for the serodiagnosis of Campylobacter pylori-associated gastritis. Scand J Gastroenterol [Suppl] 142: 53–57
7. Vaira D, D'Anastasio C, Holton J et al. (1988) Campylobacter pylori in abattoir workers: is it a zoonosis? Lancet 2: 725–726

Seroepidemiology of *Helicobacter pylori* Infection in the Armed Forces

I. Hammermeister[1], G. Janus[2], F. Schamarowski[3], M. Rudolf[4], E. Jacobs[1], and M. Kist[1]

Introduction

The mode of transmission of *Helicobacter pylori*, first isolated in 1982 [28], is still unknown, although there is increasing evidence for a person-to-person spread of infection. There are several investigations that support the idea of person-to-person transmission: a higher prevalence of *H. pylori* was found in endoscopy staff [17, 22], in institutionalized residents [2], as well as in psychiatric patients, orphanage children, and family contacts of *H. pylori*-infected patients [16, 22].

Studies on the seroepidemiology of *H. pylori* infection in the armed forces have shown an elevated risk of infection in submarine crews [7]. The results of this study are reported here. The immune response of *H. pylori*-positive patients correlates well with histologic findings in biopsy specimens [4, 10, 26, 27]. Various serologic methods have been used to detect a systemic immune response to *H. pylori*, e.g., complement fixation [3, 11, 27], passive hemagglutination [14], enzyme-linked immunosorbent assay (ELISA) systems [8, 25], and the immunoblot test [26, 27]. In the present study we used the immunoblot technique to detect an *H. pylori*-specific immune response. The aim of this prospective study was to investigate serologically the prevalence of *H. pylori* infection in well-defined risk groups, by screening over a period of 2 months to 1 year.

Materials and Methods

Western Blot Method.

We used a whole cell sonicate of a 120-kDa antigen-positive *H. pylori* strain [1]. The whole cell sonicate was separated in *sodium dodecyl sulfate–polyacrylamide gel electrophoresis* (SDS-PAGE) according to the method of Laemmli [13].

[1]Institut für Medizinische Mikrobiologie und Hygiene des Universitätsklinikums Freiburg, Hermann-Herder-Str. 11, 7800 Freiburg, FRG
[2]CHA A. Limouzin, SP 69 502 00584 Armees, France
[3]Marinesanitätsstaffel Eckernförde, 2330 Eckernförde, FRG
[4]Aufklärungsgeschwader Immelmann, 7801 Bremgarten, FRG

J.M. Pajares et al. (Eds.)
H. pylori and Gastroduodenal Pathology

Protein samples of 800 μg per gel were applied to an 8% running gel and a 4.5% stacking gel. Protein transfer from the slab gel to Immobilon membrane (Millipore, USA) was performed in a TRIS- glycin-methanol buffer by electrotransfer (30 V; 14 h). After the blotting procedure free protein binding sites were blocked with 10% skimmed milk in phosphate-buffered saline (PBS). The Immobilon membrane was cut into strips and incubated for 1 h with 3 ml patients' sera, diluted 1:150 for immunoglobulin G (IgG) and 1:60 for IgA detection. After three washes in 0.05% Tween PBS buffer for 30 min, the strips were incubated for a further hour with rabbit anti-human IgG (1:500) (Behringwerke, FRG) or rabbit anti- human IgA antibodies (1:500) (Dakopatts, FRG), respectively. After this procedure they were washed, incubated with a swine anti-rabbit antibody (1:100) (Dakopatts) for another 30 min and washed again three times. Finally, a peroxidase conjugated anti-swine antibody (1:333) (Dakopatts) was added for another 30 min. The strips were then washed three times and the reaction bands were made visible by addition of PBS buffer containing diaminobenzidine as substrate and H_2O_2 as reactant. The reaction was stopped after 10 min for IgG and 15 min for IgA by dilution with water. A clearly visible staining of the 120-kDa, 88-kDa, 86-kDa, and 82-kDa protein was assumed to indicate an *H. pylori*-specific and infection- related immune reaction, as shown in a previous study [1].

Study Groups

The following two groups presumed to be at elevated risk for person-to-person spread of infections, both by the oral-oral or fecal-oral route, were selected:

1. German submarine crews ($n = 64$, mean age: 26.2 years, range: 20–33 years). Three blood samples per individual were examined during a time period of 2 months: before, immediately after, and 3 weeks after a submarine mission. Epidemiologic characteristics of the submarine crews assumed to be risk factors were: (a) overcrowded space; (b) closed community; (c) close social contacts; (d) very limited sanitary facilities; (e) common food supply.
2. French infantry ($n = 51$, mean age: 26.5 years, range 21–39 years) consisting of regular officers having joined the military service for a minimum of 3 years. Assumed risk factors for the French infantry officers were: (a) long-lasting military service; (b) predominantly common food supply.

These potential risk groups were compared with two control groups consisting of:

1. German air force staff ($n = 74$, mean age: 23.7 years, range: 19–31 years).
2. French infantry ($n = 135$, mean age: 20.5 years, range: 19–27 years) consisting of recruits, having just started military service.

We examined three to four blood samples per individual during a time period of 1 year. The control groups were age matched to the appropriate risk groups as closely as possible. The German draft system ensures socioeconomic homogeneity between the different branches of the military forces. All study groups usually had a common food supply. Each subject was required to fill out a questionnaire to control for age, family status, gastric pain, current medical treatment, and duration of military service.

Statistical Analysis

Data were analyzed using the chi-square test and additionally by a logistic regression analysis applying the CMS SAS software program.

Results

The overall frequency of a positive antibody response to *H. pylori* in the different branches of the military forces are presented in Table 1.

The German submarine crews exhibited a markedly increased frequency in total antibody response, as well as in the IgG, IgA and combined IgG/IgA responses, when compared to the German air force staff of comparable mean age (38.1% versus 18.9%) and even to the other military branches (French infantry) enrolled in this study. This difference in frequencies for both the total antibody responses and the IgG and IgA responses in the German submarine crews and in the German air force staff was significant when analyzed by logistic regression analysis (Table 2) but was not significant when combined IgG/IgA response rates were compared.

The prevalence of *H. pylori*-specific antibodies did not differ significantly between the regular officers and the recruits, both serving in the French infantry for different time periods. The frequency of the total antibody response, IgG, IgA, and combined IgG/IgA response was comparable in both groups. There was no significant difference when logistic regression analysis was applied (Table 2). In general the prevalence of anti-*H. pylori* antibodies (especially IgG) was strongly dependent on age (Table 2). The frequency of a positive antibody response to *H. pylori* in the German submarine crews was significantly higher in comparison with the other branches of the military forces examined in this study.

Table 1. Frequency of a positive antibody response (confirmed by immunoblot) for *H. pylori* in different branches of military forces

		Age (years)					
	Subjects (*n*)	Mean	Range	Total Antibody (IgG or IgA)	IgG	IgA	IgG and IgA
German submarine crews	63	26.1	20–33	38.1	31.7	25.4	19.0
French infantry A	51	26.5	21–39	21.6	13.7	17.6	9.8
B	135	20.5	19–27	20.0	18.5	11.1	9.6
German air force	74	23.7	19–31	18.9	16.2	12.2	9.5

A, regular officers; B, recruits

Table 2. Logistic regression analysis of the specific immune response to *H. pylori* in different branches of military forces (age range 19-39 years)

	Total antibodies (IgG or IgA)			IgG			IgA			IgG and IgA		
	Chi-square	*p* value	Odds ratio	Chi-square	*p* value	Odds ratio	Chi-square	*p* value	Odds ratio	Chi-square	*p* value	Odds ratio
German submarine crews Controls: German air force	6.2	< 0.025	2.64	4.6	< 0.05	2.40	4.0	< 0.05	2.46	2.6	NS	–
French infantry A Controls: French infantry B	0.056	NS	–	0.60	NS	–	1.41	NS	–	0.001	NS	–
Age				58.8			18.6			46.0		

A. regular officers; B. recruits; NS, not significantly different from controls

Depending on the group studied, we examined in a prospective manner three to four blood samples over a time period of 2 months (German submarine crews, $n = 64$) to 1 year (French infantry recruits, $n = 135$). No seroconversion was detected during these examination periods.

Discussion

Sources and routes of *H. pylori* infection are still awaiting further elucidation, but increasing evidence for a person-to-person spread is accumulating. The oral-oral or fecal-oral mode of transmission is currently under discussion. Several reports on indirect person-to-person transmission, predominantly via endoscopes [6, 9, 21] and a higher prevalence of *H. pylori* in gastroenterologists [17], endoscopy staff [22], orphanage children [22], psychiatric patients [22], institutionalized residents, [2] and family contacts of *H. pylori*-infected patients [16, 22] confirmed the notion of a person-to-person transmission.

We examined samples from military personnel from two countries and in different branches, presumably having a differing general risk for person-to-person spread of infection. Submarine crews probably have an overall increased risk of acquiring person-to-person infections, especially during submarine missions, due to the very limited sanitary facilities which were located in a room of $1.5\,m^2$, and were used by 16–20 persons. Air force staff, as controls, have the main epidemiologic characteristics in common with the German submarine crews with the exception of the special conditions characteristic for submarine missions as mentioned above. The second presumed risk group was regular officers of the French infantry because of their already long military service, including quartering in barracks and common food supplies, compared with French infantry recruits who had just joined the military service at the beginning of the study.

The immunoblot method, with its high specifity [18, 19], permits a distinction between specific and cross-reactive antibodies. As described previously [1], the 120-kDa and 88-kDa proteins are specific for *H. pylori* and are infection-related antigens which are often combined with an 86- and 82-kDa protein band.

Considering the well-known age dependency of the *H. pylori* antibody response [12, 23, 25], the prospective study revealed a significantly higher prevalence of anti-*H. pylori*-specific antibodies in the German submarine crews in comparison to the German air force staff. There was a significant difference in the frequency of the total antibody response, the IgG antibody response, and the IgA antibody response between these two groups, whereas no significant difference in the frequency of the combined IgG/IgA response was demonstrable. This might possibly indicate that in this group the higher prevalence of anti-*H. pylori* antibodies does not indicate a higher rate of more acute, ongoing infections with *H. pylori*, which might be associated with a combined IgG/IgA response [22, 24]. On the other hand, the prevalence of combined IgG/IgA responses could be too small to be statistically significant, although it was twice as high in the German submarine crews compared to the German air force staff.

Both groups were comparable in age, sex, common food supply, and health condition when entering the army. However, in contrast to the air force staff, the submarine crews serve during their missions in an overcrowded space with extremely limited sanitary facilities, leading to very close social contacts. These conditions may facilitate oral-oral or fecal-oral routes of transmission. The submarine crews also exhibited a markedly increased frequency in their antibody response when compared to regular officers and recruits of the French infantry.

The main difference between the two study groups from the French infantry was that the regular officers had served in the army for years, whereas the recruits had just started military service. Nevertheless, there was no significant difference in the frequency of the antibody response. In fact, the regular officers of the French infantry had a distinct but not significantly lower prevalence of an IgG antibody response in comparison to the recruits; this might be influenced by a different socioeconomic status in the two study groups.

The frequency of a positive antibody response in the control groups (German air force, recruits of the French infantry) corresponds well with data from studies in which blood donors of comparable age in the two countries were examined [15, 25]. Depending on the study group in our prospective study, we examined three to four blood samples over a time period of 2 months to 1 year. During the examination period no seroconversion was detected. Several authors quote that the rate of acquisition of an *H. pylori* infection is one per 100 per year [5, 20, 23]. Although it could be shown that the German submarine crews have a significantly higher risk of infection, the study period of 2 months might be too short to detect a seroconversion. This deserves further investigation. In the case of the French infantry recruits, the sample size of the individuals examined might still be too small to guarantee a seroconversion during the observation period of 1 year.

The results of this study strongly suggest that person-to-person spread of infection, especially by the oral-oral or fecal-oral route, is the predominant mode of transmission of the *H. pylori* infection.

Conclusion

The seroprevalence of *H. pylori* infection was studied in the following branches of the armed forces: German submarine crews ($n = 64$, mean age 26.2 years) and regular officers of the French infantry ($n = 51$, mean age 26.5 years) who had served for a minimum of 3 years. The submarine crews were compared with air force staff ($n = 74$, mean age 23.7 years), and the French officers with French infantry recruits ($n = 135$, mean age 20.5 years) who had started their service at the beginning of the study. The frequency of an IgG and IgA antibody response to a 120-, 88-, 86-, and 82-kDa protein was determined by the immunoblot method. The submarine crews revealed significantly raised frequencies of the IgG and IgA response compared to air force staff. The antibody responses of French officers and recruits were not significantly different. It is concluded that submarine crews, serving during their missions in an overcrowded space with extremely limited sanitary facilities must be considered as a high-risk group for *H. pylori* infection. These results strongly suggest the person-to-person route, either oral-oral or fecal-oral, of *H. pylori* infection.

Acknowledgements. This investigation was supported by a grant from the Bundesministerium für Forschung und Technologie, Projektträgerschaften für Arbeit, Umwelt und Gesundheit (project 01KI8811/9).

References

1. Apel I, Kist M, Jacobs E, Bredt W (1988) Antibody response of patients against a 120 kDa surface protein of Campylobacter pylori. Zentralbl Bakteriol Mikrobiol Hyg [A] 268: 271–276
2. Berkowicz J, Lee A (1987) Person to person transmission of Campylobacter pylori. Lancet ii: 680–681
3. Bolton FJ, Hutchinson PM, Hinchliffe PM, Holt AV (1989) Distribution in various clinical groups of antibody to C. pylori detected by enzyme-linked immunosorbent assay, complement fixation and microagglutination tests. Serodiagn Immunother Infect Dis 3: 41–50
4. Goodwin CS, Blincow E, Peterson G, Sanderson C, Cheng W, Marshall B, Warren JR, McCulloch R (1987) Enzyme-linked immunosorbent assay for Campylobacter pyloridis correlation with presence of Campylobacter pyloridis in the gastric mucosa. J Infect Dis 155: 488–494
5. Graham DY, Adam E, Klein PD, Evans DJ, Evans DG, Hazell SL, Alpert LC, Michaletz PA, Yoshimura HH (1989) Epidemiology of Campylobacter pylori infection. Gastroenterol Clin Biol 13: 84B–88B
6. Gullini S, Boccini S, Contarini D, Macario F, Bosso O, Maini P, Bicocchi R (1988) Is transmission of Campylobacter pylori by endoscopy examination possible? Endoscopy 20: 162
7. Hammermeister I, Janus G, Schamarowski F, Rudolf M, Jacobs E, Kist M (1991) Submarine crews have an elevated risk of Helicobacter pylori infection. Eur J Clin Microbiol Infect Dis (to be published)
8. Hirschl AM, Pletschette M, Hirschl MH, Berger J, Stanek G, Rotter ML (1988) Comparison of different antigen preparations in an evaluation of the immune response to Campylobacter pylori. Eur J Clin Microbiol Infect Dis 7: 100–105
9. Hunt RH, Darkin D (1985) Epidemic hypochlorhydria. Br Med J 291: 53
10. Jones DM, Eldrigde J, Fox AJ, Sethi P, Whorwell PJ (1986) Antibody to the gastric campylobacter-like-organism ("Campylobacter pyloridis"). Clinical correlations and distribution in the normal population. J Med Microbiol 22: 57–62
11. Jones DM, Lesses AM, Eldridge J (1984) Campylobacter-like organisms on the gastric mucosa: culture, histological and serological studies. J Clin Pathol 37: 1002–1006
12. Kosunen TU, Höök J, Rautelin HI, Myllylä G (1989) Age-dependent increase of Campylobacter pylori antibodies in blood donors. Scand J Gastroenterol 24: 110–114
13. Laemmli UK (1970) Cleavage of structural proteins during the assembly of the head of bacteriophage T4. Nature 227: 680–685
14. Marshall BJ, McGechie DB, Francis GJ, Utley PJ (1984) Pyloric Campylobacter serology. Lancet ii: 281
15. Megraud F, Brassens-Rabbe MP, Denis F, Belbouri A, Duong Quynh Hoa (1989) Seroepidemiology of Campylobacter pylori infection in various populations. J Clin Microbiol 27: 1870–1873
16. Mitchell M, Bohane TD, Berkowicz J, Hazell SL, Lee A (1987) Antibody to Campylobacter pylori in families of index children with gastrointestinal illness due to Campylobacter pylori. Lancet ii: 681–682
17. Mitchell HM, Lee A, Carrick J (1989) Increased incidence of Campylobacter pylori infection in gastroenterologists: further evidence to support person-to-person transmission of C. pylori. Scand J Gastroenterol 24: 396–400

18. Newell DG (1987) Identification of the outer membrane proteins of Campylobacter pyloridis and antigenic cross-reactivity between Campylobacter pyloridis and Campylobacter jejuni. J Gen Microbiol 133:163–170
19. Perez-Perez GJ, Blaser MJ (1987) Conservation and diversity of Campylobacter pyloridis major antigens. Infect Immun 55:1256–1263
20. Perez-Perez GI, Dworkin BM, Chodos JE, Blaser MJ (1988) Campylobacter pylori antibodies in humans. Ann Intern Med 109:11–17
21. Ramsey EJ, Carey KV, Peterson WL, Jackson JJ, Murphy FK, Read NW, Taylor KB, Trier JS, Fordtran JS (1979) Epidemic gastritis with hypochlorhydria. Gastroenterology 76:1449–1457
22. Reiff A, Jacobs E, Kist M (1989) Seroepidemiological studies of the immune response to Campylobacter pylori in potential risk groups. Eur J Clin Microbiol Infect Dis 8:592–596
23. Schaub N, Stalder H, Stalder GA, Marbet UA, Vögtlin J, Affolter H, Wegmann W, Vischer WA, Zingel O, Tanner K, Dietrich FM (1988) Campylobacter pylori, Gastritis und Ulcuskrankheit. Schweiz Med Wochenschr 118:293–301
24. Vaira D, Holton J, Cairns SR, Falzon M, Polydorou A, Dowsett JF, Salmon PR (1988) Antibody titres to Campylobacter pylori after treatment for gastritis. Br Med J 297:397–398
25. Von Wulffen H, Grote HJ (1988) Enzyme-linked immunosorbent assay for detection of immunglobulin A and G antibodies to Campylobacter pylori. Eur J Clin Microbiol Infect Dis 7:559–565
26. Von Wulffen H, Grote HJ, Gatermann S, Löning T, Berger B, Buhl C (1988) Immunoblot analysis of immune response to Campylobacter pylori and its clinical associations. J Clin Pathol 41:440–446
27. Von Wulffen H, Heesemann J, Bützow GH, Löning T, Laufs R (1986) Detection of Campylobacter pyloridis in patients with antrum gastritis and peptic ulcers by culture, complement fixation test and immunoblot. J Clin Microbiol 24:716–720
28. Warren JR, Marshall BJ (1983) Unidentified curved bacilli on gastric epithelium in active chronic gastritis. Lancet i:1273–1275

Helicobacter pylori Infection Clusters in Families of Healthy Individuals

H.M. Malaty[1] and D.Y. Graham[1,2]

Introduction

Within the past decade it has been established that the *Helicobacter pylori* organism is the most common cause of chronic gastritis and *H. pylori* gastritis is strongly linked to peptic ulcer disease [6]. The epidemiology of *H. pylori* infection is beginning to be understood, but the data concerning possible reservoirs and transmission are scanty [1–3, 5, 7–9, 11–18]. Several studies have investigated the parents and siblings of children with symptomatic *H. pylori* infection and have reported a higher frequency of *H. pylori* infection among families of infected symptomatic children than among controls [3, 16]. The focus of our studies has been to attempt to avoid the bias associated with patient-based studies and to perform traditional epidemiologic studies in the healthy population. In this study, we investigated whether there was clustering of *H. pylori* infection within family units of healthy volunteers. We defined the status of the family on the basis of the results of *H. pylori* status in an index parent. We found that *H. pylori* infection was strikingly more frequent in the spouse and children of families in whom the index parent was infected, suggesting that person-to-person transmission or common source infection occurs frequently.

Methods

We investigated the frequency of *H. pylori* infection within families, defined as consisting of a husband and wife with at least one biological child, all living in the same household. Inclusion criteria required that both the parents and the children had been born in the United States, had used no antibiotic or bismuth for the previous 2 months, had no recent major illness or surgical operation, and had no symptoms referable to the upper gastrointestinal tract. *H. pylori* infection was identified using a ^{13}C-urea breath test [10] and an enzyme-linked immunosorbent assay (ELISA) for anti-*H. pylori* immunoglobulin G (IgG) [4].

[1] Department of Medicine, Veterans Affairs Medical Center, 2002 Holcombe Blud., Houston, TX 77030, USA
[2] Division of Molecular Virology, Veterans Affairs Medical Center, 2002 Holcombe Blud., Houston, TX 77030, USA

J.M. Pajares et al. (Eds.)
H. pylori and Gastroduodenal Pathology

Statistical Analyses

Families were identified on the basis of *H. pylori* status of the index parent. Two groups were defined: the negative index group in which the index parent was not infected with *H. pylori*, and the positive index group in which the parent was infected. The data were analyzed by *t* test and the Mantel-Haenszel chi-square using the SAS program (SAS Institute, Cary, NC). The *t* test or Wilcoxon rank test analyses were used to test the differences between the positive index group and the negative index group with regard to the age of the children and the parents, to the mean number in the family, and to the family income of the parents. We also examined the frequencies of *H. pylori* infection among the two groups to look for evidence of clustering of *H. pylori* infection within the positive index families. The index case was not counted in the calculations to determine the percentage infected in the family group.

Results

A total of 41 families (151 healthy individuals) residing in different geographic regions of the Houston metropolitan area were enrolled. The median number of family members was four (range three to six members). Before the results of the *H. pylori* tests were known, one parent was selected as the index case. In 22 families the index parent was not infected with *H. pylori* (negative index), but was infected (positive index) in 19 families. The two groups of families were comparable ($p > 0.1$) with regard to the number of family members, age of children and adults, and family income. The children of the uninfected index

Table 1. Summary of study populations

	Positive Index ($n = 19$)	Negative Index ($n = 22$)
Family members [a]	54	56
Number of family members (median)	4	4
Number of children	35	34
Age of parents (years) [a]		
mean	36	33
range	19–53	22–40
Age of children (years) [b]		
mean	9	6*
range	1–18	2–17
Annual family income [b]	15–25 000	15–25 000
Race		
black	12	6
white	7	16

[a] Excluding the index case.
[b] Median.
* $p = 0.05$ Wilcoxon rank test.

parent were slightly, but significantly younger than those of the infected index parents (Table 1).

Helicobacter pylori infection clustered, i.e., 68% of spouses of *H. pylori*-infected index cases were also *H. pylori* infected compared to 9% of spouses of *H. pylori*-negative index cases ($p < 0.0001$). The children of infected index parents were also more likely to be infected than children of uninfected index parents: 40% vs. 3%, respectively ($p < 0.0001$); and the results in the children were independent of whether the father or mother was the index case. Clustering of *H. pylori* infection within families suggests person-to-person transmission or common source exposure. The high frequency of *H. pylori* infection in spouses suggests that genetic factors are less important than living conditions for transmission of *H. pylori* infection.

Black families were overrepresented in the positive index group (12/19) compared to the negative index group (6/22) ($p < 0.05$). We therefore analyzed each race separately to determine if the clustering phenomenon was independent of race; the pattern of clustering was still evident.

Discussion

The design of this study was unique because it was based on healthy asymptomatic volunteers and not on the study of individuals in contact with patients with symptomatic *H. pylori* infection (many with peptic ulcers) [3, 16, 18]. Our design allowed us to examine the transmission of *H. pylori* infection within normal households not at high risk for transmission via contaminated instruments or through patient contact. The high frequency of *H. pylori* infection among the spouses and the children living in the same household with a positive index parent strongly support a process of person-to-person transmission, a common source infection, or both.

The higher prevalence of *H. pylori* infection among black families was consistent with our previous studies [5, 8] which demonstrated that the age-adjusted prevalence of *H. pylori* infection is approximately twice as great in blacks than in whites; this difference remained after adjustment for socio-economic factors. The racial difference in apparent susceptibility to *H. pylori* infection remains unexplained. The high frequency of *H. pylori* infection in spouses suggests that genetic factors are less important than living conditions for transmission of *H. pylori* infection, and this is confirmed by the finding from this study that identical twins (age 16 years) in one black family were discordant for *H. pylori* infection. Additional studies in twins are in progress.

Acknowledgements. This work was supported by the Department of Veterans Affairs; by grant DK 39919 from the National Institute of Diabetes and Digestive and Kidney Diseases; by the US Department of Agriculture/Agricultural Research Service Children's Nutrition Research Center; by the Baylor College of Medicine General Clinical Research Center at The Methodist Hospital under grant RR-00350 from the Division of Research Services, National Institutes of Health; and by the generous support of Hilda Schwartz. The investigators also acknowledge the help of Gloria Thompson, Valeria Tapscott, and Sara Sekely.

References

1. Al-Moagel MA, Evans DG, Abdulghani ME, Adams E, Evans DJ Jr, Malaty HM, Graham DY (1990) Prevalence of Helicobacter (formally Campylobacter) pylori infection in Saudia Arabia: and comparison of those with and without upper gastrointestinal symptoms. Am J Gastroenterol 85:944–948
2. Berkowicz J, Lee A (1987) Person-to-person transmission of Campylobacter pylori. Lancet ii:680–681
3. Drumm B, Perez-Perez GI, Blaser MJ, Sherman PM (1990) Intrafamilial clustering of Helicobacter pylori infection. N Engl J Med 322:359–363
4. Evans DJ Jr, Evans DG, Graham DY, Klein PD (1989) A sensitive and specific serologic test for detection of Campylobacter pylori infection. Gastroenterology 96:1004–1008
5. Fiedorek SC, Malaty HM, Evans DG, Pumphrey CL, Casteel HB, Evans DJ Jr, Graham DY (1991) Factors influencing the epidemiology of Helicobacter pylori infection in children. Pediatrics 87:578–582
6. Graham DY (1989) Campylobacter pylori and peptic ulcer disease. Gastroenterology 96 [Suppl]:615–625
7. Graham DY (1991) Helicobacter pylori in human populations: the present and predictions of the future based on the epidemiology of polio. In: Menge H, Gregor M, Tytgat GNJ, Marshall BJ, McNulty CAM (eds) Helicobacter pylori 1990: proceedings of the 2nd international symposium on Helicobacter pylori. Springer, Berlin Heidelberg New York, pp 97–102
8. Graham DY, Malaty HM, Evans DG, Evans DJ Jr, Klein PD, Adam E (1991) Epidemiology of Helicobacter pylori in an asymptomatic population in the United States: effect of age, race and socioeconomic status. Gastroenterology 100:1495–1501
9. Graham DY, Adam E, Klein PD, Evans DJ Jr, Evans DG, Hazell SL, Alpert LC, Michaletz-PA, Yoshimura HH (1989) Epidemiology of Campylobacter pylori. Gastroenterol Clin Biol 13:84B–88B
10. Graham DY, Klein PD, Evans DJ, Evans DG, Alpert LC, Opekun AR, Boutton TW (1987) Campylobacter pyloridis detected noninvasively by the 13C-urea breath test. Lancet i:1174–1177
11. Graham DY, Klein PD, Opekun AR, Boutton TW, Evans DJ Jr., Evans DG, Alpert LC, Michaletz PA, Yoshimura HH, Adam E (1988) Epidemiology of Campylobacter pylori infection: ethnic considerations. Scand J Gastroenterol 23 [Suppl 142]:9–13
12. Jones DM, Eldridge J, Whorwell PJ (1987) Antibodies to Campylobacter pyloridis in household contacts of infected patients. Br Med J 294:615
13. Klein PD, Gastrointestinal Physiology Working Group of Cayetano Herdeia and the Johns Hopkins Universities, Graham DY, Opekun AR, Sekely S (1990) Helicobacter (Campylobacter) pylori is a waterborne disease in Peruvian children. Gastroenterology 98:A69
14. Krajden S, Fuksa M, Anderson J, Kempston J, Boccia A, Petrea C, Babida C, Karmali M, Penner JL (1989) Examination of human stomach biopsies, saliva, and dental plaque for Campylobacter pylori. J Clin Microbiol 27:1397–1398
15. Langenberg W, Rauws EA, Oudbier JH, Tytgat GNJ (1990) Patient-to-patient transmission of Campylobacter pylori infection by fiberoptic gastroduodenoscopy and biopsy. J Infect Dis 161:507–511
16. Mitchell HM, Bohane TD, Berkowicz J, Hazell SL, Lee A (1987) Antibody to Campylobacter pylori in families of index children with gastrointestinal illness due to C. pylori. Lancet ii:681–682
17. Mitchell HM, Lee A, Carrick J (1989) Increased incidence of Campylobacter pylori infection in gastroenterologists: further evidence to support person-to-person transmission of C. pylori. Scand J Gastroenterol 24:396–400
18. Reiff A, Jacobs E, Kist M (1989) Seroepidemiological study of the immune response to Campylobacter pylori in potential risk groups. Eur J Clin Microbiol Infect Dis 8:592–596

Reinfection with *Helicobacter pylori* Due to Intrafamilial Clustering of the Organism

R. Collins, S. Patchett, S. Beattie, C. Keane, and C. O'Morain

Introduction

Helicobacter pylori is now well established as a pathogen in humans, being the major cause of non-autoimmune chronic gastritis [1]. There is growing evidence that it is also an important factor in duodenal ulcer (DU) relapse, with consequent implications for changes in DU therapy. However, its mode of transmission remains unclear. So far, epidemiological studies suggest possible zoonosis, or else person-to-person spread. We have studied 50 patients with endoscopically proven DU and in whom *H. pylori* was grown from antral biopsies taken at the time of diagnosis. Following appropriate treatment, at 8-week follow up all were *H.pylori* negative and their ulcers were healed. One year later, 36% had become reinfected with *H. pylori*. We set out to investigate if person-to-person spread did occur.

Methods

Patients

Eight patients with endoscopically proven DU and antral cultures positive for *H. pylori* were selected as our study group. The control group comprised eight patients with normal findings at gastroscopy and negative for *H. pylori*. The family members of each patient were screened serologically for evidence of *H. pylori* infection.

Materials

A standard enzyme-linked immunosorbent assay (ELISA) technique was used to detect *H. pylori* antibodies. Antigen was prepared from five strains of *H. pylori* cultured in the laboratory and then harvested in distilled water. The cells were disrupted by sonication to produce the antigen which coats the ELISA plates overnight. Each patient's serum was diluted with serum immune buffer and incubated for 1 h. Positive and negative controls were used for each plate.

Department of Gastroenterology, Meath/Adelaide Hospitals, Heytesbury Street, Dublin 8, Ireland

J.M. Pajares et al. (Eds.)
H. pylori and Gastroduodenal Pathology

Anti-human immunoglobulin G (IgG) conjugate was added and then the substrate (*o*-phenylenediamine). The cut-off point for active *H. pylori* infection was an optical density reading at 492 nm of > 1.0. This corresponds with an antibody titre of $\geqslant 1/3200$. We have previously shown this ELISA technique to be highly predictive of active *H. pylori* infection [2].

Results

A total of 37 family members of the *H. pylori*-negative patients and 42 family members of the *H. pylori*-positive patients were screened serologically for evidence of *H. pylori* infection. Three out of 37 (8.1%) in the former group had positive titres (i.e. $\geqslant 1/3200$, whereas 38/42 (90.5%) were positive in the latter group (Fig. 1).

Discussion

There is good evidence in support of person-to-person spread of *H. pylori* from current epidemiological data. It has clearly been shown that the prevalence of *H. pylori* infection increases with increasing age [3]. Similarly, the prevalence is much higher in underdeveloped countries where poor sanitary conditions may facilitate spread by the faecal-oral route [4]. Further evidence supporting transmission by this route is that the prevalence of antibodies to *H. pylori* among the general population is similar to the population profile of hepatitis A antibodies. One study [5] documents higher serum *H. pylori* antibody titres among patients in an institution for the mentally retarded as compared with titres in age-matched blood donor controls. Mentally retarded institutionalised patients are a known high-risk group for person-to-person spread of organisms by the faecal-oral route.

Our results clearly demonstrate intrafamilial clustering of *H. pylori.* By chance, there is a higher proportion of young people among the relatives of the *H. pylori*-positive patients. In Western societies, this is a group one would expect to be *H. pylori* negative. However, as can be seen in Fig. 2, 13 individuals in the 10–20-year-old age group were in fact *H. pylori* positive. This can be explained by intrafamilial clustering of the organism.

There have been several studies [6–8] which show serological clustering of *H. pylori* among families of index patients with endoscopic and microbiological proof of *H. pylori* infection. Our study supports these findings among the Irish population, there being no other ethnic groups involved in the study. We found no significant difference between male and female antibody titres.

In our study, all of the *H. pylori*-positive patients had DU, whereas there was a miscellany of *H. pylori*-associated conditions in the other studies. Peptic ulcer disease is characterised by relapses and remissions, and the strong association with *H. pylori* infection is well recognised [9–14]. Following successful eradication, the relapse rate is only 30% [9] and may be accounted for by intrafamilial

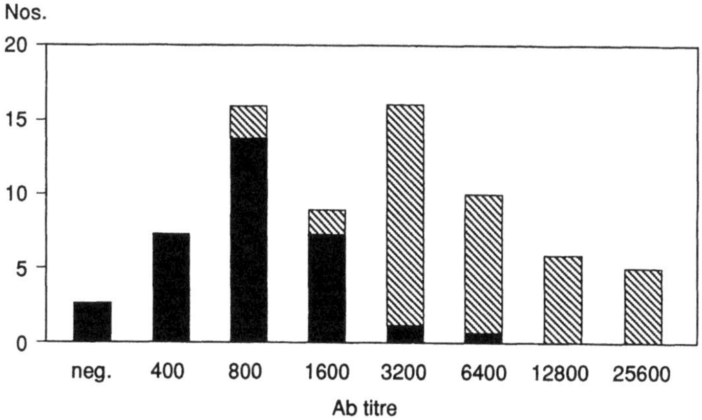

Fig. 1. Results of study to find intrafamilial clustering of *H. pylori*. *Solid columns*, *H. pylori* - negative family members; *hatched columns*, *H. pylori* - positive family members; *Ab*, antibody

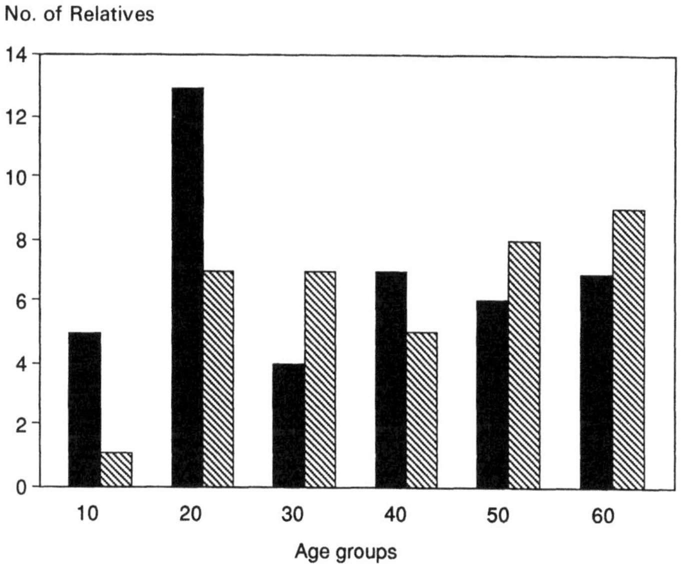

Fig. 2. Age differences. *Solid columns*, *H. pylori* - positive patients; *hatched columns*, *H. pylori* - negative patients

spread of *H. pylori*. Therefore, in cases of recurring DU despite successful healing, the optimal therapeutic approach may be to treat the patients' family contacts, thereby eradicating *H. pylori* infection and reducing the relapse rate of DU.

References

1. Marshall BT, Warren JR (1984). Unidentified curved bacilli in the stomach of patients with gastritis and peptic ulceration. Lancet i: 1311–15
2. Mathai E (1989). Campylobacter pylori: laboratory and clinical studies Thesis, University of Dublin, p 81
3. Jones DM, Eldridge J, Fox AJ, Sethi P, Whorwell PJ (1986). Antibodies to the gastric campylobacter-like organism ('Campylobacter pyloridis') - clinical correlations and distribution in the normal population. J Med Microbiol 22: 57–62
4. Graham DY, Klein PD, Opekun AR, Boutton TW et al. (1988) Epidemiology of Campylobacter pylori infection: ethnic considerations. Scand J Gastroenterol 23 [Suppl 42]: 9–13
5. Berkowicz J, Lee A (1987). Person-to-person transmission of Campylobacter pylori. Lancet ii: 680–681 (letter)
6. Drumm B, Perez-Perez G, Blaser M, Sherman P (1990) Intrafamilial clustering of Helicobacter pylori infection. N Engl J Med 322: 359–363
7. Oderda G, Ansaldi N, Boero M, Ponzetto A, Bellis D (1988). Campylobacter pylori in families of children with relapsing gastroduodenal disease due to C. pylori infection. Am J Gastroenterol 83: 1437–1438 (letter)
8. Mitchell HM, Bohane TD, Berkowicz J, Hazell SL, Lee A (1987). Antibody to Campylobacter pylori in families of index children with gastrointestinal illness due to C. pylori. Lancet ii: 681–682 (letter)
9. Patchett S, Beattie S, Leen E, Keane C, O'Morain C (1990). The role of H. pylori eradication on the natural history of duodenal ulcer disease. Enfermedades Digestivas, p 264 (abstract)
10. Marshall BJ, Warren JR, Blincow ED, Phillips M et al. (1988). Prospective double-blind trial of duodenal ulcer relapse after eradication of Campylobacter pylori. Lancet ii: 1437–1441
11. Coghlan JG, Gilligan D, Humphreys H, McKenna D, Dooley C, Sweeney E, Keane C, O'Morain C (1987). Campylobacter pylori and recurrence of duodenal ulcers-12 months follow up study. Lancet ii: 1109–1111
12. Lambert JR, Borromeo M, Korman MC, Hansky J, Eaves ER (1987) Effect of colloidal bismuth (De-Nol) on healing and relapse of duodenal ulcers - role of Campylobacter pyloridis. Gastroenterology 92: 1489
13. George LL, Borody TJ, Andrews P et al. (1990) Cure of duodenal ulcer after eradication of Helicobacter pylori. Med J Aust 153(3): 145–149
14. Rauws EAJ, Tytgat GNJ (1990). Cure of duodenal ulcer associated with eradication of Helicobacter pylori. Lancet 355 (8700): 1233–1235

Epidemiology: Summary of Workshop and Future Prospects

R.A. Feldman

In November 1990, at the European *Helicobacter pylori* study group workshop in Toledo, Spain, a 2-h epidemiology workshop reviewed three main areas: infection rates, mode of spread and reservoirs; risk factors for infection and the possible association of gastric cancer as a post-infectious complication. There was also discussion of the importance of distinguishing in treatment failure between recrudescence and re-infection.

Transmission of *H. pylori* remains poorly understood. The only available data concerning acute infection come from ingestion by adult volunteers. If humans are the source of infection, we do not know at what stage an individual is most infectious. Transmission may be most common in the first phase of the infection, regardless of the age of the person. Or perhaps children, during an acute infection, may be the most important source of infection for others, while chronic infection of adults may only rarely lead to transmission. Perhaps transmission is most likely to occur with oesophagitis and gastro-oesophageal reflux in a person of any age.

With treatment failure, there are problems distinguishing between recrudescence of an original infection, reinfection with the same organism within the family or infection with a new genotype of *H. pylori* [1, 2].

Initial Colonization

It is unclear whether *H. pylori* reaches the stomach on foods, in water or via person-to-person contact. Whatever the vehicle, *H. pylori* is prevalent worldwide in both males and females equally [3, 4]. Seropositivity to *H. pylori* develops early in developing countries [5]. In Thailand, 17.5% of children have antibodies to *H. pylori* by the age of 9 years [6], and in the Gambia 15% are positive by 20 months and 46% by 60 months [7]. By contrast, only 5% of French children less than 10 years of age are seropositive [5]. In Newcastle less than 1% of children under 5 years are seropositive and less than 6% are antibody positive in the 5–16-year age group [3]. In the United States, antibodies are rarely seen in individuals under 20 years of age [4]. Seroconversion occurs at an earlier age in lower socioeconomic groups of a population [7, 8].

Department of Epidemiology and Medical Statistics, The London Hospital Medical College at QMW, Mice End Road, London, E1 4NS, UK

J.M. Pajares et al. (Eds.)
H. pylori and Gastroduodenal Pathology

In the workshop, further data were presented. Holcombe et al. (see "*Helicobacter pylori* in children") showed a high (40%) seropositivity rate in preschool children in Maiduguri, Nigeria, and late in the plenary session Raedsch et al. showed a high positivity rate in young adults in China. Hook Nikanne et al. studied sera from a group of endoscoped patients and found history of use of alcohol and milk/butter/margarine to be factors associated with an increased rate of *H. pylori* seropositivity.

Spread by Food

Experimental studies have shown that *H. pylori* may survive in chilled foods for several days [9]. Some *H. pylori* isolated from swine have identical genotypes to those found in humans [10]. Abattoir workers are more frequently seropositive than other people working in the same environment but not having direct animal contact [11]. However, vegetarians and populations that abstain from eating either pork or beef show no difference in prevalence of seropositivity [5, 12]. In the workshop, Vaira et al. described further studies concerning an animal reservoir for *H. pylori*. Using a monoclonal antibody to human *H. pylori*, they identified *H. pylori*-like organisms in the stomachs of pigs and rabbits. At present, on balance, a zoonotic reservoir does not appear to explain the major epidemiological characteristics of *H. pylori* infection.

Spread by Water

There is inconclusive evidence concerning the spread of *H. pylori* by water. *H. pylori* has been shown to survive in water [9]. At the workshop, Graham et al. presented a report concerning Peru which showed an increased rate of seropositivity in persons from an urban area who drank from a municipal water supply, when compared with those drinking from private wells [13]. Although this may be an isolated finding, it suggests *H. pylori* reaches water from faecal contamination. Nowottny and Heilmann found no correlation between the source of drinking water and prevalence of *H. pylori* antibodies [14].

Spread by Fomites

Helicobacter pylori have been transmitted from one person to another via contact with endoscopes. Restriction enzyme analysis of bacterial DNA demonstrated that the individuals were infected with identical strains of *H. pylori*. The only relationship of these individuals to each other was that they had been serially endoscoped [15].

Spread from Person to Person?

Indirect evidence suggests that person-to-person spread of *H. pylori* may occur. Institutionalised individuals in psychiatric units and orphanages have increased antibody prevalence compared to the general population [6, 16]. Jones et al. [17], studying adults as index cases, found no increased rate of seropositivity in the patients' spouse. If person-to-person spread is a major mode of transmission, these results suggest that adults are not the major source of infection within the family. An increased rate of *H. pylori* seropositivity has been seen in gastroenterologists [18].

Several studies have demonstrated familial clustering of seropositivity to *H. pylori* when children were the index cases [19, 20]. In the workshop Malaty and Graham described family clustering of *H. pylori*, with the index adult defined by *absence of symptoms* relating to the gastrointestinal tract, absence of antimicrobial use, and a positive breath test and positive serology. When the index adult was positive by both tests, the index spouse was also frequently positive, as were the children, while when the index adult was negative, there were few other positives in the family.

At the workshop Mégraud et al. described a study in Burkina-Faso in which seropositive mothers and seropositive children were compared to seropositive mothers and seronegative children. Premastication of food was more common in the families with both mother and child positive, while water supply, sanitation and poor dental hygiene were not risk factors. De Giacomo et al. (see "*Helicobacter pylori* in Children") reported serologic studies which showed a low rate of seropositivity in young children, and family clustering of positives. Hammermeister et al. showed a higher rate of *H. pylori* seropositivity in submariners than in other military comparison groups.

If person-to-person transmission does occur, is the route faecal-oral or oral-oral? At the workshop, Blaser showed data that were compatible with a faecal-oral spread, with comparable rates of transmission of hepatitis A and other enterically transmitted organisms and the seropositivity data of *H. pylori*. Mitchell et al. [21] failed to demonstrate viable *H. pylori* in the faeces of infected patients.

In a study of 71 individuals with *H. pylori* cultured from the stomach, only one had *H. pylori* isolated from saliva [22]. In this patient, the *H. pylori* isolated both from the stomach and the saliva were indistinguishable by restriction endonuclease analysis [23]. More recently, Majmudar et al. [24] reported isolating *H. pylori* from dental plaque of 100% of 40 healthy volunteers. *H. pylori* was confirmed by smear, rapid urease testing and culture. These observations are unique.

There are now several centres using the polymerase chain reaction (PCR) to study *H. pylori* infection [25, 26]. Gobert (see "Diagnostic Progress") described use of the PCR with identification of *H. pylori* in saliva, gastric juice and biopsy specimens. Saliva was positive, in association with seropositivity, even when the biopsy was negative. Vincent et al. described a study of 20 patients in which three patients had more than one strain of *H. pylori*, when studying isolates from antrum, fundus and gastric juice using restriction enzyme analysis of strains.

Helicobacter pylori has not been isolated from nasopharyngeal, biliary, colonic nor urethral contents [27]. There are no reports suggesting sexual transmission or transmission via blood products.

At the workshop, there were three presentations concerning the relation of prior infection with *H. pylori* and subsequent gastric cancer. Forman et al. presented data on the association of positive serology to *H. pylori* and gastric cancer in China. With a large and geographically broad data set, there was an association of positive serology and gastric cancer. Talley et al. compared the frequency of *H. pylori* seropositivity in persons with gastric adenocarcinoma and other comparison groups, including those without gastrointestinal disease. The relative risk of seropositivity was high for gastric cancer, but also high for colorectal cancer. Loffeld et al. described the potential usefulness of quantitative studies of bacterial load in studying an association with pathologic findings. In addition, in the older patient group, there was an increased frequency of seropositivity in the gastric cancer patients, when compared to their healthy population.

Distinguishing Recolonization from Recrudescence

Typing of *H. pylori* by DNA restriction endonuclease analysis has allowed discrimination between *H. pylori* strains [28, 29]. Although this technique can determine whether re-isolation is due to a new or to the original strain, recrudescence as opposed to re-infection of an identical strain cannot be differentiated.

Because it is difficult to determine the time of onset of *H. pylori* infection, it is difficult to study the dynamics of the spread of infection within a family or a community. *H. pylori* immunoglobulin M (IgM) as a marker of acute infection may not help as IgM persists long after the initial contact with the organism.

Future Areas for Investigation

There are ways to answer questions relating to the source of infection and mode of spread. Following a cohort of family contacts of known cases could allow early identification of new infections and increase understanding of intra-family spread. Study of *H. pylori* infection in nursery school teachers and child minders, populations with particular exposure to children, might indicate whether children are a significant reservoir for this organism. Studies of dentists, dental technicians and their families might indicate if occupational exposure to saliva is associated with increased infection.

Having identified early infection, an assessment of whether *H. pylori* is present in saliva early in infection could be made. It may be that the organism is only present in saliva prior to, or briefly following seroconversion. Francoual et al. [30] have suggested that transmission of *H. pylori* may be associated with gastroesophageal reflux, and perhaps when reflux is present, transmission via saliva is more frequent.

Frequent exposure to *H. pylori* throughout life may explain the pattern that seropositivity increases with age. However, because infection is often chronic, an alternative hypothesis may be that most infections are acquired in childhood and the age-related increase in seroprevalence represents predominantly a cohort effect, with rates of seroprevalence in adults today predominantly the effect of infections occurring many years ago. A cohort study, using sera collected many years ago and sera collected from a similar group today could test this hypothesis.

As treatment regimens for eradication of *H. pylori* become more successful [31, 32], it is important to know if recurrence following successful treatment is the result of a genotypically new infection or recrudescence of the original infection. Molecular techniques now available can identify whether the post-treatment organism is of the same genotype as that of the original infection. These techniques cannot determine if reinfection has occurred with a genotypically identical *H. pylori* to that causing the original infection. Only understanding of how *H. pylori* spreads will solve this problem.

References

1. Bell GD (1991) Anti-Helicobacter pylori therapy: clearance, elimination, or eradication? Lancet 337 i: 310–311
2. Logan RPH, Polson RJ, Baron JH, Misiewicz JJ (1991) Follow-up after anti-Helicobacter pylori treatment. Lancet i: 562–563
3. Thomas JE, Eastham EJ, Elliott TSJ, Dobson CM, Berkeley D (1988) The prevalence of C. pylori infection in childhood and its relation to symptoms. Gut 29: A707
4. Perez-Perez GI, Dworkin BM, Chodos JE, Blaser MJ (1988) Campylobacter pylori antibodies in humans. Ann Intern Med 109: 11–17
5. Megraud F, Brassens-Rabbe MP, Denis F, Belbouri A, Hoa DQ (1989) Seroepidemiology of Campylobacter pylori infection in various populations. J Clin Microbiol 27: 1870–1873
6. Guillermo I, Perez-Perez GI, Taylor DN et al. (1990) Seroprevalence of Helicobacter pylori infections in Thailand. J Infect Dis 161: 1237–1241
7. Sullivan PB, Thomas JE, Wight DGD et al. (1990) Helicobacter pylori in Gambian children with chronic diarrhoea and malnutrition. Arch Dis Child 65: 189–191
8. The Gastrointestinal Physiology Working Group (1990) Helicobacter pylori and gastritis in Peruvian patients: relationship to socioeconomic level, age and sex. Am J Gastroenterol 85: 819–823
9. Park CE, Stankiewicz ZK, Lior H (1987) Survival of Campylobacter pylori (pyloridis) in food and water [abstract no. P-7]. In: Abstracts of the 87th annual meeting of the American Society for Microbiology. American Society for Microbiology, Washington DC p 275
10. Jones DM, Eldrige J (1987) Gastric Campylobacter-like organisms (GCLO) from man ("C. pyloridis") compared with GCLO strains from the pig, baboon and ferret [abstract no. 72]. In: Proceedings of the 4th international workshop on campylobacter infections. Goteborg
11. Vaira D, D'Anastasio C, Holton J et al. (1988) Campylobacter pylori in abattoir workers: is it a zoonosis? Lancet ii: 725–726
12. Hopkins RJ, Russell RG, O'Donnoghue JM, Wasserman SS, Lefkowitz A, Morris JG (1990) Seroprevalence of Helicobacter pylori in Seventh-Day Adventists and other groups in Maryland. Arch Intern Med 150: 2347–2348

13. Klein PD (1990) Helicobacter pylori is a waterbourne disease in Peruvian children. Gastroenterology 98: A69
14. Nowottny U, Heilmann KL (1990) Epidemiologie der Helicobacter pylori Infektion. Leber Magen Darm 20: 183–186
15. Langenberg W, Rauws EAJ, Oudbier JH, Tytgat GNJ (1990) Patient-to-patient transmission of Campylobacter pylori infection by fiberoptic gastroduodenoscopy and biopsy. J Infect Dis 161: 507–511
16. Reiff A, Jacobs E, Kist M (1989) Seroepidemiological study of the immune response to Campylobacter pylori in potential risk groups. Eur J Clin Microbiol Infect Dis 8: 592–596
17. Jones DM, Eldridge J, Whorwell PJ (1987) Antibodies to Campylobacter pyloridis in household contacts of infected patients. Br Med J 294: 615
18. Mitchell HM, Lee A, Carrick J (1989) Increased incidence of Campylobacter pylori in gastroenterologists: further evidence to support person-person transmission of C. pylori. Scand J Gastroenterol 24: 396–400
19. Mitchell HM, Bohane TD, Berkowicz J, Hazell SL, Lee A (1987) Antibody to Campylobacter pylori in families of index children with gastrointestinal illness due to C. pylori. Lancet ii: 681–682
20. Drumm B, Guillermo I, Perez-Perez GI, Blaser MJ, Sherman PM (1990) Intrafamilial clustering of Helicobacter pylori infection. N Engl J Med 322: 359–363
21. Mitchell HM, Lee A, Dick E (1990) Transmission of Helicobacter pylori. A challenge to the dogma of faecal-oral spread. [abstract no. PP957] In: Abstracts of the world congress of gastroenterology, Sydney, Australia
22. Krajden S, Fuksa M, Anderson J et al. (1989) Examination of human stomach biopsies, saliva, and dental plaque for Campylobacter pylori. J Clin Microbiol 27: 1397–1398
23. Shames B, Krajden S, Fuksa M, Babida C, Penner JL (1989) Evidence for the occurrence of the same strain of Campylobacter pylori in the stomach and dental plaque. J Clin Microbiol 27: 2849–2850
24. Majmudar P, Shah SM, Dhunjibhoy KR, Desai HG (1990) Isolation of Helicobacter pylori from dental plaques in healthy volunteers. Indian J Gastroenterol 9: 271–272
25. Gobert B. Labigne A, De Korwin JD, Conroy MC, Bene MC, Faure GC (1990) Polymerase chain reaction for Helicobacter pylori [abstract 03]. In: Third workshop of the European Helicobacter pylori study group. Rev Esp Enferm Digest 78 [Suppl 1]
26. Ho SA, Quirke P, Lewis FA et al. (1990) Helicobacter pylori detection by PCR of the gene encoding 16S ribosomal RNA in fresh and paraffin-embedded material [abstract P6]. In: Third workshop of the European Helicobacter pylori study group. Rev Esp Enferm Digest 78 [Suppl 1]
27. Blaser MJ (1987) Gastric Campylobacter-like organisms, gastritis, and peptic ulcer disease. Gastroenterology 93: 371–383
28. Morotomi M, Hoshina S, Green P et al. (1989) Oligonucleotide probe for detection and identification of Campylobacter pylori. J Clin Microbiol 27: 2652–2655
29. Simor AE, Shames B, Drumm B, Sherman P, Low DE, Penner JL (1990) Typing of Campylobacter pylori by bacterial DNA restriction endonucleases analysis and determination of plasmid profile. J Clin Microbiol 28: 83–86
30. Francoual S, Lamy PH, Le Quintre Y, Luboinski J, Petit JC (1990) Helicobacter pylori: has it a part in the lesions of the gastroesophageal reflux? J Infect Dis 162: 1414–1415
31. Rauws EAJ, Tytgat GNJ (1990) Cure of duodenal ulcer associated with eradication of Helicobacter pylori. Lancet 335: 1233–1235
32. Humphreys HS, Bourke S, Dooley C, McKenna D, Power B, Keane CT, Sweeney EC, O'Morain C (1988) Effect of treatment on Campylobacter pylori in peptic disease: a randomised prospective trial. Gut 29: 279–283

III. Pathophysiology

Superficial Components from *Helicobacter pylori* That Are Adherent to Epithelial Cells

J.L. Fauchère[1], M. Boulot Tolle[2], J.L. Beretti[2], and M.J. Blaser[3]

Introduction

The presence of *Helicobacter pylori* in the human stomach has been highly associated with gastritis and peptic ulcer disease [1–4]. Adherence of *H. pylori* to gastric cells of infected patients has been directly observed and may be important in the maintenance of infection [5–8]. The attachment of *H. pylori* to gastric cells may involve bacterial components able to bind to specific cellular receptors. The cellular receptors may be restricted to the gastric cells explaining the exclusive association of *H. pylori* with those cells, but the specificity of *H. pylori* for the gastric mucosa may also result from the particular micro-environment of the stomach; in that case, the cellular receptors binding *H. pylori* could be present on other human cells. It is thus reasonable to study the colonization factors of *H. pylori* using epithelial cell lines as has been previously described [8–10]. Using microtiter assays, we previously reported [11] a bacterial material easily extracted from the surface of *H. pylori* which included components that specifically adhered to HeLa cell membranes. This superficial adhering material (SAM) binds to a cellular receptor unrelated to *N*-acetyl-neuraminic acid and is thus different from the *N*-acetyl-neuraminyl-lactose-binding hemagglutinin previously described [12]. By gel exclusion chromatography, the adhering components of SAM copurify with urease activity in fractions which included antigens of 60, 52, 30, and 15 kDa. In this work, we further identified the adhering components of SAM and we examined the fate of viable epithelial cells coated with these adhering bacterial components.

Materials and Methods

Strains and Bacterial Procedures

The *H. pylori* strain 88D used in this work was isolated from a human gastric biopsy. It was identified and stored as previously described [11]. Bacterial cultures were grown on blood agar or Muller Hinton plus horse serum, at 37°C

[1] Microbiologie A, CHU, BP 577, 86021 Poitiers, France
[2] Faculté de Médecine Necker-Enfants Malades and Hôpital Boucicaut, Paris, France
[3] Division of Infectious Diseases, Department of Medicine, Vanderbilt University School of Medicine, Nashville, TN, USA

J.M. Pajares et al. (Eds.)
H. pylori and Gastroduodenal Pathology

under microaerobic conditions for 4 days. For viable counts, appropriate dilutions of the bacterial suspensions in 0.15 *M* NaCl were spread on blood agar and after incubation colony forming units (CFU) enumerated.

Extraction of Bacterial Surface-Exposed Material

The surface-exposed material was obtained as previously described [11, 13]. Briefly, bacterial cells of *H. pylori* cultivated for 4 days were suspended in sterile distilled water, and bacterial suspensions were enumerated by viable count; 1 ml of the suspension was saved and sonicated as total bacterial components (sonicate). The remainder of the suspension was vortex-mixed for 1 min, then centrifuged, and the supernatant was designated as the water extract (WE). The glycine extract was obtained by the method of McCoy et al. [14]. The protein contained in each extract was determined by the bicinchoninic acid method. The same number of agar plates were incubated without bacteria and treated under the same conditions to obtain a control extract (CE) devoid of bacterial components.

Assessment of Adherence of Bacterial Components on HeLa Cell Membranes

The microtiter assay for adherence is based on the immunological detection of bacterial material bound onto HeLa cell membranes as previously described [11]. This technique provides results well correlated with the microscopic reference method. Briefly, microtiter plates were coated with HeLa membranes and excess binding sites to the polystyrene were blocked with BSA. After washing of the wells, known quantities of the bacterial component were added, plates were incubated for 1 h at 37°C and then washed three times. The binding material was then quantitated by an enzyme-linked immunosorbent assay (ELISA), using rabbit anti-*H. pylori* sera. For standardization, wells were directly coated with known quantities of the studied bacterial component, and the ELISA was performed as above. The ratio of the optical density (OD) of a given assay well to the OD of the calibration well directly coated with the same quantity of bacterial component assessed the proportion of adhering material and was expressed in micrograms of adhering bacterial component per microgram of original bacterial component. This method made the results independent of both the antigenicity of the bacterial components and the affinity of the sera. In addition, using monospecific sera, it makes possible the assessment of the adherence of any particular constituent from a mixture of bacterial components.

Electrophoretic Analysis of Bacterial Extracts

The protein profiles were performed by sodium dodecyl sulphate–polyacrylamide gel phoresis (SDS-PAGE), with migration gels of 13% and stacking gels of 4% polyacrylamide. The antigenic profiles were determined by the immunoblot procedure. Two-dimensional gel electrophoresis (2DGE) was done by isoelectro-focusing (IEF) in the first dimension and SDS-PAGE in the second dimension. IEF was performed in gels containing 4% of ampholytes with a pH gradient between NaOH (20 m*M*) and H_3PO_4 (6 m*M*). 2 DGE were revealed either by silver staining or by immunoblotting.

Sera

Anti-whole-bacterial-cell serum was obtained from a rabbit immunized with sonicated bacteria. For monospecific rabbit anti-sera, antigens were obtained by preparative SDS-PAGE from 500 μg crude water extracts from the 88D strain. After separation, antigens were electro-eluted and electro-dialyzed against SDS-free buffer. Antigens were given subcutaneously to rabbits with Freund complete adjuvant. After two booster shots, the rabbits were bled 6 weeks after the first injection. Titers of the sera were checked by ELISA, and specificity was checked by immunoblotting SDS-PAGE or 2DGE of homologous WE against each serum. Three sera (A, B, C) to specified antigens were prepared.

Cytotoxic Activity of *H. pylori* SAM

A standard HeLa cell monolayer was incubated at 37°C for 1 h with various quantities of SAM in a 96-well microtiter plate. After elimination of non-adherent material by washing, HeLa cells were incubated 12 h at 37°C in the presence of 20 m*M* urea in a buffered medium, pH 7.4. Cell monolayers were then washed to remove the killed detached cells, and the viable cells still adherent on the polystyrene were assessed by Giemsa staining, elution of the stain in 0.1% SDS, and measurement of the OD 620 of the eluate. A HeLa cell monolayer treated in the same conditions with minimum essential medium (MEM), replacing SAM and urea, represented the 100% cell viability control.

Results

Protein and Antigenic Profiles of the Surface-Exposed Material

The WE and glycine extract (GE) from *H. pylori* were found to contain 20% and 6%, respectively, of the total protein present in the sonicated whole bacterial cells. No difference was observed between protein and antigenic profiles of WE and GE, with major components of approximately 60, 50, 40, 30, 25, and 15 kDa.

Two-dimensional gel electrophoresis of WE were silver stained or were immunoblotted with anti-*H. pylori* monospecific sera. The protein pattern of WE looked complex (Fig. 1) with several antigens exhibiting different pH_i and similar M/W. The antigenic patterns obtained with the three monospecific sera showed the following antigens (Fig. 2). With serum A: antigen 1 (Agl), mass =

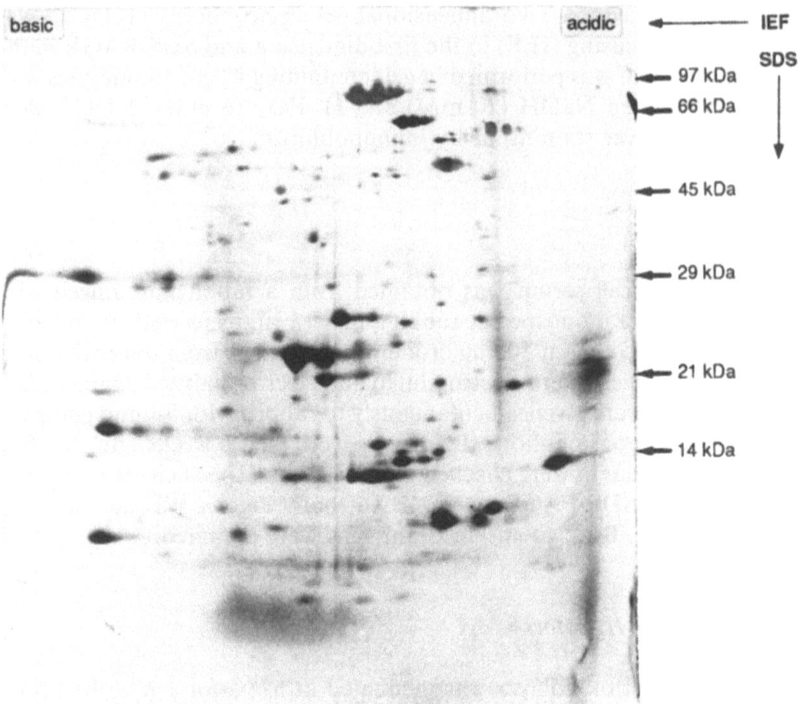

Fig. 1. Protein profile pattern of water extract from *H. pylori* strain 88D. Water extract underwent 2DGE with IEF in the first dimension (pH gradient between NaOH 20 m*M* and H_3PO_4 6 m*M*) and SDS-PAGE in the second dimension (13% polyacrylamide)

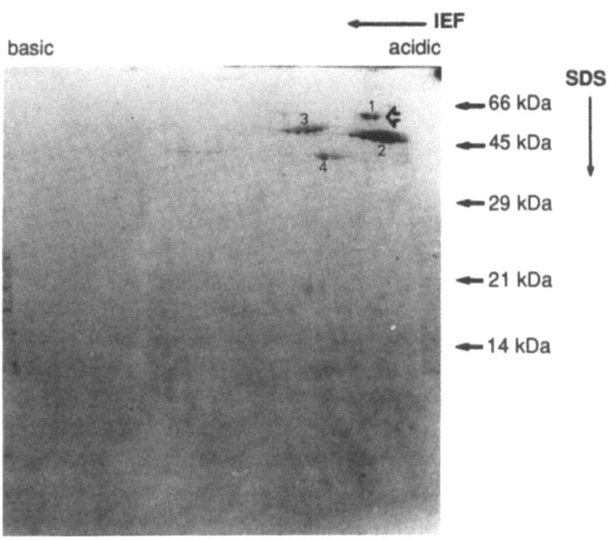

Fig. 2a (caption see p. 69)

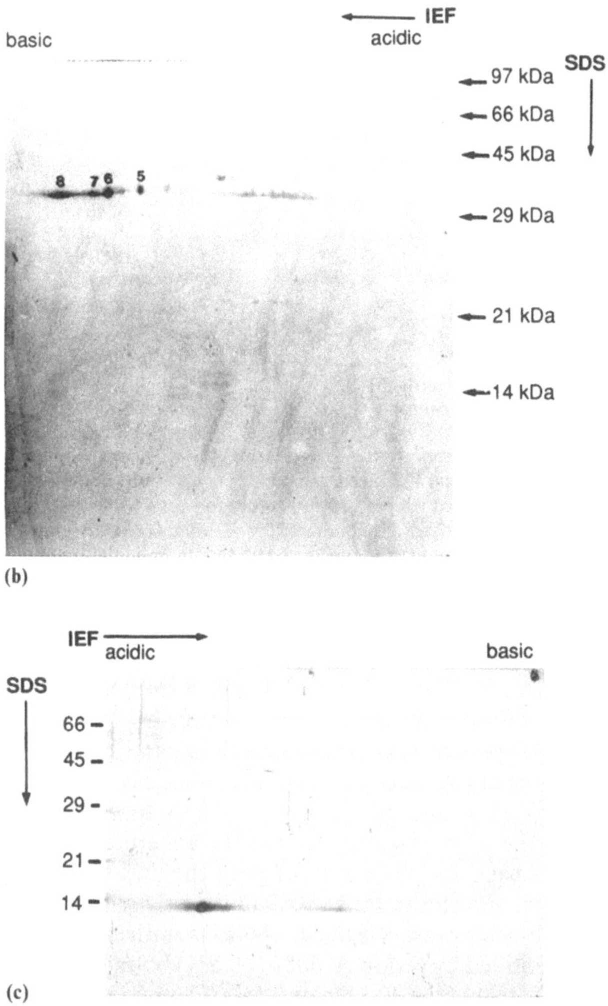

Fig. 2a–c. Antigenic patterns of water extract from strain 88D that underwent 2DGE (see legend to Fig. 1) and immunoblotted with monospecific sera A, B or C (see text)

66 kDa, pH_i = 4.1–4.3; Ag 2, mass = 48 kDa, pH_i = 3.9–4.6; Ag 3, mass = 50 kDa, pH_i = 5.0–5.5; Ag 4, mass = 42 kDa, pH_i = 4.7–5.1. With serum B: Ag 5–8, mass = 31 kDa and pH_i = 9.1, 8.8, 8.4, and 7.8, respectively. With serum C: Ag 9, mass = 15 kDa. The CE contained no detectable protein and no *H. pylori* antigen.

Identification of the Adhering Components of SAM

After gel exclusion chromatography, the fractions exhibiting the highest adherence activity were immunoblotted with an antiserum raised against whole

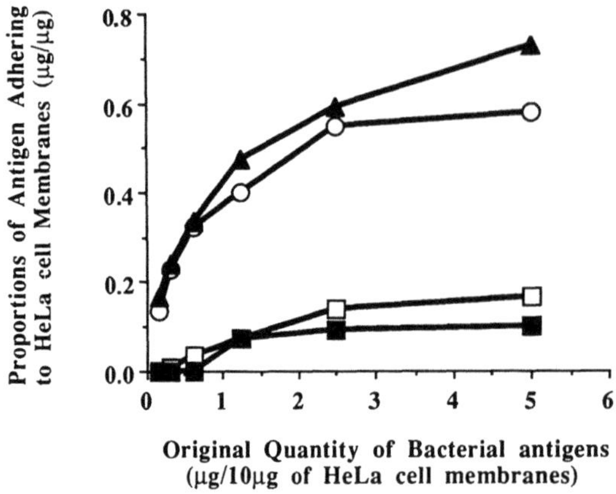

Fig. 3. Proportion of antigens of water extract from strain 88D, adhering to HeLa cell membranes. ELISA adherence assay was performed with 10 µg HeLa cell membranes and various quantities of water extract. The adhering material was revealed with either anti-serum to whole bacterial cell or monospecific sera A, B or C. Serum A recognized antigens between 42 and 66 kDa, serum B recognized antigens of 30 kDa and serum C recognized antigens of 15 kDa. *Triangles*, all antigens; *circles*, 66-kDa antigens; *open squares*, 30-kDa antigens; *solid squares*, 15-kDa antigens

bacterial cells. These fractions included a major antigenic band of 60 kDa and minor bands at about 52, 30, and 15 kDa [11]. These bands were not (or were only slightly) present in the fractions from the nonadherent peaks.

To identify the single subcomponent that is the bacterial ligand binding to the host cell receptor, we used the above A, B, C sera in the adherence ELISA to assess the proportion of a given group of antigens binding to HeLa cell membranes. We assessed both the original quantities and the adhering fractions of each group of antigens with each specific serum or with the anti-whole-cell serum (Fig. 3). Anti-whole-cell serum showed that 80% of the original bacterial material adhered on HeLa cell membranes. Of the 42–66-kDa antigens present in the original material and identified by serum A, 60% (± 5%) bound to HeLa cell membranes versus 12% (± 2%) of the 30-kDa antigens (identified by serum B) and less than 10% for the 15-kDa antigens (identified by serum C). These results suggest that 42–66-kDa antigens include the bacterial ligand binding to the cellular receptor.

Fate of Host Cells Coated with *H. pylori* SAM

We then determined the fate of viable HeLa cells coated with *H. pylori* SAM. In the absence of urea, fewer than 10% of the cells were killed regardless of the quantity of SAM coating the cells. In the presence of urea, the percentage of killed cells increased with the quantity of SAM coating the cells (Fig. 4). This effect was inhibited by aceto-hydroxamic acid (1%), a well-known inhibitor of urease (Fig. 4). In absence of SAM coating the cells, viability of HeLa cells was 88% after 12 h of incubation with urea 100 m*M*. Thus, SAM exhibits a dose-related cytotoxicity related with urease activity, presumably due to released

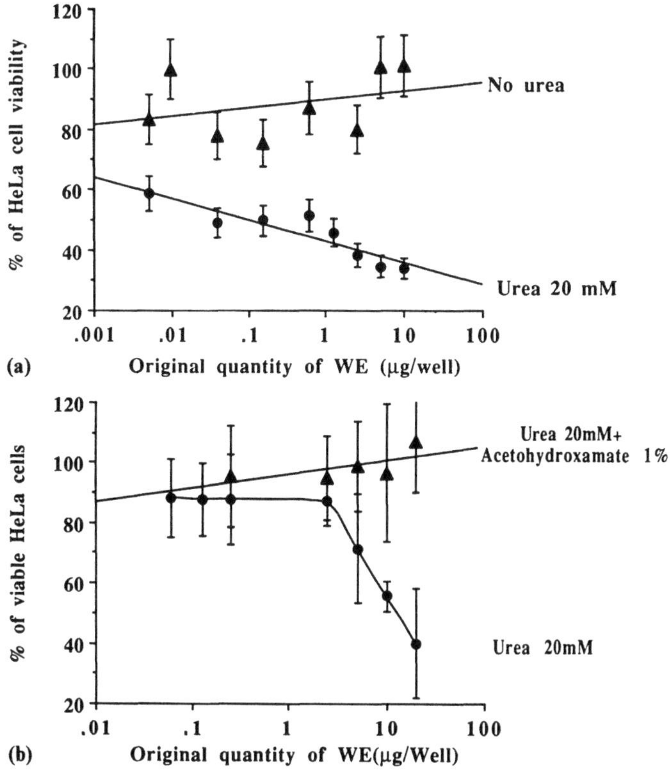

Fig. 4a–b. Cytotoxic activity of superficial material from *H. pylori*, fixed to viable HeLa cells. HeLa cells were coated with SAM, washed, and incubated for 12 h with urea. The killed cells were eliminated, and the remaining viable cells were assessed colorimetrically. The results show a urea-dependant cytotoxicity related to the quantity of SAM coating the cells (**a**). This cytotoxicity was essentially completely inhibited by acetohydroxamic acid, a known urease inhibitor (**b**). In the absence of SAM coating the cells, 88% of HeLa cells retained their viability after 12-h incubation with 100 m*M* urea

ammonia. There was a dose-related cytotoxic effect of ammonium sulfate on HeLa cells with 100% killed in the presence of ammonium sulfate 0.05 m*M*, further suggesting the toxicity of ammonium ions to these cells.

Conclusions

Our data indicate that *H. pylori* exhibits an abundant surface-exposed material, easily extracted and containing both urease and adherence activities. These two activities are present in components of similar molecular weights, including a high-molecular-weight protein complex and three subunits of 42–66, 30, and 15 kDa, respectively. The bacterial ligand which binds to the cellular receptor is associated with the antigens between 42 and 66 kDa. Fixation of SAM to the target cell leads to a urease-dependant cytotoxic effect.

It is therefore possible that the adhering component may be a component complexed with urease in the surface-exposed material. The adhering component may be the urease itself acting as an adhesin while it is attached to the bacterial cell or adhering directly when exported from the cell. Urease that is fixed to gastric cells may be ideally located to efficiently hydrolyze urea extracted from the blood stream. The urease activity concentrated at the membrane level of the host cell may participate in cell damage. Both the previously described hemagglutinins and SAM may be relevent in *H. pylori* colonization of the human gastric mucosa. In favor of this hypothesis is that 75% of colonized patients develop an antibody response against the hemagglutinin and 98% have antibodies against urease [15], demonstrating the close association of these two components with the colonized mucosa.

Acknowledgements. This work was partly supported by University Paris V, INSERM, Phillipe Foundation, and the IPSEN Institut.

References

1. Marshall BJ, McGechie DB, Rogers PA, Glancy RJ (1985) Pyloric Campylobacter infection and gastroduodenal disease. Med J Aust 142:439–444
2. McNulty CAM, Watson DM (1984) Spiral bacteria of the gastric antrum. Lancet 1:1068–1069
3. Rathbone BJ, Wyatt JI, Heatley RV (1986) Campylobacter pylori—a new factor in peptic ulcer disease. Gut 17:635–641
4. Warren JR, Marshall BJ (1983) Unidentified curved bacilli on the gastric epithelium in active chronic gastritis. Lancet 1:1273–1275
5. Hazell SL, Lee A, Brady L, Hennessy W (1986) Campylobacter pyloridis and gastritis: association with intercellular spaces and adaptation to an environment of mucus as important factors in colonization of the gastric epithelium. J Infect Dis 153:658–663
6. Chen XG, Correa P, Offerhaus J, Rodriguez E, Janney F et al. (1986) Ultrastructure of the gastric mucosa harboring Campylobacter-like organisms. Am J Clin Pathol 86:575–582
7. Van Spreeuwel JP, Duursma GC, Meijer CJ, Bax R, Rosekrans PC et al. (1985) Campylobacter colitis: histological, immunohistochemical and ultrastructural findings. Gut 26:945–951
8. Evans DG, Evans DJ Jr, Graham DY (1989) Receptor-mediated adherence of Campylobacter pylori to mouse Y-1 adrenal cell monolayers. Infect Immun 57:2272–2278
9. Neman-Simha V, Megraud F (1988) In vitro model for Campylobacter pylori adherence properties. Infect Immun 56:3329–3333
10. Huang J, Smyth CJ, Kennedy NP, Arbuthnott JP, Napoleon Keeling PW (1988) Haemagglutinating activity of Campylobacter pylori. FEMS Microbiol Lett 56:109–112
11. Fauchere JL, Blaser MJ (1990) Adherence of Helicobacter pylori and its surface components to HeLa cell membranes. Microb Pathog 9:427–439
12. Evans DG, Doyle JE, Jr, Moulds JJ, Graham DY (1988) N-acetylneuraminyl-lactose-binding fibrillar hemagglutinin of Campylobacter pylori: a putative colonization factor antigen. Infect Immun 56:2896–2906
13. Pei, ZH, Ellison PT, Lewis RV, Blaser MJ (1988) Purification and characterization of a family of high molecular weight surface-array proteins from Campylobacter fetus. J Biol Chem 263:6416–6420
14. McCoy EC, Doyle D, Burda K, Corbeil LB, Winter AJ (1975) Superficial antigens of Campylobacter (Vibrio) fetus: characterization of an antiphagocytic component. Infect Immun 11:517–525
15. Evans DJ, Evans EG, Smith KE, Graham DY (1989) Serum antibody responses to the N-acetylneuraminyl-lactose-binding hemagglutinin of Campylobacter pylori. Infect Immun 57:664–667

Bismuth Bioavailability from ^{205}Bi-Labelled Bismuth Compounds Used in Suppression of *Helicobacter pylori* and Treatment of Peptic Ulcer

B. Dresow, R. Fischer, E.E. Gabbe, and H.C. Heinrich

Introduction

Bismuth compounds have been used for over 2 centuries mainly for the treatment of gastrointestinal disorders. Long duration of the therapy periods with high dosages (up to 20 g per day) had occasionally led to serious side effects in the past, including encephalopathies and nephropathies [1, 2].

Attention has once again been drawn to the use of bismuth preparations when these compounds were shown to be effective in the suppression of *Helicobacter pylori* infections [3]. These bacteria were found to be present on gastric mucosa of patients with gastric and/or duodenal ulcers and are discussed as an important factor in the etiology of peptic ulcer disease [4].

Despite the examples of bismuth intoxications described in the past and the widespread use of pharmaceutical bismuth preparations nowadays, there are only a few studies in the literature concerning the absorption and pharmacokinetic data of bismuth from oral dosages in human. Postabsorptive serum bismuth concentrations were measured in patients receiving oral colloidal bismuth subcitrate [5]. Bierer [6] found that, from orally administered bismuth subsalicylate, more than 99% of the ingested bismuth was excreted with the feces and only about 0.003% was excreted in urine during the first 8 days following. administration. Patients receiving colloidal bismuth subcitrate had higher plasma bismuth levels than those receiving basic bismuth salicylate [7]. Further studies, especially with different bismuth preparations in comparison, are not known from the literature.

The aim of the present work was to study the bioavailability of bismuth from various single dose ^{205}Bi-labelled pharmaceutical bismuth preparations used in peptic ulcer therapy in humans.

Material and Methods

Tracer ^{205}Bi-nitrate ($t_{1/2}$ = 15.3 days) was isolated from proton-irradiated ^{206}Pb-enriched lead target as described previously [8].

Abteilung Medizinische Biochemie, Universitätskrankenhaus Eppendorf, Martinistr. 52, 2000 Hamburg 20, FRG

J.M. Pajares et al. (Eds.)
H. pylori and Gastroduodenal Pathology

^{205}Bi-labelled basic bismuth salicylate, basic bismuth gallate, basic bismuth nitrate, and bismuth aluminate were synthesized according to Dresow et al. [9]. ^{205}Bi-colloidal bismuth subcitrate was prepared by exchange labelling of the commercial preparation (Byk Gulden, Konstanz) [9].

The absorption of radiobismuth was studied in five male and one female volunteers (30–60 years) who gave their free and informed consent to their participation in this study. The ^{205}Bi-preparations (200 mg bismuth, 72–370 kBq) were administered together with 200 ml water after an overnight fast. After administration the volunteers fasted for a further 2 h. Urine and feces were collected for 4–20 days following administration. According to ICRP 30 (assuming a bismuth absorption of 5% from various bismuth compounds) the weighted committed dose equivalent was calculated to account for 0.08 and 0.4 mSv for the administered ^{205}Bi activities which is equivalent to 0.06% –0.34% of the natural radiation burden.

Total body radioactivity was measured in a 4π geometry liquid scintillation whole body counter in the γ-energy range of 200–1800 keV [10] and simultaneously by a fixed-position Germanium body detection system (two 93% p-HPGe type systems [11]) (Fig. 1). Retention measurements were performed 20 min after oral application (= 100% reference value) and on days 7, 11, 14, and 21 post oral intake. Whole body count rates were normalized to an aliquot of the administered ^{205}Bi charge inside a body phantom. The lower detection limit for the 4π whole body counter was 40 Bq (measuring time = 300 s), equivalent to a retention value of 0.05%–0.02%. The simultaneous measurement with the HPGe body counter system, detection limit 50 Bq (measuring

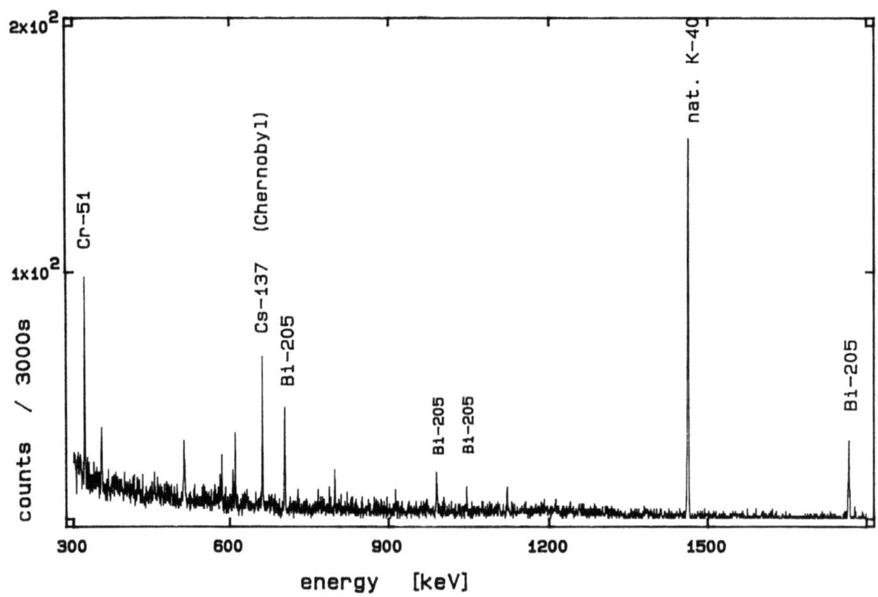

Fig. 1. Total body γ-spectrum (RF) 7 days after oral intake of 200 mg (160 kBq) colloidal bismuth subcitrate and "nonabsorbable" $^{51}CrCL_3$ tracer dose (400 kBq), recorded by two HPGe detectors (E_{rel} = 93%; dE = 2keV)

time = 3000 s), demonstrated that any change in total body activity was caused unambiguously by ^{205}Bi.

Radioactivities of aliquots from the original oral doses and of urine and feces were spectroscopied in a shielded Germanium detector (36% n-HPGe type, dE = 1.9 keV; Canberra GmbH, Frankfurt) showing ^{205}Bi, ^{134}Cs, and ^{137}Cs (from the Chernobyl fallout), and ^{40}K (natural) isotopes. Excreta were measured in a standard 1.2 l Marinelli beaker geometry with a detection limit of 0.4 Bq equivalent to 0.0005%–0.0001% of oral dose for ^{205}Bi and data acquisition times up to 24 h.

Compartment Model Analyses

Experimental urinary excretion data, analyzed by a two- or three-compartment model, resulted in more precise uptake parameters (Table 1) independently of the different collection periods (3–20 days). Originally, this model was developed for similar experiments in rats [9]. The exponential function for accumulated urine data, $U(t)$, is characterized by rate constants, k_i, and compartmental uptake parameters, A_i:

$$U(t) = \sum_i A_i[1 - \exp(-k_i t)];\ i = 1, 2, N \cdots$$

Data were fitted by a nonlinear least square method (Marquardt algorithm). Total uptake was calculated as $\Sigma_i A_i$, assuming urinary excretion as the only elimination pathway for absorbed bismuth. This procedure, performed for basic bismuth salicylate and colloidal bismuth subcitrate in five subjects in intra-individual comparison, is documented in Fig. 2. The almost ten times higher

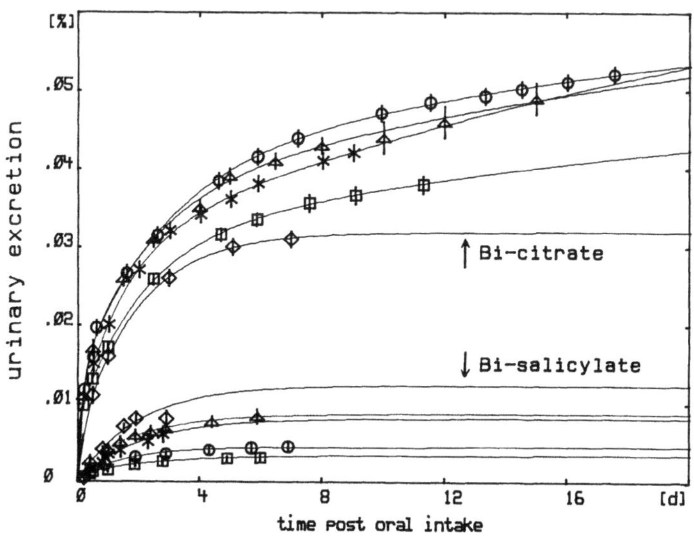

Fig. 2. Fitted individual ^{205}Bi model curves from accumulated urinary excretion after oral intake of colloidal bismuth subcitrate or basic bismuth salicylate (200 mg bismuth, 70-300 kBq ^{205}Bi) in five subjects. Uptake was calculated from asymptotic function values

Table 1. Experimental ^{205}Bi 7-day whole body retention and accumulated urinary excretion from oral ^{205}Bi labelled compounds (200 mg bismuth, 72-370 kBq) as a percentage of dose (mean ± S.D.) and compartmental uptake parameters

^{205}Bi preparation	(n)	Experimental values: Seven-day WBR (%)	Experimental values: Urinary excretion (%)	Two-compartment model: Uptake (%)	Three-compartment model: Uptake (%)	Three-compartment model: A3 (%)
Colloidal bismuth subcitrate	5	0.10 ± 0.12	0.042 ± 0.008	0.043 ± 0.008	0.067 ± 0.009	0.033 ± 0.011
Basic bismuth gallate	3	0.12 ± 0.04	0.038 ± 0.010	0.039 ± 0.010		
Basic bismuth nitrate	3	0.02 ± 0.02	0.004 ± 0.001	0.004 ± 0.001		
Basic bismuth salicylate	5	0.03 ± 0.03	0.005 ± 0.003	0.005 ± 0.002		
Bismuth aluminate	3	< 0.02	0.002 ± 0.001	0.003 ± 0.002		
Compartmental kinetic parameters (days):						
				$T2(k_1)$: 0.12 ± 0.08	0.04 ± 0.03	
				$T2(k_2)$: 1.50 ± 0.42	1.41 ± 0.37	
				$T2(k_3)$: 21[a]		

WBR, whole body retention. [a]Fixed biological half-life of 21 days from [5].

absorption of bismuth from the colloidal subcitrate compared to the basic salicylate is one of the results summarized in Table 1.

The accumulated ^{205}Bi urine excretion data could be described adequately (chi-square $\leqslant 1$) by a two-compartment model with mean biological half-lives of 0.12 and 1.5 days. There were no significant differences in the kinetic parameters among the five different bismuth preparations used.

In addition, a three-compartment model with a long-term biological half-life fixed to 21 days according to Froomes et al. [5] was applied to the urinary excretion data from ^{105}Bi-colloidal bismuth subcitrate ingestion. Neither a significant improvement in fit residuals nor any distinct change in kinetic parameters was evaluated, whereas nearly 50% of bismuth may be eliminated by the third compartment with a rate constant of 0.033 day^{-1}.

Results

The absorption of bismuth from five ^{205}Bi-labelled pharmaceutical bismuth preparations was studied in humans. From single oral doses (200 mg bismuth, 72–370 kBq ^{205}Bi) of all five compounds under study only minimal amounts of bismuth (< 0.1% of dose) were absorbed and excreted with the urine. As judged from the accumulated urinary excretion, significantly higher amounts of radiobismuth were absorbed from the colloidal bismuth subcitrate and the basic bismuth gallate as compared to the bismuth aluminate, basic bismuth salicylate or the basic bismuth nitrate (Table 1).

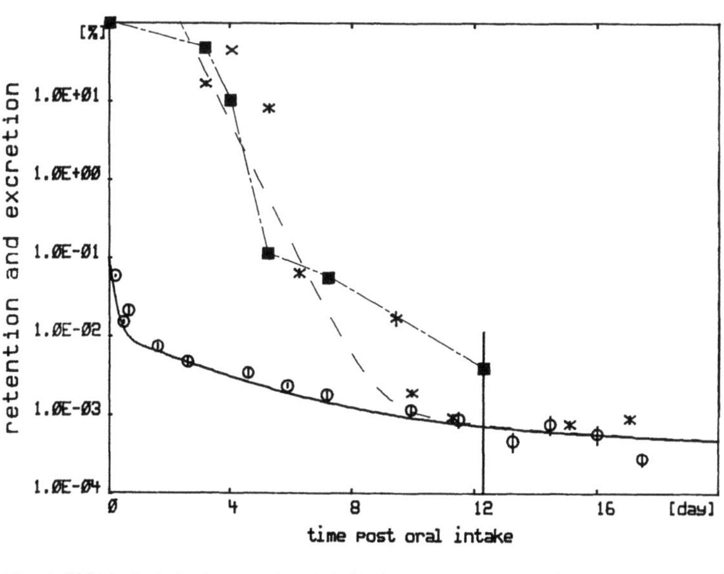

Fig. 3. ^{205}Bi whole body retention (HPGe detector; *squares*), urinary (per day; *circles*) and fecal (per day; *crosses*) excretion rates (percentage of dose) after oral intake of 200 mg colloidal bismuth subcitrate (160 kBq) with three-compartment model fit for 24-h urine loss

The radiobismuth whole body retention measured 7 days after administration indicated that $\leqslant 0.1\%$ of the administered dose of bismuth accumulated in the body. In parallel with the urine data, a fivefold higher whole body radioactivity was observed for the ^{205}Bi-colloidal bismuth subcitrate and the ^{205}Bi basic bismuth gallate compared to the ^{205}Bi-labelled salicylate, nitrate, and aluminate (Table 1). Ten to twelve days after ingestion of a single dose of the labelled compounds, the whole body retention was below the detection limit.

Calculated from the whole body retention and from accumulated urinary excretion, more than 99.9% of the oral bismuth is excreted with the feces 7 days after administration. In order to determine the ratio of bile:urine excretion of absorbed bismuth, additional ^{205}Bi measurements of feces were performed in five cases. From day 12 after ingestion radiobismuth activities in feces (bile excretion) and urine of the same interval were in the ratio of 1 : 1 (Fig. 3).

Discussion

From oral doses of five V^{205} Bi-labelled pharmaceutical bismuth preparations, the radiobismuth absorption and excretion with the urine was found to be extremely low for all compounds tested. Differences in the urinary excretion rate indicate an approximately ten times higher bismuth absorption rate from colloidal bismuth subcitrate and basic bismuth gallate than from the other compounds involved in this study. These observations are in good agreement with recent results [7] that higher plasma bismuth concentrations were measured in patients receiving colloidal bismuth subcitrate (DeNo 1; Gist Brocades, The Netherlands) over a period of 4 weeks than in a group of patients on basic bismuth salicylate for the same period. Also comparable with our results are data published by Bierer [6]. He found that from 4.6 g orally administered bismuth subsalicylate (basic bismuth salicylate) more than 99% of the ingested bismuth was excreted with the feces while only about 0.003% was found in the urine during the first 8 days following administration.

We have shown recently in rats [9] that the intestinal absorption of bismuth from various ^{205}Bi-labelled pharmaceutical bismuth preparations was in general small: Calculated from accumulated ^{205}Bi-urinary excretion and ^{205}Bi whole body retention in rats, a significantly higher uptake was observed from colloidal bismuth subcitrate (0.35% of dose) as compared to gallate (0.11%), salicylate (0.08%), nitrate (0.07%) and aluminate (0.04%).

More than 99.9% of the ingested radiobismuth is excreted with the feces. Monitoring the rate of fecal to urinary ^{205}Bi-excretion, a continuous decay up to 10–12 days following administration is observed, indicating a long intestinal passage time of the nonabsorbed bismuth. Beyond day 12 the ratio of radiobismuth excreted with feces (bile excretion) and urine is 1:1. In studies in rats Vienet et al. [12] and Dresow et al. [9] showed a bile:urine excretion rate of absorbed bismuth of 1:4. A bismuth bile:urine ratio of 1:2 was recently determined in patients directly after 4 days of bismuth subcitrate therapy [13]. Therefore, more realistic bismuth absorption values would be achieved in multiplying the uptake parameters of Table 1 by a factor of 1.5.

The very low whole body radioactivities measured 7 days after administration gave no hint of any accumulation of bismuth in the total body. Owing to the relatively long gastrointestinal passage of the high fraction of nonabsorbed bismuth competing with the short biological half-life of the absorbed ^{205}Bi, the 7-day whole body retention even seem to be too high.

Biological half-lives of 0.12 (fast transit?) and 1.5 days were evaluated from compartment model analyses of urine excretion data. Most biological half-lives of bismuth were determined from elimination in blood and urine after discontinuing multiple dose therapy with colloidal bismuth subcitrate over several weeks. These data are dominated in the elimination as well as in the accumulation phase by long-term components of 10–30 days [5]. A short half-life of 1.4 days was registered in the single dose experiment of Bierer [6].

Model Simulation for Typical Therapy Regimens

Theoretically the pharmacokinetic parameters calculated from ^{205}Bi-urine data (Table 1) enable the prediction of total body content from the continuous bismuth intake rate (multiple dosing). Assuming a constant ratio between bismuth plasma levels and total body bismuth, the mean plasma concentrations in the accumulation phase of the multiple dose experiment of Froomes et al. [5] were reanalyzed with our three-compartment model parameters of Table 1. Compartmental uptake values, A_i, were increased by a factor of 1.5 [13] due to additional bile (feces) excretion. A bismuth plasma concentration of 10 μg/l corresponds to 1 mg total body content (Fig. 4).

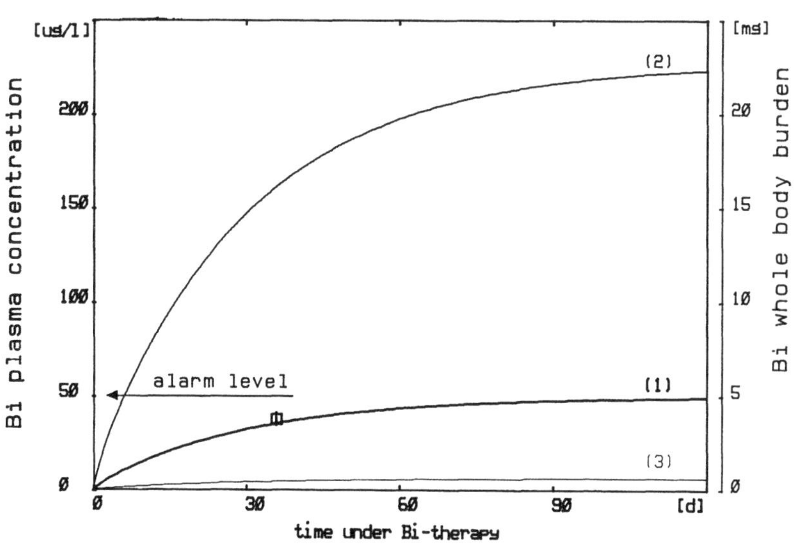

Fig. 4. Model simulation of plasma bismuth levels for present therapy schemes: therapy with 430 mg/day bismuth subcitrate (1), 860 mg/day bismuth subcitrate and increased uptake of 0.23% (2), and with 720 mg/day bismuth salicylate (3). (Model parameters are derived from the experimental data of Froomes et al. [5] and this work)

Curve 1, calculated for a recommended colloidal bismuth subcitrate dose of 430 mg/day, intersects the maximal plasma concentration (33.7 μg/l) of the Froomes experiment at 36 days after the beginning of therapy. Under steady-state conditions beyond 2 months of continuous therapy, the alarm level of 50 μg/l [2] will be easily achieved. Curve 2 was calculated with twice the multiple dose (860 mg/day) and for an increased uptake value of 0.23%. Plasma levels of 880 μg/l, as recently reported by Playford et al. [14] will not be achieved under normal circumstances. Curve 3, calculated for basic bismuth salicylate and a therapy dose of 720 mg/day, will probably remain far below the alarm level.

Conclusion

An evaluation of bismuth absorption from oral ^{205}Bi-labelled compounds was only possible from urinary excretion data. The accumulated radiobismuth urine excretion determined at least the lower limit of bismuth uptake. Part of the absorbed bismuth could have been excreted with the feces (bile excretion). Following a single oral dose, an accumulation of bismuth in the body was not found as demonstrated by whole body retention measurements.

From the five bismuth compounds compared in this study, considerable bismuth amounts were absorbed only from the colloidal bismuth subcitrate and the basic bismuth gallate. The relatively higher bioavailability of bismuth from these salts may be one explanation for the reported bismuth intoxications. In view of higher dosages and/or longer therapy intervals, e.g., in the treatment of *H. pylori* infections, the differences in absorption among the different bismuth compounds should be kept in mind.

Acknowledgement. This research was sponsored by the Kernforschungszentrum Karlsruhe GmbH.

References

1. Buge A, Rancurel G, Poisson M, Dechy H (1974) Encephalopathies myocloniques par les sels de bismuth. Nouv Presse Med 3:2300–2315
2. Hillemand P, Palliere M, Laquais B, Bouvet P (1977) Traitement bismuthique et bismuthemie. Semin Paris 53:1663–1669
3. McNulty CAM, Dent J, Wise R (1985) Susceptibility of clinical isolates of campylobacter pyloridis to 11 antimicrobial agents. Antimicrob Agents Chemother 28:837–841
4. Marshall BJ, Mcgechie DB, Rogers PA, Glancy RJ (1985) Pyloric campylobacter infection and gastroduodenal disease. Med J Aust 142:439–444
5. Froomes PRA, Wan AT, Keech AC, McNeil JJ, McLean AJ (1989) Absorption and elimination of bismuth from oral doses of tripotassium dicitrato bismuthate. Eur J Clin Pharmacacol 37:533–536
6. Bierer DW (1990) Bismuth subsalycilate: history, chemistry, and safety. Rev Infect Dis 12 [Suppl 1]:3–8
7. Raedsch R, Walter-Sack I, Weber E, Blessing J (1990) Pharmakokinetik von Wismut-Präraten bei Patienten mit Gastritis und Ulkuskrankheiten Klin Wochenschr 68:488

8. Fischer R, Dresow B, Wendel J, Bechthold V, Heinrich HC (1991) Bi-205/Bi-206 cyclotron production from Pb-isotopes for absorption studies in man. Int J Appl Radiat Isot (submitted)
9. Dresow B, Nielson P, Fischer R, Wendel J, Gabbe EE, Heinrich HC (1991) Bioavailability of bismuth from ^{205}Bi-labelled pharmaceutical oral Bi-preparations in rats. Arch Toxicol (65: 646–650)
10. Heinrich HC, Gabbe EE, Whang DH (1965) Empfindlichkeits- und Gütekenngrößen des Hamburger 4π-Großraum-Radioaktivitätsdetektors mit flüssigem organischen Szintillator. Atompraxis 11:430–439
11. Fischer R, Heinrich HC (1990) Identifizierung und Quantifizierung inkorporierter Radionuklide im menschlichen Körper. BUNR-Schlußbericht St Sch 1024
12. Vienet R, Bouvet P, Istin M (1983) Cinétique et distribution du ^{206}Bi chez le rat et le lapin: un modèle. Int J Appl Radiat Isot 34:747–753
13. McLean AJ, Islam S, Lambert JR (1990) Anomalous short plasma elimination half life in a patient intoxicated with bismuth subcitrate. Gut 31:1086
14. Playford RJ, Matthews CH, Campbell MJ, Delves HT, Hla KK, Hodgon HJF, Calam J (1990) Bismuth induced encephalopathy caused by tripotassium dicitrato bismuthate in a patient with chronic renal failure. Gut 31:359–360

Optical and Electronic Findings in *Helicobacter pylori* Infection of Antral Mucosa

J.D. Delgado Bellido, F. Rivera Hueto, J.J. Rios Martin,
and J.M. Herrerias Gutierrez

Introduction

Helicobacter pylori (HP) infection is etiologically related to type B chronic gastritis, which is the underlying condition in most cases of peptic ulcer, especially duodenal ulcer (DU). Chronic gastritis affects the antral mucosa principally, with focal involvement of the fundus, showing a close association between the presence of HP colonization and the increase of polymorphonuclear leukocytes (PMN) [1].

Histopathological studies, specially those made with electron microscopy (EM), can contribute to elucidate the pathogenic mechanisms and the spatial distribution of the HP with relation to the gastroduodenal mucosa [2–9]. The HP is often associated with the intercellular junctions of gastric and duodenal epithelial cells [6, 10]. The localization of HP, close to intercellular junctions, is thought to be due to the presence of adequate matters for growth, such as hemin and urea [10]. The HP induces several ultrastructural alterations: cell necrosis, phagolysosomes (vacuolization), alterations of the intercellular complexes, luminal bulging of mucosal cells, edema, degeneration of microvilli, depletion of mucous granules, and others [3–5, 7, 9]. Some histochemical studies demonstrated an increase of glycoproteins rich in sialic acid inside the apical cytoplasm of the gastric mucosal cells, related to the HP attachment [7].

The EM studies show the presence of attachment structures, called pedestals, similar to those observed in the enteropathogenic *Escherichia coli* [11]. The adherence of different HP strains, studied in vitro by means of immunofluorescence and both scanning and transmission EM, shows a close association between bacteria and two out of the four cell lines studied. Cup-like attachment structures, different from those of pedestals, have also been observed [12].

The attachment to mucosal surfaces is the initial event in the pathogenesis of most infectious diseases caused by bacteria in animals and humans. Bacterial adherence is a selective phenomenon mediated by specific adhesive molecules on the bacteria surface, called adhesins, as well as those on host cell membranes, or receptors [13]. The different appearances of bacterial adhesins have been classified into three groups: fimbriae or pili, fibrillae, and afimbriae or nonfimbrial. The latter is beyond resolution of EM, and the HP adhesin is supposed to have a nonfimbrial structure with specific physical and biochemical properties: antigenic, heat sensitive, destroyed by pronase and papain, but resistant to

C/. LEVIES 7, 41004 Sevilla, Spain

J.M. Pajares et al. (Eds.)
H. pylori and Gastroduodenal Pathology

pepsin and trypsin [14]. Besides, it has been demonstrated a sialoconjugate structure, both in the adhesin (*N*-acetylneuraminyllactose) [14], and in the receptor by means of specific hemagglutination-inhibition assays [14, 15].

The HP is a vigorous urease producer, and it is precisely this property that is being used for its detection [16]. There is a good correlation between the results of the urea test and the histopathological findings [17]. Moreover, an inverse relationship between the time for a positive reaction and the density of HP colonization has been suggested [18].

Our study is an attempt to observe the mucosal changes both at optical and ultrastructural levels, in selected antral biopsy samples, and to examine the spatial relationship between the bacteria and gastric mucosal cells.

Material and Methods

We performed a histopathological study of antral biopsies, taken within 5 cm from the pylorus, in 32 successive patients. At least two samples from each patient were taken at the beginning of gastroscopy.

The first sample was used in a rapid urea test, commercially outlined for HP detection (CLOtest, Delta West Ltd., Western Australia). The biopsy specimen was inserted into the CLOtest at room temperature and evaluated after 5 and 20 min: the presence of HP was estimated to be positive if a red or pink coloration appeared surrounding the biopsy specimen. In case of a negative test, a further re-evaluation was made after 3 h. The positive tests from 1–3 h were considered slightly positive because of scanty colonization.

The second biopsy specimen, embedded in formaldehyde, was processed for optic microscopy (OM). Two different stains were used: hematoxylin and eosin (HE) and Warthin-Starry (WS).

In order to detect profuse colonized samples of antral mucosa for the ultrastructural study, a third antral sample was obtained from a selected group of patients. The rules outlined for this selection are: (a) endoscopic diagnosis of active DU; and (b) positive CLOtest within the first 5 min.

The selected samples were fixed for 4 h into a 2.5% glutaraldehyde solution, buffered with a 0.1-*M* sodium cacodylate solution at pH 7.2. The fragments were dehydrated with ethanol at increasing concentrations (30%, 50%, 70%, 90%, 100 × 2), 15 min in each step, and later on into oxide of propylene. The inclusion and fulfillment of the blocks into polyethylene capsules was made into Spurr resin, according to the usual practice of an EM laboratory. The sections, randomly chosen, were observed and photographed with a Zeiss EM900 electron microscope.

Results

The CLOtest results were the following: 22 positive, three slightly positive, three and seven negative. With HE stain, bacillary structures were identified in 20 cases (19 CLOtest positive, one CLOtest slightly positive), whereas no bacteria were detected in the 12 remaining cases (CLOtest: three positive, two slightly positive, and seven negative). In nine cases with an endoscopic diagnosis of DU, the CLOtest came out positive before 5 min. These cases were selected for EM study.

The histopathological diagnoses have been grouped into three classes according to the type, degree, and depth of infiltrate. We have used the same terminology and histopathological concepts employed by Barwick [19]. Chronic active gastritis–superficial with PMN infiltrate–was diagnosed in 13 cases; chronic gastritis–superficial without PMN infiltrate–in 11 cases; chronic atrophic gastritis–implication of glandular region with or without acute infiltrate–in one case; no changes in six cases; and carcinoma in one case.

A good correlation of the CLOtest results with both histopathological findings and HE stain was observed ($p < 0.0001$) (Figs. 1, 2). No differences were found in the detection of HP with WS or HE if a suitable objective (of oil immersion) and enlargement (× 1000) were used.

Bacillary structures of curved appearance were identified with OM and EM. The optical changes observed in the HP antral infection were: a raised PMN infiltrate in relation to HP presence; HP accumulation either close or adhered to mucous membrane; an intercellular junction predilection; presence of epithelial "microerosions"; transformation from a cylindrical to a cuboidal form of the infected epithelial cells. Focal areas of intestinal metaplasia were detected, but none of them presented HP colonization.

The ultrastructural changes observed in the nine cases selected for EM study could be grouped into two stages (Table 1) according to the spatial distribution between the HP and the glycocalyxes of mucosal cells. These stages have different changes and HP density, and could be simultaneous in neighboring areas. Our observations show an inverse relationship between the degree of

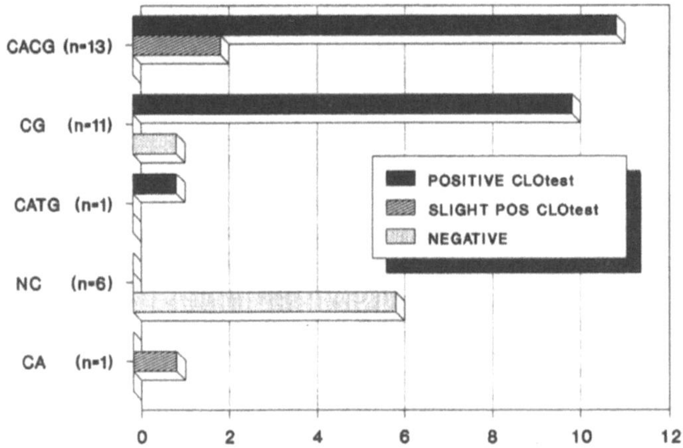

Fig. 1. CLOtest–histopathology correlation ($n = 32$) ($p < 0.001$). *Solid columns*, positive CLOtest; *hatched columns*, slightly positive CLOtest; *dotted columns*, negative CLOtest; *CACG*, chronic active gastritis; *CG*, chronic gastritis; *CATG*, chronic atrophic gastritis; *NC*, no change; *CA*, carcinoma

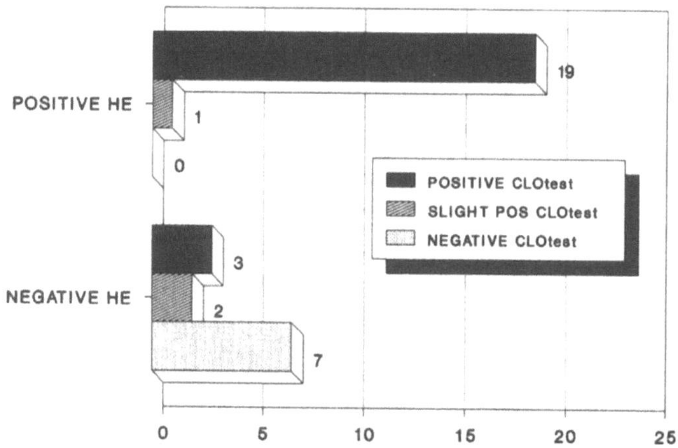

Fig. 2. CLOtest–hematoxylin and eosin stain (*HE*) correlation ($n = 32$) ($p < 0.001$). *Solid columns*, positive CLOtest; *hatched columns*, slightly positive CLOtest; *dotted columns, negative CLOtest*

Table 1. Stages of *Helicobacter pylori* attachment

	Mucosal Cell		
Stage	Glycocalix	Cytoplasm	*H pylori*
Approximation	Preserved microvilli	Normal vacuoles	Numerous Intact Free into mucus
Adhesion			
Early phase	Altered microvilli Bulging	Normal vacuoles	Afimbriae network Blurred HP edge Predilection to intercellular junctions
Later phase	Pedestals Cup-like Microvilli loss	Altered electronic density of vacuoles	Bacterial destruction

damage to the glycocalyx and the germ density. The first stage, or approximation, with a high HP density, presents undamaged bacteria, a preserved glycocalyx and microvilli system, together with normal mucous vacuoles (Fig. 3). The second stage, or adhesion proper, has two phases: the early phase characterized by an afimbrial network between the bacteria and microvilli. In this phase, the glycocalyx could be altered, or slightly distorted, and presents luminal bulging (Fig. 4). Bacteria are concentrated close to intercellular junctions, and some of them which are nearer to the glycocalyx could show blurred those edges (Fig. 5).

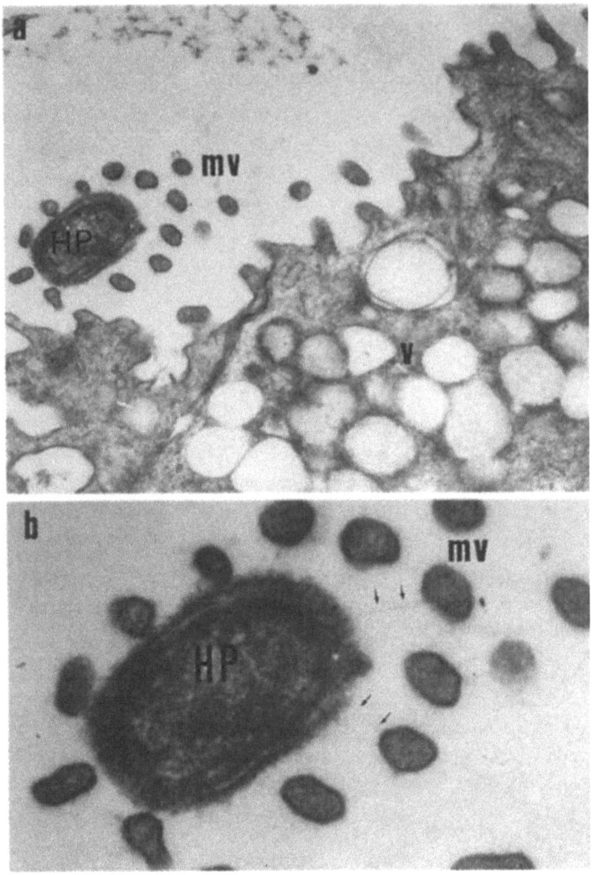

Fig. 3. a Microvilli (mv) system surrounding HP, with preserved bacterial edges. ×12000. **b** Enlargement of previous **a** with a thin network extended between HP capsule and microvilli (*arrows*)

In the late phase of adhesion, it is possible to observe a low density of bacteria, in most cases attached, with "mature" attachment implements: pedestals or cup-like structure. There is a complete loss of the microvilli system and an abnormal electronic density of mucous apical vacuoles. Some cells take on a metaplastic appearance. At this stage, it is common to observe bacterial destruction.

Discussion

The urease tests are suitable tools for a quick detection of HP, with high sensitivity and specificity [17, 18]. These tests could be used during the endoscopic procedure, allowing a rapid diagnosis of antral infection. Also, a

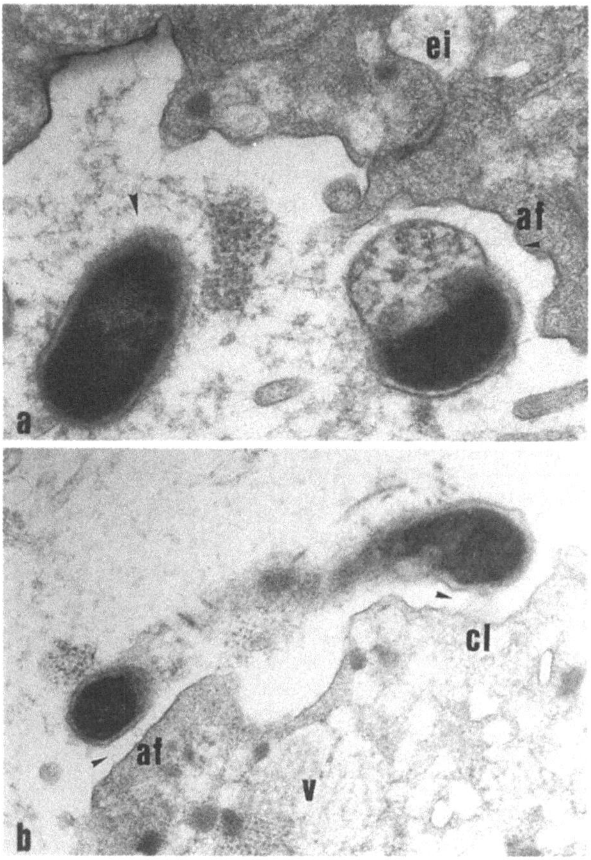

Fig. 4. a Close to intercellular junction (*ei*), early phase of adhesion, with bacterial edge slightly blurred (*arrow*) ; (af), **b** a fimbrial network. Concurrence of adhesion structures: afimbriae (af) and cup-like (cl) structures. (arrows)

good correlation between the positive CLOtest and histopathological findings of chronic gastritis has been observed in this study. Besides, the inverse relationship between time and density suggested by other authors [18] is confirmed, provided there is a good selection of samples for EM.

The OM observations suggest morphological transformations in mucous epithelial cells of the antrum in the context of type B of chronic gastritis. Our OM study supports the association of the HP infection with the well-known findings of type B chronic gastritis [1, 19].

In our opinion, the most interesting aspect of EM study is the attachment process and the ultrastructural alterations induced by HP into glycocalyx. The concept of stages in relation to bacterial attachment to epithelial surfaces has been used to define the spatial distribution of other bacteria such as *E. coli* [20] and the HP itself [9] to glycocalyx. The HP attachment to the glycocalyx of mucosal antral cells takes place after the complete loss of the microvilli system. First, the bacterial afimbriae are in contact with microvilli. Then cup-like and

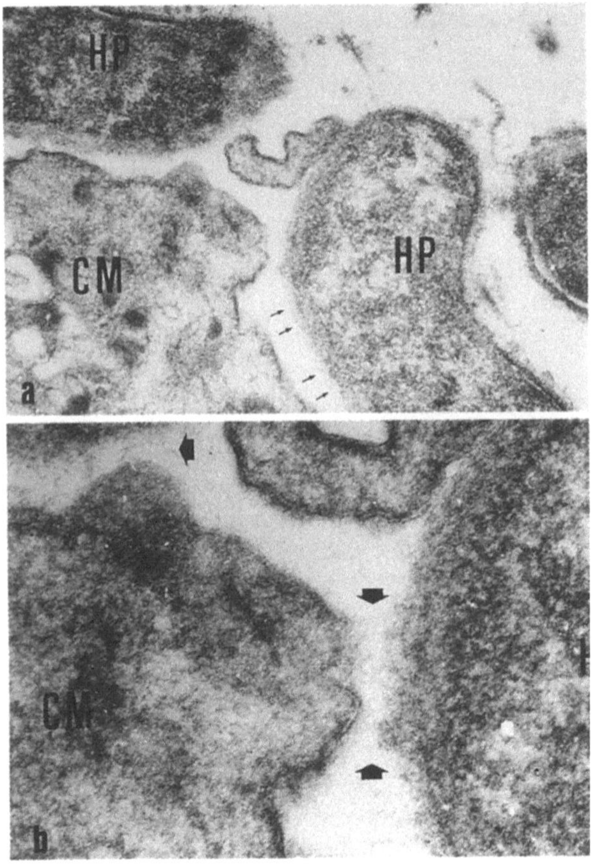

Fig. 5. a Glycocalyx slimming, microprotrusions, and blurring of the bacterial edge nearest to glycocalyx. **b** Enlargement of the closest areas. *CM*,

pedestal-like extrusions of mucosal cell glycocalyxes appear. These changes have been observed in most of the Gram-negative bacteria [13] and are attributed to the interaction between polysaccharides of the microvillar glycocalyx and antigenic saccharides of the bacteria [13, 14, 20]. The increase of sialic residues, detected with *Limax flavus* agglutinin in the presence of attached HP [7], is consistent with the proposed biochemical nature of both fibrillar hemagglutinin of HP [14] and its mucosal receptors [14, 15]. Moreover, although the sialic acid is not a saccharide abounding in normal gastric mucins, in which the fucose predominates, it is prevalent in intestinal metaplasia areas. To this effect, the metaplastic appearance taken on by antral mucosa in the late phase of bacterial attachment, and observed in this study, might have a histochemical correspondence.

Our belief is that an accurate knowledge of the ways of HP attachment could contribute to the development of new medicaments against bacterial and/or glycocalyx surface components involved in the adhesion.

References

1. Graham DY (1989) Campylobacter pylori and peptic ulcer disease. Gastroenterology 96:615–625
2. Price AB, Levi J, Dolby JM, Dunscombe PL, Smith A, Clark J, Stephenson ML (1985) Campylobacter pyloridis in peptic ulcer disease: microbiology, pathology, and scanning electron microscopy. Gut 26:1183–1188
3. Tricottet V, Bruneval P, Vire O et al. (1986) Campylobacter-like organisms and surface epithelium abnormalities in active, chronic gastritis in humans: an ultrastructural study. Ultrastruct Pathol 10:113–122
4. Chen XG, Correa P, Offerhaus J et al. (1986) Ultrastructure of the gastric mucosa harboring Campylobacter-like organisms. Am J Clin Pathol 86:575–582
5. Fiocca R, Villani L, Turpini F, Turpini R, Solcia E (1987) High incidence of Campylobacter-like organisms in endoscopic biopsies from patients with gastritis, with or without peptic ulcer. Digestion 38:234–244
6. Bode G, Malfertheiner P, Ditschuneit H (1987) Invasion of Campylobacter-like organism in the duodenal mucosa in patients with active duodenal ulcer. Klin Wochenschr 63:144–146
7. Bode G, Malfertheiner P, Ditschuneit H (1988) Pathogenic implications of ultrastructural findings in Campylobacter pylori related to gastroduodenal disease. Scand J Gastroenterol 23 [Suppl 142] :25–39
8. Caselli M, Bovolenta MR, Aleotti A, Trevisani L, Stabellini G, Ricci N (1988) Epithelial morphology of duodenal bulb and Campylobacter-like organisms. J Submicrosc Cytol Pathol 20:237–242
9. Caselli M, Trevisani L, Pazzi P, Putinati S, Stabellini G (1988) Campylobacter pylori et muqueuse gastrique. Gastroenterol Clin Biol 12:586–587
10. Hazell SL, Lee A, Brady L, Hennessy W (1986) Campylobacter pyloridis and gastritis: association with intercellular spaces and adaptation to an environment of mucus as important factors in colonization of the gastric epithelium. J Infect Dis 153:658–663
11. Goodwin CS, Armstrong JA, Marshall BJ (1986) Campylobacter pyloridis, gastritis, and peptic ulceration. J Clin Pathol 39:353–365
12. Neman-Simha V, Megraud F (1988) In vitro model for Campylobacter pylori adherence properties. Infect Immun 56:3329–3333
13. Beachey EH (1981) Bacterial adherence: adhesin-receptor interations mediating the attachment of bacteria to mucosal surfaces. J Infect Dis 143:325–345
14. Evans DG, Evans DJ, Moulds JJ, Graham DY (1988) N-acetylneuraminyllactose-binding fibrillar hemagglutinin of Campylobacter pylori: a putative colonization factor antigen. Infect Immun, 56:2896–2906
15. Ligwood CA, Pellizzari A, Law H, Sherman P, Drumm B (1989) Gastric glycerolipid as a receptor for Campylobacter pylori. Lancet 2:238–239
16. Langenberg ML, Tygat GN, Schipper ME, Rietra PJ, Zanen HC (1984) Campylobacter-like organisms in the stomach of patients and healthy individuals. Lancet 1:1348–1349
17. Vaira D, Holton J, Cairns SR et al. (1988) Four rapid urease test (RUT) for the detection of Campylobacter pylori (CP) : is it reliable enough to start therapy? J Clin Pathol 41:355–356
18. Borromeo M, Lambert JR, Pinkard KJ (1987) Evaluation of "CLO-test" to detect Campylobacter pyloridis in gastric mucosa. J Clin Pathol 40:462–468
19. Barwick KW (1988) Chronic gastritis. The pathologist's role. Pathol Annu 23:223–251
20. Cantey JR, Lushbaugh WB, Inman LR (1981) Attachment of bacteria to intestinal epithelial cells in diarrhea caused by Escherichia coli strain 'RDEC-1 in the rabbit: stages and role of capsule. J Infect Dis 143:219–230

Bacteria and Gastrin Release

J. Calam[1], R.A. Goodlad[2], C.Y. Lee[2], B. Ratcliffe[3], M.E. Coates[4], G.W.H. Stamp[2], and N.A. Wright[2]

Introduction

Expansion of Scientific knowledge and techniques has put pressure on researchers to narrow their interest to quite specific topics. Apart from the clinical benefits that it is bringing, work on *Helicobacter pylori* is refreshing because it forces us to think laterally and consider interactions between changes observed in different disciplines such as biochemistry, physiology, immunology and bacteriology.

H. pylori and Gastrin Release

We considered that *H. pylori* might increase gastrin release by making the antral micro-environment alkaline through urease-derived ammonia [23]. Our results, and those of others [15, 24, 27], showed that *H. pylori* does indeed increase gastrin release. It has also been confirmed that the antral mucous layer is more alkaline when *H. pylori* is present [18], although it is not clear how much the rather small difference observed, pH 7.0 versus pH 6.4, would affect gastrin release after 2–3 and 24 h, respectively. Inhibition of urease by acetohydroxamic amic acid [13] or De-Nol plus antibiotics [8] had no significant effect on gastrin release after 2–3 and 24 h, respectively. Inhibition of urease by acetohydroxamic acid was not complete, but their treatment with De-Nol plus antibiotics did more or less abolish urease activity as measured by the ^{13}C-urea breath test. This treatment also diminished the number of acute, but not the number of chronic, inflammatory cells in the antral mucosa without affecting gastrin release.

Teichman et al. [35] showed that certain cytokines which are released during local immune responses can release gastrin. Subsequently Wyatt et

[1]Department of Medicine, Royal Postgraduate Medical School, Hammersmith Hospital, Du Cane Road, London W12 ONN, UK
[2]Department of Histopathology, Imperial Cancer Research Fund, Lincoln's Inn Fields, London WC2A 3PN, UK
[3]Department of Chemistry and Life Sciences, The Polytechnic of North London, Holloway Road, London, W7 8DB, UK
[4]Department of Biochemistry, University of Surrey, Guildford, Surrey GU2 5XH, UK

J.M. Pajares et al. (Eds.)
H. pylori and Gastroduodenal Pathology

al. [38] found elevated basal plasma gastrin in a few patients who had antral gastritis without *H. pylori* and suggested that this bacteria increases gastrin release by causing inflammation rather than by changing pH. Crabtree et al. [10] have shown that the release of the cytokine, tumour necrosis factor, is greatly increased by infection with *H. pylori* (see J.E. Crabtree et al., this volume). We previously showed that antral "G-cell hyperfunction" in patients with duodenal ulcer disease is associated with fasting duodeno-gastric reflux which may well have caused antral gastritis [5]. In these patients the antrum seemed to be abnormally sensitive to weak stimulants of gastrin release [9]. In summary, it has been confirmed that *H. pylori* increases release of gastrin, but inhibition of urease for about 24 h does not return plasma gastrin concentrations to normal. The consensus at Toledo was that *H. pylori*-associated gastrin release is likely to be largely driven by local inflammation. However, this may be difficult to prove without using specific inhibitors of the inflammatory mediator(s) thought to be responsible.

H. pylori and Gastric Acid Secretion in Duodenal Ulcer Disease

It is well known that patients with duodenal ulcer disease tend to secrete more acid than normal individuals [1, 19], but is this due to infection with *H. pylori*? We found that maximal acid output in the majority of patients with duodenal ulcer who have a positive biopsy urease test was significantly higher than in the small minority who have a negative urease test [23]. McColl et al. [27] reported a significant elevation of postprandial intragastric pH in patients with duodenal ulcer after eradication of *H. pylori* but, on the basis of further studies, concluded that the bacteria does not significantly affect intra-gastric acidity [28]. In future it may be necessary to consider together the variety of factors which might affect gastric acid secretion in *H. pylori*-associated duodenal ulcer disease. Patients with duodenal ulcer disease have a remarkable lack of atrophic gastritis in the body of the stomach [17]. This might be the cause of their increased numbers of parietal cells [19], notwithstanding any trophic effect of *H. pylori*-associated rises in gastrin [16]. *H. pylori* colonises the body as well as the antrum of the stomach and might have local effects on parietal cell function. Cave and Vargas [7] have described a protein produced by *H. pylori* which inhibits gastric acid secretion, so that in this respect its eradication might tend to increase acid secretion. The bacteria can also produce superficial gastritis in the acid-secreting region of the stomach, which might possibly also affect acid secretion.

Hypergastrinaemia Associated with Achlorhydria

The work described here [6] also considered physiology in relation to bacteria and inflammation, but in relation to the marked rise in gastrin which occurs in the absence of gastric acid secretion. This phenomenon is important because it occasionally leads to gastric carcinoid tumours in patients with gastric atrophy [2, 32] and frequently produces such tumours in rats during pharmacological suppression of acid secretion [4, 22]. It is still not clear how a lack of acid increases gastrin release. There is evidence that a physiological reflex is involved in rapid changes in gastric release which occur, for example, on acidification of the lumen during meal stimulation [37]. However, other responses such as the rise in gastrin on gastric neutralisation in humans do not start for about 5 h [31, 37], which seems somewhat slow for a simple physiological reflex.

Saliva contains about 1000000 organisms per millilitre [12] and about 1 l is swallowed per day. Gastric acid is strongly antiseptic but bacteria proliferate in the stomach if it is neutral. Initially the bacteria found are aerobes from saliva, but later the stomach contains anaerobes and facultative anaerobes which are normally present in the colon. Gastrointestinal bacteria have many effects on epithelial function, including the stimulation of gastrin release [23]. They also stimulate immunological activity [10, 26] (see J.E. Crabtree et al., this volume). We asked whether bacteria contribute to the rise in gastrin which occurs on suppression of acid secretion and tested this by comparing the effect of the H2 receptor antagonist loxtidine [4] on gastrin release in germ-free and conventional rats.

Methods

We studied matched germ-free and conventional Lister hooded rats of the same strain. They received either saline, or loxtidine 10 mg/kg per day or 70 mg/kg per day s.c. by osmotic mini-pumps. It was estimated that these doses would give partial and complete suppression of acid secretion, respectively [4]. Each treatment was given to eight germ-free rats, mean weight 306 g, six male and two female, and eight sex-matched conventional rats with a mean weight of 276 g. The greater weight of the germ-free rats was due to the distended caecum found in such animals. The animals fed and drank water ad lib. After 1 week they were anaesthetised with pentobarbital and killed by exsanguination from the heart. Gastrin was measured in ethylenediaminotetraacetate (EDTA) plasma by radioimmunoassay using antiserum G179 provided by Prof. Bloom [23].

The animals' stomachs were collected for histological examination. Longitudinal sections of the posterior gastric wall were fixed in neutral buffered formalin and processed to paraffin wax. Sections were stained by haematoxylin and eosin, peroxidase, chromotrope 2R and toluidine blue. Examinations were performed blind. Where appropriate, cell populations were assessed formally by examination of the basal zones and inter-glandular compartments. For each specimen 10–22 high-power fields were examined, using a 10 × 10 graticule.

Results

Gastrin

One-way analysis of variance (ANOVAR) showed a significant effect of loxtidine on gastrin in both germ-free ($p = 0.001$) and conventional ($p < 0.001$) rats (Table 1). On two-way ANOVAR the difference in gastrin between germ-free and conventional rats was significant ($p < 0.05$), and there was a significant interaction between bacterial status and dose ($p = 0.02$). This indicated that bacterial status effected the change in gastrin produced by the drug. Further analysis by t tests showed that the difference in gastrin between germ-free and conventional rats was significant only after the high dose of loxtidine ($p < 0.05$). This was because the high dose produced a significant rise in gastrin above low dose ($p < 0.05$) in conventional, but not in germ-free rats.

Histology

There was no evident difference between the groups in mucosal thickness or the numbers of neutrophils, lymphocytes, mononuclear cells or mast cells, but differences in the numbers of eosinophils were noted.

One-way ANOVAR showed no significant effect of loxtidine dose on eosinophil counts in the whole field, at the base of the gland or between glands in germ-free rats. However, in the conventional groups loxtidine significantly increased the number of eosinophils in all the regions scored ($p < 0.001$). Two-way ANOVAR showed a highly significant effect of dose ($p < 0.001$) but no effect of microbial status. There was, however, a significant interaction between bacterial status and dose on eosinophils ($p < 0.05$). A similar effect was seen for eosinophils at the base (effect of drug, $p < 0.001$, interaction $p < 0.02$) and for eosinophils between the glands (effect of drug $p = 0.002$), except that there was no significant interaction term for the latter.

The results of t tests are shown in Table 2. In conventional rats the drug produced a highly significant rise ($p < 0.0001$) in total eosinophil count from 13.1 (control) to 37 ± 4 (high dose) cells per field. Germ-free rats showed a smaller rise from 21 ± 3 to 28 ± 3 cells per field which was not significant

Table 1. Plasma gastrin concentrations in germ-free and conventional rats after control and loxtidine infusions

Drug	Dose (mg/kg per day)	Plasma gastrin concentration (mean ± SEM; pmol/l)		
		Germ-free	Conventional	p
Saline	—	59 ± 11	36 ± 8	NS
Loxtidine	10	153 ± 30	181 ± 27	NS
Loxtidine	70	178 ± 11	278 ± 26	< 0.005

NS, not significant

Table 2. Numbers of eosinophils per microscope field in germ-free and conventional rats after control and loxtidine infusions

Drug	Dose (mg/kg per day)	Eosinophil count (per 40x field; mean ± SEM)		
		Germ-free	Conventional	*p*
Total				
Saline	–	21 ± 3	13 ± 1	< 0.05
Loxtidine	10	31 ± 4	32 ± 5	NS
Loxtidine	70	28 ± 3	37 ± 4	< 0.05
At base				
Saline	–	16 ± 2	11 ± 0.1	< 0.02
Loxtidine	10	31 ± 4	25 ± 4	NS
Loxtidine	70	21 ± 2	31 ± 3	< 0.01
Between crypts				
Saline	–	5 ± 1	3 ± 0.4	< 0.05
Loxtidine	10	9 ± 2	7 ± 1	NS
Loxtidine	70	6 ± 1	7 ± 1	NS

NS, not significant

($p = 0.052$). However, germ-free rats had more eosinophils after control infusions than conventional rats ($p < 0.05$ for total count).

Discussion

The findings show for the first time that the microbial flora contributes to the rise in gastrin which occurs when gastric acid secretion is inhibited [20–22, 25, 33]. However, plasma gastrin concentrations were similar in conventional and germ-free rats in the absence of the drug and during administration of loxtidine at a dose which partially inhibits acid secretion. This suggests that the difference seen during complete suppression of acid secretion is specific, rather than due to a generalised tendency of germ-free rats to release less gastrin or to respond less to loxtidine.

The idea that bacteria might be involved in the rise in gastrin which occurs on suppressing acid is consistent with known temporal aspects of this response including its slow onset in humans [31]. In humans it is necessary to have continuous neutralisation for the rise in gastrin to become fully established. Patients with atrophic gastritis who secrete no acid at all have plasma gastrin levels about 30 times higher than the norm, whereas patients on long-term omeprazole, which keeps the stomach near neutral for all but about 3 h per day, have gastrin levels which are only about four times higher than the norm [20, 21, 25]. These slow effects contrast with other effects of pH on gastrin release, such as that of acidification on basal or meal-stimulated gastrin release, which occur within 30 min [37]. as expected for a simple physiological reflex.

Work on reflexes within the gastric epithelium suggests that somatostatin is involved in a variety of inhibitory effects, including the inhibition of gastrin release by a low intragastric pH [3, 29, 36]. Somatostatin does inhibit gastrin release, and gastrin-releasing cells (G cells) are physically surrounded by the projections of somatostatin-releasing cells. Vascular perfusion of rat stomach with antibodies against somatostatin increased both basal and stimulated gastrin release [36]. Studies have shown that pH-driven changes in gastrin release are accompanied by reciprocal changes in indicators of somatostatin synthesis or release [3]. Thus it seems likely that antral somatostatin cells are involved in the regulation of gastrin release by lumenal pH.

Our results appear to distinguish two ways in which suppression of acid secretion affects gastrin release on the basis of whether bacteria are needed or not. The low dose of loxtidine was chosen to allow some acid secretion. This elevated plasma gastrin levels in germ-free as well as conventional animals, therefore, presumably, by a physiological reflex. This might correspond to the rapid response seen in humans [37]. The additional rise in gastrin produced by the high dose of loxtidine required the presence of bacteria, and bacteria might contribute to the slow response seen in humans. This may explain its slow onset as well as the need for continuous neutralisation. Bacteria do survive during periods when the pH is above 4 during treatment with cimetidine, but overgrowth does not become established because the gastric lumen is sterilised by secretion of acid during the night [30].

Bacteria can affect many aspects of gastro-intestinal function. Some produce specific bio-active factors such as cholera toxin. Others alter gut function by metabolising substances in the lumen, for example, bacterial digestion of fibre in the hind gut affects growth of colonocytes [14]. Bacteria also affect the gut by provoking the release on inflammatory mediators [2, 10, 34, 35]. Histological examination of rats' gastric mucosa in the present study showed a dose-dependent increase in eosinophils in the conventional rats, but not in the germ-free rats, and a significant interaction between the effects of bacterial status and dose on eosinophil count. Conventional rats had significantly more eosinophils than germ-free rats during high-dose, but not during low-dose loxtidine. Eosinophils are known to release the cytokine interleukin I [11]. Two other cytokines, interleukin II and gamma interferon, have been shown to release gastrin from the isolated perfused canine stomach [35]. Therefore it is possible that intra-epithelial eosinophils contributed to the effect of loxtidine on gastrin release in the presence of bacteria. This is consistent with the idea that gastric inflammatory cells modulate gastrin release [35, 38]. We are currently extending this work to examine the possible role of intragastric bacteria in the greatly increased gastrin release seen in some patients with achlorhydria.

Hopefully the bringing together of various scientific disciplines that has been stimulated by work on *H. pylori* will encourage re-examination of the mechanisms underlying a number of related disease states.

Acknowledgements. We thank Viv Ronaasen for skilled technical assistance, Glaxo for providing loxtidine and advice on its use, Prof. Bloom for radioimmunoassay reagents and the editor of Clinical Science for permission to describe work in press [6].

References

1. Blair AJI, Feldman M, Barnett C, Walsh JH, Richardson T (1987) Detailed comparison of basal and food-stimulated gastric acid secretion rates and serum gastrin concentrations in duodenal ulcer patients and normal subjects. J Clin Invest 79:582–587
2. Borch K, Renvall H, Leidberg GV (1985) Gastric endocrine cell hyperplasia and carcinoid tumours in pernicious anaemia. Gastroenterology 88:638–642
3. Brand SJ, Stone D (1988) Reciprocal regulation of antral gastrin and somatostatin gene expression by Omeprazole-induced achlorhydria. J Clin Invest 82:1059–1066
4. Brittain RT, Jack D, Reeves JJ, Stables R (1985) Pharmacological basis for the induction of gastric carcinoid tumours in the rat by loxtidine, an unsurmountable histamine H2-receptor blocking drug. Br J Pharmacol 85:843–847
5. Calam J, Tracy HJ (1980) Pyloric reflux and gastrin cell hyperfunction. Lancet 2:918
6. Calam J, Goodlad RA, Lee CY, Ratcliffe B, Coates ME, Stamp GWH, Wright NA (1991) Achlorhydria, hypergastrinaemia : the role of bacteria. Clin Sci 80:281–284
7. Cave DR, Vargas M (1989) Effect of a Campylobacter pylori protein on acid secretion by parietal cells. Lancet 2:187–189
8. Chiltajallu RS, Dorian CA, McColl KEL (1990) Is Helicobacter pylori-associated hypergastrinaemia due to the bacterium's urease or the antral gastritis? Gut 31:A1175
9. Cooper RG, Dockray GJ, Calam J, Walker RJ (1985) Acid and gastrin responses during intragastric titration in normal subjects and duodenal ulcer patients with G-cell hyperfunction. Gut 26:232–236
10. Crabtree JE, Shallcross TM, Wyatt JI, Heatley RV (1990) Tumour necrosis factor alpha secretion by Helicobacter pylori colonised gastric mucosa. Gut 31:A600–601
11. Del Pozo V, De Andres B, Martin E, Maruri N, Zubeldia JM, Palomino P, Lahoz C (1990) Murine eosinophils and IL-1: aIL-1 mRNA detected by in situ hybridisation. J Immunol 144:3117–3122
12. Drasar BS, Shiner J, McLeod GM (1969) Studies of the flora of the gastrointestinal tract in healthy and achlorhydric persons. Gastroenterology 56:71–79
13. El Nujumi AM, Dorain CA, Chittajallu RS, Neithercut WD, McColl KEL (1991) Effect of inhibition of H. pylori urease activity by acetohydroxamic acid on serum gastrin in duodenal ulcer subjects. Gut 32:866–870
14. Goodlad RA, Ratcliffe BR, Fordham JP, Wright NA (1989) Does dietary fiber stimulate epithelial cell proliferation in germ-free rats? Gut 30:820–825
15. Graham DY, Opekun A, Lew GM, Evans DJJ, Klein PD, Evans DG (1990) Ablation of exaggerated meal-stimulated gastrin release in duodenal ulcer patients after clearance of Helicobacter (Campylobacter) pylori infection. Am J Gastroenterol 85:394–398
16. Johnson LR (1976) The trophic effect of gastrointestinal hormones. Gastroenterology 70:278–283
17. Kekki M, Sipponen P, Siurula M (1984) Progression of antral and body gastritis in active and healed duodenal ulcer and duodenitis. Scand J Gastroenterol 19:382–388
18. Kelly SM, Crampton J, Hunter JO (1990) Helicobacter pylori increases the pH of the gastric mucosa in vivo. Gut 31:A1177–1178
19. Lam Sk (1984) Pathogenesis and pathophysiology of duodenal ulcer. Clin Gastroenterol 13:447–472
20. Lanzon-Miller S, Pounder REMR, Chronos NAF, Ball S, Mercieca JE, Plausson M, Cederberg C (1987) Twenty-four-hour intragastric acidity and plasma gastrin concentration in healthy subjects and patients with duodenal or gastric ulcer, or pernicious anaemia. Aliment Pharmacol Ther 1:225–237
21. Lanzon-Miller S, Pounder RE, Hamilton MR, Ball S, Chronos NAF, Raymond F, Olausson M, Cederberg C (1987) Twenty-four hour intragastric acidity and plasma gastrin concentration before and during treatment with either ranitidine on omeprazole. Aliment Pharmacol Ther 1:239–251
22. Larsson H, Carlsson E, Mattsson H, Lundell L, Sundler F, Sundell G, Wallmark B, Watanabe T, Hakanson R (1986) Plasma gastrin and gastric enterochromaffin like cell activation and proliferation. Studies with omeprazole and ranitidine in intact and antrectomised rats. Gastroenterology 90:391–399

23. Levi S, Beardshall K, Haddad G, Playford R, Ghosh P, Calam J (1989) Campylobacter and duodenal ulcers: the gastrin link. Lancet 1:1167–1168
24. Levi S, Beardshall K, Swift I, Foulks W, Playford R, Ghosh P, Calam J (1989) Antral Campylobacter pylori, hypergastrinaemia, and duodenal ulcers: Effect of eradicating the organism. Br Med J 299:1504–1505
25. Lind T, Cederberg C, Forssell H, Olausson M, Olbe L (1988) Relationship between reduction of gastric acid secretion and plasma gastrin concentration during omeprazole treatment. Scand J Gastroenterol 23:1259
26. Mai UEH, Perez-Perez GI, Wahl LM, Wahl SM, laser MJ, Smith PD (1990) Inflammatory and cytoprotective responses by human monocytes are induced by Helicobacter pylori: possible role in the pathogenesis of type B gastritis (abstract). Gastroenterology 981:A662
27. McColl KE, Fullarton GM, ElNujumi AM, Macdonald AM, Brown IL (1989) Lowered acidity gastrin and gastric acidity after eradication of Campylobacter pylori in duodenal ulcer. Lancet 2:499–500
28. McColl KEL, Fullarton GM, El NuJumi AM, Macdonald AMI, Dahil S, Hilditch TE (1990) Serum gastrin and gastric acid status one and seven months after eradication of Helicobacter pylori in duodenal ulcer patients. Gut 31:A601
29. Moss S, Calam J (1990) Pathophysiology of gastrointestinal hormones. Curr Opin Gastroenterol 6:877–881
30. Muscroft TJ, Burdon DW, Youngs D, Keighley MRB (1981) Cimetidine is unlikely to increase formation of intragastric N-nitroso-compounds in patients taking normal diet. Lancet 1:408–410
31. Peters M, Feldman M, Walsh J, Richardson C, (1983) Effect of gastric alkalinization on serum gastrin concentrations in humans. Gastroenterology 85:35–39
32. Richards AT, Hinder RA, Harrison AC (1987) Gastric carcinoid tumours associated with hypergastrinaemia and pernicious anaemia - regression of tumours by antrectomy. S Afr Med J 72:51–54
33. Ryberg B, Mattsson H, Carlsson E (1988) Effects of Omeprazole and Ranitidine on gastric acid secretion, blood gastrin levels and [3H] thymidine incorporation in the oxyntic mucosa from dogs and rats. Digestion 39:91–99
34. Sjogren RW, Sherman PM, Boedeker EC (1989) Altered intestinal motility precedes diarrhoea during Escherichia coli infection. Am J Physiol 257:G725–G731
35. Teichman RK, Pratschke E, Grab J, Hammer C, Brendel W (1986) Gastrin release by interleukin-2 and gamma-interferon in vitro. Can J Physiol Pharmacol (suppl) 64:62
36. Walsh JH (1987) Gastrointestinal hormones. In: Johnson LR (ed) Physiology of the gastrointestinal tract. Raven, New York, pp 181–254
37. Walsh JH, Grossman M (1975) Medical progress: gastrin. N Engl J Med 292:1324–1332
38. Wyatt JI, Rathbone BJ, Green DM, Primrose J (1989) Raised fasting serum gastrin in chronic gastritis is independent of Campylobacter pylori status and duodenal ulceration. Gut 30:A1483

Helicobacter pylori, Gastritis, Duodenal Ulcer, and Gastric Histamine Content

D.M.M. Queiroz[1], E.N. Mendes[1], G.A. Rocha[1], J.R. Cunha-Melo[2], A.J.A. Barbosa[3], G.F. Lima, Jr[2], and C.A. Oliveira[4]

Introduction

Helicobacter pylori has been observed in the gastric mucosa of up to 100% of duodenal ulcer (DU) patients [2]; however, the role of this microorganism in duodenal ulcer pathogenesis is still poorly understood.

It has recently been proposed that *H. pylori* breaks urea into ammonia which raises the antral surface pH, leading to an increase of gastrin release and therefore gastric acid secretion [3].

It has also been demonstrated that the serum gastrin concentration of adults with DU [6] and of children with upper gastrointestinal disorders [7] decreases after *H. pylori* eradication with antimicrobial agents.

Since histamine has been suggested as the final common mediator for all gastric secretagogues [1, 4, 11], including gastrin, we determined the histamine concentration in the oxyntic mucosa of *H. pylori*-positive patients both with and without Du and of *H. pylori*-negative subjects.

Materials and Methods

This project was approved by the Ethics Committee of the Hospital das Clinicas, Universidade Federal de Minas Gerais, Brazil, and consent was obtained from all the patients studied.

A total of 43 patients who underwent endoscopy for investigation of dyspeptic symptoms were studied. Except for ten (eight men, mean age 46.2 years) *H. pylori*-positive patients whose DU healed after treatment with H2 antagonists (800 mg/day for 6 weeks), no other patient received nonsteroidal anti-inflammatory drugs, H2 receptor antagonists, antibiotics, or any medication for at least 30 days before the study. Eleven patients (eight men, mean age

[1]Laboratory of Research in Bacteriology, Faculdade de Medicina, Universidade Federal de Minas Gerais, Av. Alfredo Balena 190/464, CEP 30130, Belo Horizonte, Minas Gerais, Brazil
[2]Department of Surgery, Faculdade de Medicina, Universidade Federal de Minas Gerais, Av. Alfredo Balena 190/464, CEP 30130, Belo Horizonte, Minas Gerais, Brazil
[3]Department of Pathology, Faculdade de Medicina, Universidade Federal de Minas Gerais, Av. Alfredo Balena 190/464, CEP 30130, Belo Horizonte, Minas Gerais, Brazil
[4]Department of Medicine, Faculdade de Medicina, Universidade Federal de Minas Gerais, Av. Alfredo Balena 190/464, CEP 30130, Belo Horizonte, Minas Gerais, Brazil

J.M. Pajares et al. (Eds.)
H. pylori and Gastroduodenal Pathology

40.8 years) were *H. pylori* positive with endoscopically proven DU; nine (six men, mean age 41.9 years) were *H. pylori* positive with histological gastritis without DU; six (five men, mean age 39.3 years) were *H. pylori* positive and presented with healed DU after highly selective vagotomy, and seven (six men, mean age 42.2 years) were *H. pylori* negative without DU.

Biopsy specimens from the antral and oxyntic mucosa were employed for histology and *H. pylori* identification (culture, urease test, and carbolfuchsin-stained smear).

Three fragments from the oxyntic mucosa were collected to determine the histamine content. They were immediately frozen and stored in liquid nitrogen until processing. Histamine was extracted [9] in a plastic tube containing 0.1 *M* phosphate buffer pH 7.9. The fragments were homogenized with a Teflon pestle tissue homogenizer and heated for 10 min in boiling water. After that the tubes were centrifuged at 5000 rpm at 4°C and the supernatant was lyophilized. Histamine concentration was determined after resuspension of dry residue by enzymatic isotopic assay [8].

Statistical analyses were performed by the two-tailed t test and differences were taken as significant when p was less than 0.05.

Results and Discussion

The gastric oxyntic mucosa histamine concentration of all the groups studied is shown in Table 1. In the present study significant differences were observed when the histamine content of *H. pylori*-negative subjects was compared to *H. pylori*-positive patients with DU ($p = 0.0008$) and *H. pylori*-positive patients with gastritis without DU ($p = 0.0060$); however, there were no statistical differences between the latter two groups. These *H. pylori*-positive patients, who had not previously been treated, presented low endogenous histamine which could be due to an increased acid secretion induced by *H. pylori.* On the other hand, *H. pylori*-positive patients after vagotomy as well as those who were treated with H2 antagonists did not present lower gastric histamine content ($p = 0.066$ and $p = 0.2816$, respectively). Some studies have demonstrated that vagotomy, like H2 antagonists, significantly increases gastric histamine contents in patients with DU [5, 10, 11]. According to Man et al. [5], this fact may be

Table 1. Gastric oxyntic mucosa histamine content in *H. pylori*-positive and -negative patients.

Patients	Concentration (μg/g)
HP negative	50.29 ± 14.06
HP positive	
With DU	27.76 ± 9.17
With gastritis without DU	29.60 ± 11.64
After vagotomy	63.42 ± 7.55
After H2 antagonist treatment	56.68 ± 9.62

explained by an inhibition of gastric secretion, and less histamine may be lost into the gastric juice resulting in an increased content of endogenous histamine stores in the gastric mucosa.

The results of this study suggest that *H. pylori*-positive patients, except those after vagotomy and those who were treated with H2 antagonists, present a decrease of the "stored" histamine which could be due to increased acid secretion.

References

1. Code CF (1977) Reflections on histamine gastric secretion and H2 receptor. N Engl J Med 296:1459–1462
2. Graham DY (1989) Campylobacter pylori and peptic ulcer disease. Gastroenterology 96:615–625
3. Levi S, Beardshall K, Haddad G, Playford R, Ghosh P, Calam J (1989) Campylobacter pylori and duodenal ulcers: the gastrin link. Lancet 1:1167–1168
4. Man WK, Ingoldby CJH, Spencer J (1984) Is pentagastrin-stimulated secretion mediated by histamine? Gut 25:965–970
5. Man WK, Saunders JH, Ingoldby C, Spencer J (1981) Effect of cimetidine on the amounts of histamine in the gastric mucosa of patients with gastric or duodenal ulcers. Gut 22:923–926
6. McColl KEL, Fullarton GM, EL Nujumi AM, MacDonald AM, Brown IL, Hilditch TE (1989) Lowered gastrin and gastric acidity after eradication of Campylobacter pylori in duodenal ulcer. Lancet 2:490–500
7. Oderda G, Vaira D, Holton J, Ainley C, Altare F, Ansaldi N (1989) Amoxycillin plus tinidazole for Campylobacter pylori gastritis in children: assessment by serum IgG antibody, pepsinogen I, and gastrin levels. Lancet 1:690–692
8. Shaff RE, Beaven MA (1979) Increased sensitivity of the enzymatic isotopic assay of histamine measurement of histamine in plasma and serum. Anal Biochem 94:425–430
9. Snyder SH, Baldessarini RJ, Axelrod J (1966) A sensitive and specific enzymatic isotopic assay for tissue histamine. J Pharmacol Exp Ther 153:544–549
10. Troidl H, Rohde H, Lorenz W, Hafner G, Hamelmann H (1978) Effect of seletive gastric vagotomy on histamine concentration in gastric mucosa of patients with duodenal ulcer. Br J Surg 65:10–16
11. Waldum HL, Sandvik AK (1989) Histamine and the stomach. Scand J Gastroenterol 24:130–139

Detection of Phospholipases and Cytotoxic Activities of *Helicobacter pylori*

M.A. Daw[1], R. O'Moore[2], and C. O'Morain[2]

Introduction

The production of phospholipases and cytotoxins is an important virulence factor in the pathogenesis of enteric bacteria. Phospholipases A have been detected in a large number of bacteria, including *Escherichia coli* and *Vibrio parahaemolyticus* [1], which may have a biological role to play. Phospholipases C are the most frequently studied phospholipases of bacterial origin and are believed to be toxic for the host cell.

The exact pathogenic mechanism by which *Helicobacter pylori* causes disease is not fully understood. Culture filtrates of *H. pylori* cause cytopathic effects like vacuolation and rounding in vitro. Chen et al. [2] have described the intracellular damage of mucus-secreting cells in association with *H. pylori.* However, when antral biopsies were taken from patients with *H. pylori* gastritis, there was a 95% association between the presence of *H. pylori* and damaged mucosal cells [3]. Leunk et al. [4] also found that culture filtrates of *H. pylori* can cause non-lethal cytopathic effects in vitro in seven of nine mammalian cell lines tested. Although no single toxin capable of causing these effects has been demonstrated, a variety of enzymes capable of causing cytotoxic effects directly or indirectly have been identified [5].

The substrates of these enzymes are phospholipids, particularly phosphotidylcholine (also called lecithin) which is almost exclusively found in mammalian membrane. Phospholipases can therefore weaken the cell membrane and lead to cell damage. Since *H. pylori* remains in close contact with the mucosal cells, even small amounts of phospholipases produced by this organism could be harmful. In addition to this direct effect, phospholipases acting on lecithin can produce lysolecithin which is cytotoxic [6].

In ulcer patients there is an increased amounts of lysolecithin in the gastric juice compared to the normal individuals which may be related to the action of pancreatic juice [7]. However, if *H. pylori* is capable of producing phospholipases, then lysolecithin could be a product of this enzymatic activity.

The aims of the present study were to determine the ability of *H. pylori* to produce phospholipase A and phosphalipase C and to investigate the cytotoxic activity of the sonicates containing these enzymes using a tissue culture cell line. Such enzymes and their cytotoxic activity may contribute to the pathogenesis of *H. pylori*-associated duodenal ulcer and gastritis.

[1] Department of Medical Microbiology, Faculty of Medicine, Tripoli, Libya
[2] Departments of Biochemistry, St. James's Hospital, and Gastroenterology, Meath/Adelaide Hospital, Trinity College, Dublin, Ireland

J.M. Pajares et al. (Eds.)
H. pylori and Gastroduodenal Pathology

Materials and Methods

Bacterial Strains

Twenty-three isolates of *H. pylori* were isolated from antral biopsies of different patients with duodenal ulcer and gastritis. Bacterial isolates were grown on chocolate agar (5% v/v heated horse blood) and incubated under microaerophilic conditions at 37°C for 5–7 days. Each strain was scraped off and suspended in *Tris*-HCl buffer (pH 7.4) and sonicated for 5 min, centrifuged at 1200 rpm (4°C) for 45 min and then each supernatant was investigated for a phospholipases A and C, and cytotoxic activity.

Phospholipase A

Phospholipase A activity of *H. pylori* was studied by incubating culture supernatants with lecithin containing labelled ^{14}C fatty acids. The products of enzymatic activity were separated in thin-layer chromatography (TLC) and subjected to autoradiography. The TLC plates were also stained with iodine and the radioactivity in each band was assessed. ^{14}C-lecithin (NEC-764 Med Labs), which contains toluene, was dried in a vortex dryer and then made up to 50 ml in a mixture of chloroform and methanol (2:1 v/v) containing 3 m*M* cold lecithin.

Hot lecithin (2.2 × 106 dpm; 5 ml) was added to the cold lecithin (10 mg) and the mixture was dried down in a vortex drier under vacuum. To this was added 10 ml 10 m*M* Tris buffer (pH 7.4) and glass beads. The hot and cold lecithin mixture was brought into suspension in the aqueous phase by sonicator at maximal settings in a Dawe Soniprobe sonicator for 2 min; 500 μl of this lecithin mixture was mixed with 500 μl culture supernate. This was incubated at 37°C for 4 h in initial experiments and overnight in the later experiment. The products were extracted using 4 ml choloroform-methanol (2:1), then the chloroform phase was separated and evaporated under vacuum. The residue was redissolved in 100 μl chloroform methanol; 50 μl of this mixture was introduced into the tank for the run. The plates were left in the tank until the solvent (acetic acid, *d*-isopropylether, *n*-heptane) reached a level of a few millimetres from the top of the plate. They were dried and exposed for autoradiography (Hyperfilm MP, Amersham). The plates were also stained with iodine vapour and the bands from the corresponding areas of active and heated mixture were scraped from the silica and the radioactivity was measured in a scintillation counter.

Phospholipase C

Quantitative phospholipase C assays were performed by measuring the cleavage of *p*-nitrophenylphosphorylcholine (NPPC; Sigma Chemical, UK), chromogenic phospholipase C substrate as described by Kurioka and Matsuda [8]. Immediately after sonication, buffer was added to give a final concentration of 0.25 *M* Tris-HCl, pH 7.2, 1 m*M* ZnCl. For phospholipase C assays, 50 μl samples were serially diluted (twofold dilution) in 0.25 *M* Tris-HCl, pH 7.2, and 150 μl freshly prepared substrate reagent was added to each well to give the following final concentration of reagents: 10 m*M* NPPC, 1 m*M* ZnCl, 60% (v/v) glycerol, 0.25 *M* Tris-HCl, pH 7.2. The plates were incubated at 37 °C for 4 h, after which colour development was monitored by measuring the absorbence at 410 nm using an enzyme-linked immunosorbent assay (ELISA) plate reader. Phospholipase C activity was also tested after mixing the supernatant with 10^{-3} *M* EDTA, 0.25% Sodium Laurylsulfate (SLS), 60% v/v sorbitol instead of glycerol and after boiling the supernatant at 100°C for 15 min.

Cytotoxic Activity

Cultured cells used for the detection of cytotoxic activity were HeLa cells (tissue culture cell line). They were used as described by Daw [9]. The cell line was maintained in Earle's medium, counted and distributed in small plastic pijou (Trac) tubes with coverslips. One millilitre of each supernatant was added to each coverslip and incubated at 37 °C for 48 h, cells were examined microscopically for the presence of intracellular vacuolisation and the ability of HeLa cells to take up the stain. The tests were scored as positive if > 50% of the cells showed vacuolisation and damage.

Results

Phospholipases activity of *H. pylori* filtrates was investigated using both phosphotidylcholine-containing ^{14}C-labelled fatty acids (neutral system) as a substrate for phospholipase A and NPPC as chromogenic phospholipase C substrate. The 23 strains studied showed a phospholipase A activity when incubated with radiolabelled phosphotidylcholine (4 × 10 rpm). All the strains split phosphotidycholine up to 20% when incubated for 4 h. This activity was increased by an overnight incubation.

Phospholipase C activity was detected in all strains tested. This was determined by measuring the optical density (OD) at 410 nm. The mean OD was 0.150 ± 0.025 for 10 (43%) strains and 0.270 ± 0.28 for 13 (57%) strains.

HeLa cell tissue culture cell line was used to screen the cytotoxic activity produced by *H. pylori*. The results revealed that some of the culture filtrates induced alteration in the appearence of the cell line. Of the 23 strains studied, only 15 (65%) showed damage to the cell monolayer. The response consisted of intracellular vacuolisation and release in the cytoplasm after 48 h of incubation with *H. pylori* filtrates.

Consideration was given to the possibility that a large amount of phospholipases, particularly phospholipase C, might be responsible for the cytotoxic activity. All the culture filtrates which induced cytotoxic activity on HeLa cells contained phospholipases (Table 1), although the filtrates were not concentrated, and the cytotoxicity was not quantitatively determined. In 11 (48%) strains there was a correlation between the phospholipase C level and the

Table 1. Phospholipases and cytotoxic activity of *H. pylori*

Isolates studied	Phospholipase activity		Cytotoxic activity on HeLa cells
	A	C	
1	+ +	+ + +	+ + +
5	+ +	+ +	−
3	+ +	+ + +	+ +
4	+ +	+ + +	+
7	+	+	+
3	+	+	−

Symbols: − , no activity; + , weak activity; + + , moderate activity; + + + , strong activity.

Table 2. Characteristics of phospholipase C (PLC) activity of *H. pylori*

Property tests[a]	Phospholipase C activity
PLC in the supernatant alone	+ +
PLC in the supernatant with	
$ZnCl_2$ (10^{-4} M)	+ + +
Heat (100°C)	−
PH (7.2)	+ + +
EDTA (10^{-3} M)	−
SLS (0.25%)	−
Glycerol (60%)	+ +
Sorbitol (60%)	+ + +

[a] The tests were carried on the supernatant of the isolate who showed the highest level of PLC activity.

cytotoxic damage induced in the HeLa cell monolayer. Phospholipase C activity as shown in Table 2 was inhibited by EDTA, SLS and boiling while it was maximised in presence of sorbitol, $ZnCl_2$ and glycerol at neutral pH.

Discussion

Helicobator pylori strains were found to possess enzymatic ability both by producing phospholipase A which splits phosphotidylcholine and phospholipase C which breaks NPPC to *p*-nitrophenol. They can also cause cytotoxic damage to the HeLa cell monolayer. This enzymatic activity may play a significant role in the hydrolysis of phospholipid layer in the mucosal surface in *H. pylori*-associated duodenal ulcer and gastritis. Close examination of biopsy specimens reveals that *H. pylori* can be consistently seen within the gastric mucosa and in close proximity to epithelial cells [10]. This close contact, may allow the phospholipases and other toxic factors to produce damage to the epithelial cells. In other studies [11, 12] we have shown that *H. pylori* can produce different phospholipases which may be capable of rapid degradation of the gastric mucosa.

In biopsy specimens, *H. pylori* gastritis may be associated with a varying percentage of damaged mucosal cells, which may be due to a varying percentage of contents in the micro-environment [13]. Phospholipase C in this study was variable among different strains of *H. pylori* rather than phospholipase A. This may be used as a marker in scoring the pathogenicity of the organism. *H. pylori* strains isolated from patients with duodenal ulcer have a higher level of phospholipase C than those isolated from patients with gastritis [14].

Cytotoxic damage of *H. pylori* to the tissue culture cell line have been described by different investigators. Leunk et al. [4] have found that *H. pylori* can produce toxin which causes vacuolisation of several tissue culture cell lines, similar results were also found by Xu et al. [13]. This is, in accord with our results. Of our strains 65% showed damage to the HeLa cell monolayer, although there is no evidence in our results that the toxin was cytolethal but it may have been cytolytic (formidable) where the cytoplasm was released from the cells and subsequently may lead to their death–some of the strains were found to be haemolytic, (M.A. Daw, unpublished data).

The phospholipases and cytotoxic effect were found to be active at neutral pH. This may suggest that the urease adopts a neutral pH surrounding the bacteria (micro-environment) which allows it to survive in an acidic environment, and it also permits these enzymes to operate at a maximal efficiency [15]. Urease-negative strains of *H. pylori* were also found to be incapable of producing intracellular vacuolisation [13]. Attempts were, made to correlate pospholipase C activities with the cytotoxic effect. This remained coupled in only 48% of the strains, but the specificity of this toxicity is still to be determined. Phospholipases together with other enzymes may cause direct or indirect damage within the mucosal layer and that could make the mucosal surface appear more hydrophilic. However, this has been found in duodenal ulcer patients, which may account for our results. Spychal et al. [16] showed that the presence of *H. pylori* infection was associated with an increase in surface hydrophilicity. The role of these enzymes and their effects on the surface activity of the gastric mucosa is still to be determined.

Acknowledgements. Dr. M.A. Daw would like to thank the staff of the Endoscopy Unit, Meath/Adelaide Hospitals, the staff of the Clinical Microbiology Department, St. James's Hospital, and Ms. Jane Murray for preparing the manuscript.

References

1. Mollby R (1978) Bacterial phospholipases. In: Jeljaszewicz J, Wadstrom T (ed) Bacterial toxins and cell membrane. Academic, London, pp 367–424
2. Chen XG, Correa P, Offerhaus J, Rodriguez E, Janney F, Hoffmann E, Fox J, Hunter F, Diavolitsis S (1986) Ultrastructure of the gastric mucosa harboring campylobacter-like organisms. Am J Clin Pathol 86 : 575–582

3. Bayerdorffer E, Oertel H, Lehn N, Kasper G, Mannes GA, Sauerbruch T, Stolte M (1989) Campylobacter pylori–primary pathogen in active chronic gastritis? In: Megraud F, Lamouliatte H (ed) Gastroduodenal pathology and Campylobacter pylori. Elsevier, Amsterdam, pp 249–256
4. Leunk RD, Johoanso: PT, David BC, Kraft WG, Morgan DR (1988) Cytotoxic activity in broth-culture filtrates of Campylobacter pylori. J Med Microbiol 26 : 93–99
5. Megraud F (1989) Campylobacter pylori: enzymes. In: Rathbone BJ, Heatley RV (ed) Campylobacter pylori and gastroduodenal disease. Blackwell, Oxford, pp 39–47
6. Lehninger AL (1980) Lipids, liproprotein and membranes. Worth, New York pp 279–308
7. Rhodes J, Bernardo OE, Phillips SF, Rovelstad RA, Hoffmann AF (1969) Increased refulx of bile into the stomach in patients with gastric ulcer. Gastroenterology 57 : 241–252
8. Kuriko S, Matsuda M (1976) Phospholipase C assay using P-nitrophenylphosphorylcholine together with sorbitol and its application to studying the metal and detergent requirement of the enzyme. Anal Biochem 75 : 281–289
9. Daw MA (1989) Epidemiology of Gram-negative bacilli in neutropenic patients with particular reference to Enterobacter cloacae. Ph D thesis, Dublin University
10. Hazell L, Lee A, Brady L, Hennessy W (1986) Campylobacter pylori and gastritis: association with intracellular spaces and adpition to an environment of mucus as important factors in colonisation of the gastric epithelium. J Infect Dis 153 : 658–663
11. O'Morain C, Mathi E, Daw MA, Cafferky, M, Keane C, O'Moore R (1990) Lipolytic activity of Helicobacter pylori. Gastroenterology 98 : A101
12. Daw MA, Cotter L, Healy M, O'Moore R, Keane C, O'Morain C (1990) Phospholipases and cytotoxic activity of Helicobacter pylori. Rev Esp Enferm Dig 78 [Suppl] : 78–79
13. Xu J, Goodwin S, Cooper M, Robinson J (1990) Intracellular vacuolization caused by the urease of Helicobacter pylori. J Infect Dis 161 : 1302–1304
14. Daw MA, Cotter L, Healy M, O'Moore R, Keane C, O'Morain C (1991) Are Helicobacter pylori strains isolated from duodenal ulcer patients more virulent than those isolated from gastritis? Gastroentrology 100:A 572
15. Daw MA, Deegan P, Leen E, O'Morain C (1991). Effect of Omeprazole on Helicobacter pylori and associated gastritis. Aliment Pharmacol Therap 5:435–439
16. Spychal RT, Goggin PM, Marrero JM, Saverymuttu SH, Yu CW, Corbishley CM, Maxwell JD, Northfield TC (1990) Surface hydrophobicity of gastric mucosa in peptic ulcer disease. Relationship to gastritis and Campylobacter pylori infection. Gastroenterology 98 : 1250–1254

IV. Pathology

Antral Follicular Gastritis Is Characteristic of *Helicobacter pylori* Infection

F.J. Sancho, S. Sainz, J. Mones, B. Mirelis, E. Oliva, and F. Vilardell

Introduction

Chronic gastritis is an inflammatory process of the stomach mucosa which is thought to be caused by a wide variety of mechanisms. In spite of several efforts to systematize the morphological types of gastritis [1–3], a definite aetiological classification has not yet been established.

The aim of this study was to evaluate the inflammatory response of the gastric mucosa to infection by *Helicobacter pylori* (HP) and to establish whether a specific variety of gastritis was produced.

Patients and Methods

We randomly studied 114 patients who were referred to the digestive endoscopy unit of our hospital with symptoms suggesting gastroduodenal ulcers; 43 of them were diagnosed on endoscopy as having duodenal ulcer (DU) and the remaining 71 as having non-ulcer dyspepsia (NUD) [4]. Four antral biopsies were taken from all patients. Two biopsies were fixed in formalin and sent to the pathology department where they were routinely processed and evaluated by the same pathologist. The histological changes and the existence of germs morphologically compatible with HP were assessed.

The urease test was performed on another biopsy using the following methodology: the biopsy was placed in Christensen's medium at room temperature and was observed every hour for the first 4 hours and subsequently after 24, 48 and 72 h. The test was considered to be positive when the colour change took place within the first 24 h.

The last biopsy was sent to the microbiology department in a sterile test tube and was processed in the following way: it was crushed with a mortar and some fragments were used for an extension and Gram stain. The other fragments were innoculated in two different culture media: agar-blood and agar-Skirrow (selective medium with antibiotics, Oxoid SR69). They were incubated at 37°C in a microaerophilic atmosphere (Anaerocult C, Merck) for 7 days and were assessed

Hospital de la Sta. Creu i Sant Paux, Servicios de Anatomia Patológica, Patologia Digestivay Microbiologia, Avda. San Antonio M. Claret 167, Barcelona 08025, Spain

J.M. Pajares et al. (Eds.)
H. pylori and Gastroduodenal Pathology

daily for growth of the characteristic colonies. The colonies were identified using a urease test, oxidase test and Gram stain.

Criteria for infection were either a positive culture or a positive Gram stain with a positive urease test and with germs morphologically characteristic of HP in the gastric mucosa. The morphological changes were evaluated using Whitehead's criteria [2]. Statistical analysis was performed using the chi-square test with Yates' correction.

Results

A total of 84 (74.5%) of the 114 patients fulfilled the criteria for infection by HP, 36 (83.7%) of the 43 patients with DU and 48 (67.6%) of the 71 with NUD. Only three patients were considered infected in spite of a negative culture because they fulfilled the three conditions previously described.

The mucosa of the 84 infected patients was pathological in all cases, and in 80 the morphologically characteristic bacilli in contact with the surface epithelium were visualized. In 69 (82.1%) of these 84 infected patients, a diffuse infiltration of the lamina propia by lymphocytes and plasma cells which separated the glands was observed. In general there was no significant decrease in their number nor was atrophy present (Fig. 1). Both in the lamina propia and intra-epithelially there were numerous polymorphonuclear leucocytes (Fig. 2). One lymphoid follicle with a germinal centre was found in at least one of the two samples in these 69 cases. This could be designated chronic active gastritis with lymph nodes (CAG + LN) or antral follicular gastritis.

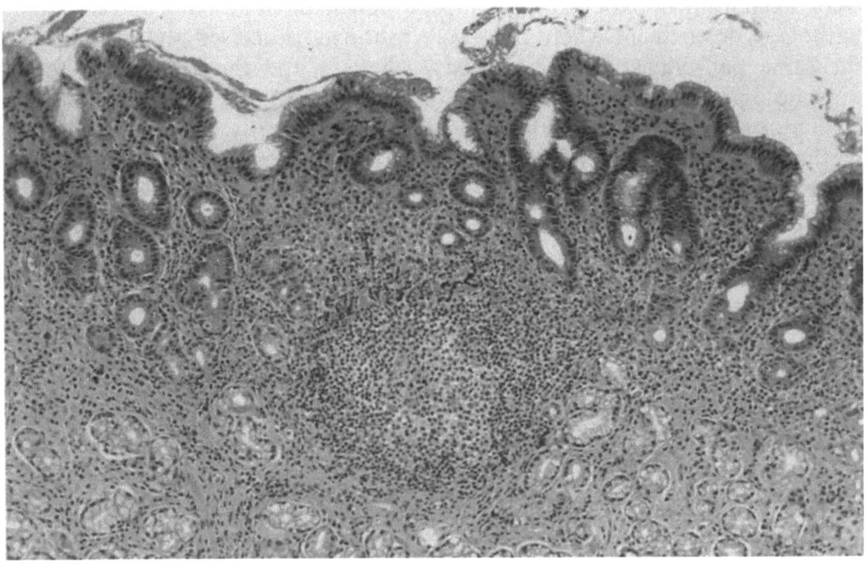

Fig. 1. Antral mucosa. Chronic active gastritis with lymph nodes. Haemotoxylin and eosin; ×40

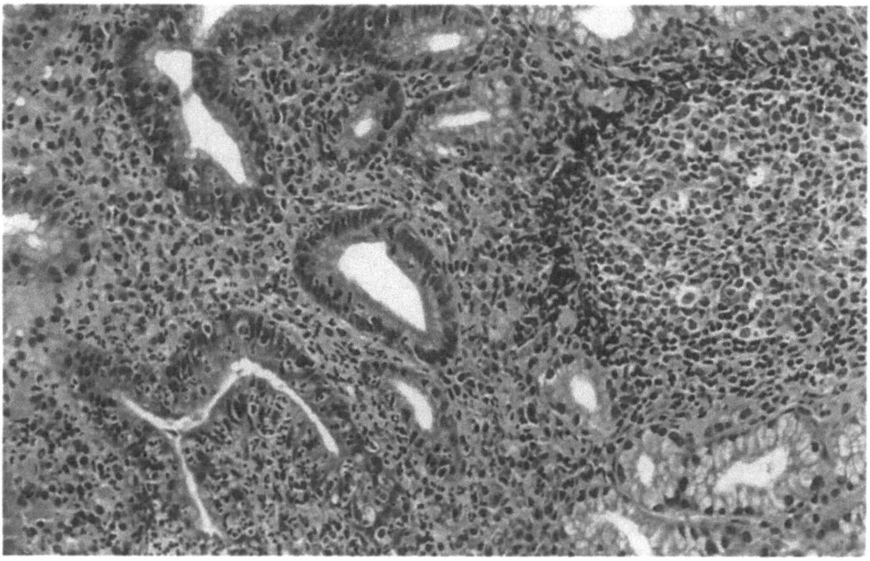

Fig. 2. Numerous intra-epithelial polymorphonuclear leucocytes. Lymph node with germinal centre at the bottom. Haemotoxylin and eosin; ×200

Out of the remaining 15 infected patients, in 11 cases CAG was observed without the presence of lymph nodes in either of the two samples, and in the other four patients chronic superficial gastritis (SG) was diagnosed.

The morphology in the 30 patients who were not infected was the following: in 13 cases the mucosa was strictly normal, in four SG changes were observed, in six CAG and in seven CAG + NL. Germs were not visualized in any of the patients with normal mucosa or SG, nor in four of the six with CAG, but were found in all the biopsies in which the morphology was CAG + NL.

The association between infection by HP and the morphology of CAG + LN was found to be statistically significant ($p > 0.0001$).

Discussion

The pathogenicity of HP was demonstrated in Marshall's [5] and Morris's [6] experiments in which Koch's postulates were fulfilled. Both authors ingested pure cultures of HP after having observed their undamaged mucosa using endoscopy and biopsy and they subsequently developed acute gastritis. In Morris's case the gastritis became chronic and persisted, though mildly, after eradication of the germ.

Furthermore, it has been observed that the presence of HP produces an immune response which is both local (specific antibodies in gastric juice) and systemic (antibodies in serum) [7]. If the close association between chronic gastritis and the presence of HP observed in numerous studies [8, 9] is taken into consideration, it is logical to assume that the inflammation observed in the gastric mucosa of patients with HP can be considered, at least in part, to be an immune reaction to infection. The study which demonstrated this most effectively is that of Rauws et al. [10], which evaluated in a semi-quantitative way the degree of inflammation of the gastric mucosa in a large series of patients infected by HP. After eradicating the germs using various types of treatment, they verified the progressive decrease of inflammation and suggested the cause–effect relationship between infection by HP and CAG. This causal relationship was substantiated by the fact that in our study no patient with normal mucosa presented infection by HP.

In spite of the known association between chronic gastritis and HP, the existing morphological descriptions are not very detailed. Wyatt et al. [11], in a retrospective study, identified lymphoid follicles in 27.4% of HP-associated gastritis. In our series, analysing two samples of antral mucosa in each patient, this percentage rose to 82.1% of infected patients. Although they were also observed in seven patients not considered to be infected, in all these cases numerous germs morphologically compatible with HP were visualized in contact with the surface epithelium. The negativity of the culture and the Gram stain in these cases could be attributed to a sample error, given that these techniques were performed on the same fragment of mucosa which was different from those used for the histological study.

The highly significant relationship observed in our series between infection by HP and CAG + LN (antral follicular gastritis) shows quite clearly that these changes can be explained as a local response to the immune stimulation by the bacterial antigens. As a result, this type of gastritis should be accepted as being aetiologically related to HP infection and is sufficiently different from the rest of inflammatory processes of the gastric mucosa which are still of unknown aetiology.

References

1. Lambert R (1972) Chronic gastritis. Digestion 7:83–128
2. Whitehead R (1985) Mucosal biopsy of the gastrointestinal tract, 3rd edn. In: Major problems in pathology, vol 3. Saunders, Philadelphia
3. Correa P (1988) Chronic gastritis: a clinico-pathological classification. Am J Gastroenterol 83:504–509
4. Talley NJ, Phillips SF (1988) Non-ulcer dyspepsia: potential causes and pathophysiology. Ann Intern Med 108:865–879
5. Marshall BJ, Amstrong JA, McGechie DB, Glancy RJ (1985) Attempt to fulfill Koch's postulates for pyloric Campylobacter. Med J Aust 142:436–439
6. Morris A, Nicholson G (1987) Ingestion of Campylobacter pylori causes gastritis and raised fasting gastric pH. Am J Gastroenterol 82:192–199
7. Rathbone BJ, Wyleatt JI, Worsely BW et al. (1986) Systemic and local antibody response to gastric Campylobacter pyloris in non ulcer dyspepsia. Gut 27:642–647
8. Blaser MJ (1987) Gastric Campylobacter-like organisms, gastritis, and peptic ulcer disease. Gastroenterology 93:371–383

9. Rathbone BJ, Wyatt JI, Heatley RV (1986) Campylobacter pyloridis. A new factor in peptic ulcer disease. Gut 23:635–641
10. Rauws EAJ, Langenberg W, Houthoff HJ, Zanen HC, Tytgat GNJ (1988) Campylobacter pyloridis- associated chronic active antral gastritis. A prospective study of its prevalence and the effects of antibacterial and antiulcer treatment. Gastroenterology 94:33–40
11. Wyatt JI, Rathbone BJ (1988) Immune response of the gastric mucosa to Campylobacter pylori. Scand J Gastroenterol 23 [Suppl 142]:44–99

Defining the Role of *Helicobacter pylori* in Relationship to Relapse of Duodenal Ulcer

R. Logan, R. Polson, M. Walker, P. Gummett, H. Baron, and G. Misiewicz

Introduction

Helicobacter pylori is the most common cause of histological gastritis [1], and in addition several studies have shown that the recurrence of duodenal ulcer disease (DU) can be prevented by eradication of *H. pylori* [2–4], so implicating *H. pylori* in the aetiology of recurrent DU [5]. These studies have also been interpreted as providing proof that *H. pylori* is the cause of DU, and although this may be true, sceptics argue that to show a causal relationship, a temporal sequence between the recurrence of *H. pylori* and DU has to be demonstrated. Thus the aim of this study was to demonstrate whether the recurrence of *H. pylori* preceded the recurrence of DU.

Patients and Methods

Patients between the ages of 18 and 70 years with an endoscopically confirmed DU (minimal diameter > 5 mm) were considered for entry into the study. Ethical committee approval and patients' written informed consent were obtained. Previous resective gastric surgery, recent (within 2 months) use of bismuth salts or antibiotic combinations known to be active in vivo against *H. pylori* were exclusion criteria.

At endoscopy (Olympus GIF1T20) DU size and site were recorded, and biopsies near ($n = 2$) and away ($n = 2$) from the ulcer were taken for histology. In addition, *H. pylori* status was assessed by four antral biopsies, two for histology haematoxylin and eosin and Gimenez stains) and two for culture (Oxoid SR147 with microaerophilic conditions for up to 10 days). Biopsy forceps (Olympus FB13K) were sterilised by autoclaving and endoscopes disinfected between patients by using an automatic washing machine (KeyMed EW20) [6].

Non-invasive assessments of *H. pylori* status were made with the ^{13}C-urea breath test (^{13}C-UBT) using the European standard protocol [7] with pooled breath collection from $t = 10$ to $t = 40$ min. A positive result was recorded if the excess δ^{13} CO_2 excretion was > 5 per mil (standard delta notation).

Parkside Helicobacter Study Group, Central Middlesex Hospital, Acton Lane, Park Royal, London NW10 7NS, UK

J.M. Pajares et al. (Eds.)
H. pylori and Gastroduodenal Pathology

All patients had endoscopy, biopsy and ^{13}C-UBT performed before starting colloidal bismuth subcitrate (CBS) 120 mg q.d.s. for 28 days. Immediately after finishing treatment with CBS patients were re-endoscoped and breath tested: patients with unhealed ulcers were given a further 1 month of treatment with CBS, whilst patients with healed ulcers were entered for follow up.

Follow Up

After ulcer healing the ^{13}C-UBT was repeated at weekly intervals for 1 month, and then at 3, 6 and 12 months after finishing CBS. The gastroduodenoscopy (OGD) was repeated immediately if the ^{13}C-UBT became positive (excess $\delta^{13}CO_2$ excretion > 5 per millilitre) and at 1, 3, 6 and 12 months after finishing CBS. Ulcer recurrence (minimal diameter > 5 mm) was taken as the end point.

Results

Twenty patients (11 men, 9 women; median age 38 years, range 20–68 years) were entered into the study. All patients had *H. pylori* detected by histology, culture and ^{13}C-UBT before starting treatment with CBS. The mean ($\pm$ SEM) pre-treatment excess δ^{13} CO_2 excretion was 25.6 ($\pm$ 9.6) per mil.

After 4 weeks of treatment *H. pylori* was not detected by histology, culture or ^{13}C-UBT in 19/20 (95%) patients all of whom had healed DUs. An unhealed DU was seen in one patient in whom histology and ^{13}C-UBT (excess $\delta^{13}CO_2$ = 7.8 per mil vs. 28.6 per mil pre-treatment) were still positive for *H. pylori.* Ulcer healing and clearance of *H. pylori* were achieved after a further 4 weeks of treatment with CBS.

Recurrence of *H. pylori*

^{13}C-UBT follow up at 1 week showed a recurrence of *H. pylori* in 14/20 patients, which was confirmed on endoscopy (median time to repeat OGD was 10 days after finishing CBS), at which stage only one patient had a recurrent DU (asymptomatic; Fig. 1). Three out of six of the remaining patients had a recurrence of *H. pylori* within 1 month of finishing CBS.

Three out of 20 (15%) patients had successfully eradicated *H. pylori* with no *H. pylori* detected by histology, culture or ^{13}C-UBT 1 month after finishing CBS. All three patients have remained well, with normal endoscopy and histology, and a negative ^{13}C-UBT at 3, 6, and 12 months.

DU Recurrence

DU recurrence occurred after *H. pylori* recurrence in 16/17 patients, 50% recurring after 6 weeks. The remaining patient had an asymptomatic DU

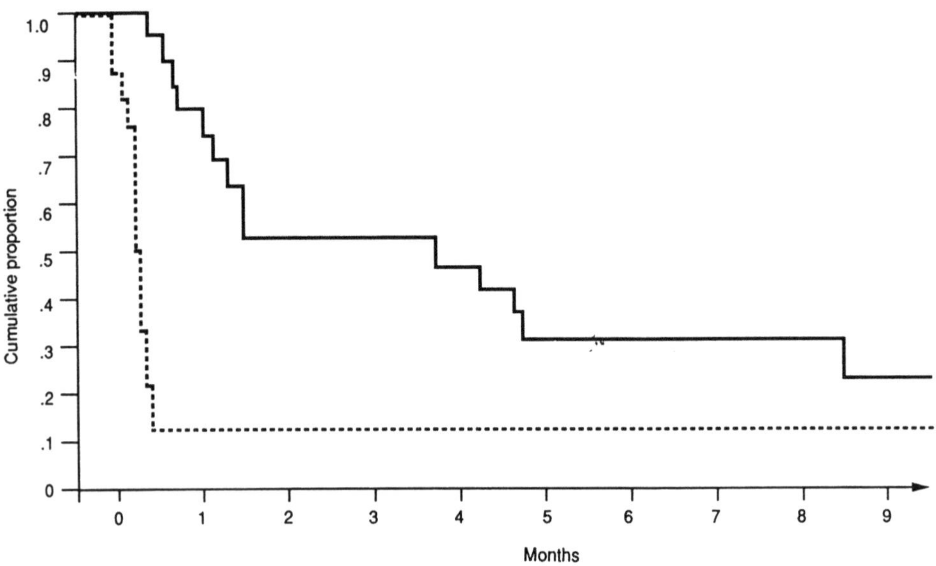

Fig. 1. A life table analysis of the recurrence of *H. pylori* (– – –) in relation to recurrence of duodenal ulcer (———). *H. pylori* recurs within days of finishing tripotassium dicitratobismuthate and always precedes recurrence of duodenal ulcer. In three patients the infection is eradicated and they remain ulcer free

recurrence within 10 days of finishing CBS. Three of 20 patients in whom *H. pylori* was eradicated have had no ulcer recurrence at 1-year follow up.

Discussion

The results from this study show that the recurrence of *H. pylori* precedes the recurrence of DU. This study also confirms previous experience that eradication of *H. pylori* can produce prolonged periods of ulcer remission and hopefully cure. Whilst awaiting the development of suitable animal models, this temporal relationship between the recurrence of *H. pylori* and DU shows *H. pylori* to be the major factor in the aetiology of DU.

If used alone, CBS 120 mg q.d.s. for 4 weeks will eradicate *H. pylori* infection in 15% of cases, whilst in the remaining 75% *H. pylori* will recur within days of finishing treatment. This suggests that prolonged courses of CBS are of little benefit, indeed recent studies have shown that a similar rate of *H. pylori* recurrence is seen after only 1 week of CBS. Therefore, to cure DU by *H. pylori* eradication, antibiotic/bismuth combination therapies will have to be developed.

References

1. Rauws EA, Langenberg W, Houthoff HJ, Zanen HC, Tytgat GNJ (1988) Campylobacter pyloridis-associated chronic active antral gastritis. A prospective study of its prevalence and the effects of antibacterial and antiulcer treatment. Gastroenterology 94:33–40
2. Marshall BJ, Goodwin CS, Warren JR et al. (1988) A prospective double-blind trial of duodenal ulcer relapse after eradication of Campylobacter pylori. Lancet ii:1437–1442
3. Coghlan JG, Humphries H, Dooley C et al. (1987) Campylobacter pylori and recurrence of duodenal ulcers–a 12-month follow-up study. Lancet ii:1109–1111
4. Rauws EAJ, Tytgat GNJ (1990) Eradication of H. pylori cures duodenal ulcer. Lancet 335:1233–1235
5. Tytgat GNJ, Axon ATR, Dixon MF, Graham DY, Lee A, Marshall BJ (1991) Helicobacter pylori: causal agent in peptic ulcer disease? In: World congress of gastroenterology working party report. J Gastroenterol Hepatol 6:103–137
6. Weller IVD, Williams CB, Jeffries DJ et al. (1988) Cleaning and disinfection of equipment for gastrointestinal flexible endoscopy: interim recommendations of a Working Party of the British Society of Gastroenterology. Gut 29:1134–1151
7. Logan RPH, Dill S, Walker MM et al. (1990) Evaluation of the European standard ^{13}C-urea breath test for the detection of Heliocbacter pylori. Gut 30:A1177

Gastritis and *Helicobacter pylori* Prevalence in Patients with Heterotopic Gastric Mucosa or Gastric Metaplasia in the Duodenum

H. K. Koch and S. Boemke

Introduction

Helicobacter pylori colonization of gastric mucosa is strongly correlated to duodenal ulcer [4]. Since *H. pylori* can only be found on the columnar surface epithelia of gastric pits and not on intestinal epithelia, it has been suggested that gastric metaplasia of the duodenal mucosa is a prerequisite for *H. pylori* colonization of the duodenum and this colonization may lead to ulceration [7, 8]. Beside gastric metaplasia of the duodenal mucosa, there is another condition–heterotopic gastric mucosa in the duodenum–in which columnar surface epithelia are present in the duodenum possibly facilitating the growth of *H. pylori* in the duodenum. It is known that gastric metaplasia in the duodenum results from damage to the normal duodenal epithelium, e.g., by increased acid secretion of the stomach [2, 5]. The assumption that gastric metaplasia and colonization with *H. pylori* produce duodenal ulcer is therefore difficult to accept since the presence of gastric metaplasia already indicates a noxious effect to duodenal mucosa, which is presumably not caused by *H. pylori*. To study the influence of coexisting *H. pylori* infection and the presence of gastric surface epithelia in the duodenum we therefore evaluated the prevalence of gastric metaplasia and *H. pylori* infection in four predefined groups of patients. In addition, cases with heterotopic gastric mucosa were collected from our files and analyzed with respect to *H. pylori* infection and mucosal pathology.

Material and Methods

Four groups were defined as follows:

1. No pathologic features by endoscopic study
2. Endoscopic diagnosis of gastritis
3. Duodenal ulcer
4. Gastric ulcer

Patients were sampled for the four groups over 3 months, the inclusion criteria being: (a) Biopsies of duodenum, and the pyloric and fundic areas of the

Institute of Pathology, Rosastr. 9, 7800 Freiburg im Breisgau, FRG

J.M. Pajares et al. (Eds.)
H. pylori and Gastroduodenal Pathology

Table 1. Number of patients, sex distribution, and mean age in the five groups investigated

Group	Total (*n*)	Male (*n*)	Female (*n*)	Mean age (years)
1. Normal endoscopic picture	100	46	54	44.2
2. Gastritis (endoscopic diagnosis)	113	65	48	53.1
3. Duodenal ulcer	89	63	26	49.6
4. Gastric ulcer	32	15	17	61.4
5. Heterotopic fundic mucosa	69	36	33	58.1

stomach; (b) sufficient clinical information for unequivocal allocation to one of the four groups.

In addition, 69 of 126 patients with heterotopic gastric mucosa in the duodenum had duodenal as well as gastric biopsies and were investigated for *H. pylori* infection and gastric mucosal pathology (Table 1).

The histologic examination was performed on routine step sections stained by hematoxylin and eosin, and periodic-acid-schiff reaction. Special attention was given to *H. pylori* organisms by screening the slides with a high-power dry objective (×63) or additional examination with oil immersion (×100) in difficult cases. The presence or absence of gastric metaplasia in the duodenum was ascertained, and in the gastric biopsies the type of gastritis was determined using the classification proposed by Whitehead [6]. Since patients with gastric ulcer often only had biopsies of the gastric mucosa, only 32 patients could be recruited for this study.

Results

Of 334 patients in groups 1–4, 139 (42%) were found to have gastric metaplasia in the duodenum. *H. pylori* prevalence in patients with gastric metaplasia (79%) was 20% higher than in patients without (58%). The active type of gastritis in the pyloric region was more frequent in patients with gastric metaplasia (71% vs. 58%). In patients both with and without gastric metaplasia *H. pylori* was prefentially seen with active gastritis. Patients with gastric metaplasia generally had a higher rate of *H. pylori* infection with both active or inactive gastritis and both in pyloric or fundic regions (Table 2).

The greatest difference was seen in cases of inactive antral (60% vs. 31%) or fundic gastritis (79% vs. 67%).

Analyzing the group-specific prevalence of gastral metaplasia, *H. pylori* prevalence, and the type of gastritis, the following results were obtained. Gastric metaplasia was seen in 87% of patients with duodenal ulcer, one third of patients with endoscopic diagnosis of gastritis or gastric ulcer, and 17% of patients whose mucosa appeared normal at endoscopy. In groups 1–4 *H. pylori* prevalence in patients with or without gastric metaplasia no longer differed significantly (Table 3). There was also no difference with respect to the type of gastritis in patients with or without gastric metaplasia when the four groups

Table 2. Type of gastritis in antral or fundic mucosa and *H. pylori* prevalence in patients with or without duodenal gastric metaplasia

		With gastric metaplasia		Without gastric metaplasia		Total	
Histology	Location	($n = 139$)	HP+($n = 110$)	($n = 195$)	HP+($n = 114$)	($n = 334$)	HP+($n = 224$)
ISG	Antral	2	(1)	14	(2)	16	(3)
	Fundic	37	(30)	46	(33)	84	(63)
ICG	Antral	18	(11)	21	(9)	39	(20)
	Fundic	1	–	5	(1)	6	(1)
ASG	Antral	3	(3)	3	–	6	(3)
	Fundic	53	(51)	54	(51)	107	(102)
ACG	Antral	96	(93)	110	(101)	206	(195)
	Fundic	14	(13)	32	(23)	46	(36)
Normal[a]	Antral	15	–	35	–	40	–
	Fundic	25	(10)	49	(4)	74	(14)

[a] The sum of the columns for antral or fundic biopsies does not give the total of 139 cases with gastric metaplasia or 195 cases without gastric metaplasia since in patients with gastric ulcer of either the antral or fundic region it was sometimes impossible to determine the type of gastritis in this region.
HP+ , positive for *H. pylori*; IsG, inactive surface gastritis; ICG, inactive chronic gastritis; ASG, active surface gastritis; ACG, active chronic gastritis.

Table 3. Group-specific prevalence of gastric metaplasia, *H. pylori*, infection and type of gastritis in patients with or without gastric metaplasia

		Group 1		Group 2		Group 3		Group 4	
		+ GM ($n = 17$)	− GM ($n = 83$)	+ GM ($n = 34$)	− GM ($n = 79$)	+ GM ($n = 77$)	− GM ($n = 12$)	+ GM ($n = 11$)	− GM ($n = 21$)
HP + (%) (HP duod.)		5(29) –	38(45)	23(67) (3)	52(66)	74(96) (19)	11(92)	8(72) (1)	13(62)
ISG	Antral	1	9	1	5	–	–	–	–
	Fundic	2	17	13	24	20	1	2	4
ICG	Antral	3	8	8	10	6	1	1	2
	Fundic	1	2	–	3	–	–	–	–
ASG	Antral	–	1	–	–	3	–	–	2
	Fundic	2	15	6	24	40	9	5	6
ACG	Antral	2	35	20	53	66	10	8	12
	Fundic	1	13	6	12	6	1	1	6

GM, gastric metaplasia. See Table 2 for other abbreviations.

Table 4. Type of gastritis in antral or fundic region in patients with heterotopic fundic mucosa in the duodenum ($n = 69$)

Histology	Antral biopsies (n)	Fundic biopsies (n)
ISG	12	11
ICG	12	1
ASG	–	–
ACG	7	2
Normal	31	30
fundic gland cyst	–	7
Other	7	16

See Table 2 for abbreviations.

were analyzed separately. If gastric metaplasia was present in patients without duodenal ulcer, gastritis was not necessarily associated. In 23 of 139 patients with gastric metaplasia (16%) *H. pylori* was found in the duodenal biopsy. Nineteen of these patients had duodenal ulcer, 22 had active antral gastritis, and 16 had active fundic gastritis as well. Only one had normal fundic mucosa.

In 69 patients with heterotopic fundic mucosa in the duodenum only nine had *H. pylori* infection. With regard to the mean age of this group of patients, a surprisingly small percentage with active gastritis was seen (Table 4). Only nine patients had polypoid foveolar hyperplasia and stromal edema similar to the lesion known as reflux gastritis [1]. Within the heterotopic mucosa two patients showed active granulocytic inflammation with some epithelial defects. The others, including three patients in whom *H. pylori* was found on the heterotopic mucosa in the duodenum, had only mild and nonspecific inactive inflammation.

Discussion

As could be expected, gastric metaplasia was most frequent in patients with manifest duodenal ulcer. It was also seen in one third of patients with endoscopic diagnosis of gastritis or with gastric ulcer. None of these patients had had a history of duodenal ulcer. Whereas the overall analysis implied a higher incidence of associated *H. pylori* infection and active type of gastritis in patients with gastric metaplasia, this no longer held true when the analysis was performed in groups 1–4 separately. The impression of a higher incidence of associated *H. pylori* infection or active gastritis in patients with gastric metaplasia was clearly caused by the high percentage of patients with duodenal ulcer, which contributed more than half of all patients with gastric metaplasia. In all studies conducted so far and in our own experience duodenal ulcer is most frequently associated with *H. pylori* infection and active gastritis [3, 4]. The few cases of duodenal ulcer without detectable gastric metaplasia still had a high rate of *H. pylori* infection and active gastritis. It is therefore not evident that *H. pylori* infection and active gastritis should directly contribute to the evolution of duodenal ulcer by inducing gastric metaplasia, microbial colonization, and eventual ulceration. This is even more unlikely since patients without doudenal ulcer showed no significant difference of *H. pylori* infection rate or incidence of active gastritis. Although the number of patients with gastric metaplasia, *H. pylori* infection, and gastritis but without duodenal ulcer is small, it should be expected that some should have a history of previous duodenal ulcer if there really were a significant link between the triad of gastritis, *H. pylori* infection, and gastric metaplasia and the risk of duodenal ulcer. Unless such a relationship can be shown in a prospective study of patients with this triad, the hypothetical explanation for the possible pathogenetic mechanism of *H. pylori* in duodenal ulcer offered by the gastric metaplasia, bacterial colonization, and ulceration sequence should be questioned.

The evaluation of patients with heterotopic fundic mucosa in the duodenum demonstrated that this lesion is not only morphologically different from gastric metaplasia; it is a coincidental finding endoscopically, in most instances identified as a small polypoid lesion evoking the macroscopic differential diagnosis of Brunner gland hyperplasia or heterotopic mucosa. Considering the age of these patients, the low incidence of associated active gastritis is even more significant when compared to patients with gastric metaplasia. The ulceration of this heterotopic mucosa is a rare phenomenon which was not observed in our patients, even in those who had *H. pylori* infection and gastritis or in the three patients in whom *H. pylori* was found on the heterotopic mucosa. This observation lends no further argument to the hypothesis of a gastric metaplasia, microbial colonization, and ulceration sequence [7, 8].

Gastric metaplasia in the duodenum was a well-known phenomenon before *H. pylori* infection and its implications were realized. Since it is most often observed in ulcerated or eroded mucosa or mucosa recovering from an injury, it was simply accepted as evidence of a disturbance in terminal differentiation of regenerating epithelia. In our opinion, more and convincing data are needed before this view is changed in favor of a causal relationship to mucosal damage.

References

1. Dixon MF, O'Connor HJ, Axon ATR, King RFJG, Johnston D (1986) Reflux gastritis: distinct histopathological entity? J Clin Pathol 39:524–530
2. James AH (1964) Gastric epithelium in the duodenum. Gut 5:285–294
3. Koch HK, Baumert B, Koch U, Oehlert M, Oehlert W (1990) Prevalence of campylobacter pylori as demonstrated by histology or CLO-test in different types of gastritis. A study in 5 clinically predefined groups of patients. Pathol Res Pract 186:154–158
4. Mc Kinlay AW, Upadhyay R, Gemmell CG, Russell RI (1990) Helicobacter pylori: bridging the credibility gap. Gut 31:940–945
5. Patrick WJA, Denham D, Forrest APM (1974) Mucous change in the human duodenum: a light and electron microscopic study and correlation with disease and gastric acid secretion. Gut 15:767–776
6. Whitehead R (1985) Mucosal biopsy of the gastrointestinal tract. In: Bennington JL (ed) Major problems in pathology, vol 3, 3rd edn. Saunders, Philadelphia, pp 41–58
7. Wyatt JI (1989) Campylobacter pylori, duodenitis and duodenal ulceration. In: Rathbone BJ, Heatley RV (eds) Campylobacter pylori and gastroduodenal disease. Blackwell, Oxford, pp 117–124
8. Wyatt JI, Rathbone BJ, Dixon MF, Heatley RV, Axon ATR (1988) Campylobacter pylori and development of duodenal ulcer. Lancet I:118–119

Evolution of Gastritis After *Helicobacter pylori* Eradication

P. Vincent[1], H. Lamouliatte[2], I. Darchis[3], M. Lecomte[4], F. Mégraud[5], and A. Mascarel[6]

Introduction

Since the first description of *Helicobacter pylori* [8], many studies have confirmed the presence of the bacteria in different locations of the stomach [4, 5, 9]. It is well recognized that *H. pylori* can survive in gastric juice at low pH [2] with the help of its urease [1] by ammonia production from the urea of both gastric juice and mucosa [5]. Also, *H. pylori* can colonise a formerly normal gastric mucosa, and its presence is closely related to histological signs of polymorphonuclear cell invasion and atrophic glandular lesions, even without clinical symptomatology.

However, the pathogenic potential of *H. pylori* remains unclear. High immunoglobulin G (IgG) levels to *H. pylori* have been reported in patients without type A gastritis [3]. The relation to lymphocytic gastritis is not constant [3, 7], and some cases of a spontaneous disappearance of *H. pylori* have been reported [7]. Finally, more data are needed to prove the aetiological role of the bacteria in gastritis. Koch's postulates, for instance, imply that the cure of causal agent must be followed by a disappearance of the disease.

Since the clinical evolution of the non-ulcer dyspepsia syndrome is difficult to measure and a reduction of pain is sometimes seen in spite of ineffective treatment (placebo effect), the aim of this study was to follow up the evolution of histological lesions of gastritis when *H. pylori* was eradicated.

Patients and Methods

We studied 161 antral and fundic biopsies from 32 patients (treated or not), at 1 and 4 months after the first gastroscopy, and compared the histological evolution of gastric mucosa according to the bacteriological persistence or eradication of *H. pylori*. A total of 32 patients, [25 men aged 23–66 (mean

[1] Department of Microbiology, Faculty of Medicine, 1 pl. de Verdun, 59045 Lille Cedex, France
[2] Department of Gastroenterology, CHR, Bordeaux, France
[3] Department of Gastroenterology, CHR, Lille, France
[4] Department of Histopathology, CHR, Lille, France
[5] Department of Microbiology, CHR, Bordeaux, France
[6] Department of Histopathology, CHR, Bordeaux, France

J.M. Pajares et al. (Eds.)
H. pylori and Gastroduodenal Pathology

= 38.5 ± 13.1) and seven women aged 24–59 (mean = 37.7 ± 12.4)] with histologically and bacteriologically proven *H. pylori* gastritis were randomly treated for 1 month either with amoxicillin (2 g per day) and tinidazole (1.5 g per day) ($n = 20$; M/F ratio = 9; mean age = 40.4 ± 13.9) or by placebo ($n = 12$; M/F ratio = 1.4; mean age = 34.8 ± 10.4). Patients were followed up, and a control gastroduodenoscopy was made at 1 month and 4 months after the treatment ($n = 31$ and $n = 20$ patients, respectively).

Before each endoscopy, the fibreoptic gastroduodenoscope was disinfected in a 2% glutaraldehyde solution for 10 min and then rinsed with distilled sterilised water. Biopsies (one from the gastric antrum and one from the fundus) were harvested for both histological and bacteriological examination. Haematoxylin and eosin staining was used for histological observation. The gastritis grade according to Whitehead, presence or not of polynuclear cells and presence or not of *H. pylori* were noted. To confirm the viability of *H. pylori*, samples were plated onto selective and unselective media, incubated under microaerophilic conditions for, up to 5 days for bacterial identification (colony and cell morphology, Gram staining and urease).

From a bacteriological point of view, biopsies were classified as follows: (a) persistence: when *H. pylori* was cultured from the biopsy (no treatment or therapeutic failure); (b) clearance: no *H. pylori* after 1 month; and (c) eradication: no *H. pylori* after 4 months. The chi-square test was used for statistical comparisons.

Results

At the first endoscopy, all the patients had viable *H. pylori* in at least one biopsy. In the antrum, 93.8% of the 32 biopsies had *H. pylori*, and all of these also had polymorphonuclear cells. Whitehead graduation of the gastritis was as follows: 1, 31.3%; 2, 46.9%; 3, 18.7%; and 4, 3.1%. In the fundus, 75.0% of the 32 biopsies had *H. pylori*, and from these 79.2% also had polymorphonuclear cells. Whitehead graduation of fundic lesions was as follows: 0, 12.5%; 1, 59.4%; 2, 28.1%; and 3 or 4, 0%.

Evolution of Polymorphonuclear Cells in the Whole Biopsies. After 1 month 88% of both antral and fundic biopsies with *H. pylori* still had polymorphonuclear cells, versus 18.4% of the biopsies where *H. pylori* had been cleared (Table 1). After 4 months, there were, respectively, 84.2% and 14.2%. These differences were highly significant ($p < 0.001$ at 1 month and $p < 0.001$ at 4 months). According to the location of lesions, whatever the time, the antrum had 96.3% of biopsies with polymorphonuclear when *H. pylori* was persistent versus 24.0% when *H. pylori* was cleared or eradicated. The fundus had, respectively, 70.6% and 11.7% ($p < 0.001$ and $p < 0.01$ for antrum and fundus respectively).

Evolution of Atrophic Lesions. After 1 month 78.2% of the biopsies with *H. pylori* had a stable or increasing atrophic grade versus 55.1% when *H. pylori* was cleared (Table 2). After 4 months, there were, respectively, 84.2% and 21%.

Table 1. Evolution of polymorphonuclear cells in biopsies according to the microbiological status of H. pylori (P.)

	1 month		4 months	
	Antrum	Fundus	Antrum	Fundus
H. pylori positive				
PN +	14	8	12	4
PN −	1	2	0	3
H. pylori negative				
PN −	12	19	7	11
PN +	5	2	1	2

Figures show number of biopsies affected. PN +, polymorphonuclear cells present; PN −, polymorphonuclear cells not present.

Table 2. Evolution of atrophic lesions according to H. pylori microbiological status

Follow up	*H. pylori* microbiological status	Histological Whitehead grade ANTRUM			FUNDUS		
		+	0	−	+	0	−
1 month	Clearance	7	7	3	6	5	1
	Persistence	3	6	4	2	7	1
4 months	Eradication	5	3		6	3	2
	Persistence	2	6	4	1	6	−

Figures show number of biopsies affected; +, decreasing; 0, stable; −, increasing glandular atrophy.

Furthermore, in the antrum, no patient showed increasing atrophic lesions when *H. pylori* had been eradicated for 4 months.

Discussion

These results confirm the close relationship between *H. pylori* and gastritis lesions. Polymorphonuclear cell reaction was stopped early when *H. pylori* was cleared even if a bacteriological recrudescence followed. The glandular atrophy was improved more slowly in a smaller but significant proportion of the cases and no increase in the Whitehead grade was observed in the antrum after eradication.

The parallel between *H. pylori* and mucosal gastritis lesions is a very strong indication of *H. pylori* being a causal agent of the gastritis. However, longer follow up is needed to prove the full disappearance of gastritis when *H. pylori* is definitely eradicated and the reappearance of lesions if there is reinfection.

References

1. Cox DM, McLaren A, Snowden MA (1990) Isolation and characterisation of urease negative variants of H. pylori. Rev Esp Enferm Dig 78 (S1): 29
2. Delmee M, Debongnie JC, Warzee P, Mainguet P (1987) Une méthode originale de prélèvement bactériologique pour l'étude du Campylobacter pylori dans l'ulcère duodénal. Gastroenterol Clin Biol 11: 550–553
3. Dixon MF, Wyatt JI, Burke DA, Rathbone BJ (1988) Lymphocytic gastritis relationship to Campylobacter pylori infection. J Pathol 154: 125–132
4. Goodwin CS, Armstrong JA, Marshall BJ (1986) Campylobacter pyloridis gastritis and peptic ulceration. J Clin Pathol 39: 353–356
5. Graham DY, Klein PD (1987) Campylobacter pyloridis gastritis: the past the present, and speculations about the future. Am J Gastrolenterol 82: 283–286
6. Hazell SL, Lee A (1986) Campylobacter pyloridis, urease, hydrogen-ion back diffusion and gastric ulcers. Lancet 1: 15–17
7. Jones EA, Flejou JF, Potet F, Muzeau F, Molas G, Rotenberg A, Goldfain D. (1990) Lymphocytic gastritis: a clinicopathological study of 32 patients. Eur J Gastroenterol Hepatol 2: 367–372
8. Marshall BJ (1983) Unidentified curved bacilli on gastric epithelium in active chronic gastritis. Lancet 1: 1273–1275
9. Marshall BJ (1986) Campylobacter pyloridis and gastritis. J Infect Dis 82: 283–286

Histological Improvements in Gastritis After Eradication of *Helicobacter pylori* Infection

R. Petrino, D. Bellis, M. Boero, A. Ponzetto, L. Chiandussi, and A. Di Napoli

Introduction

It is worldwide accepted that *Helicobacter pylori* (HP) gastric infection is one of the main causes of chronic antral gastritis: starting from the lamina propria lymphocites and polymorphs infiltrate the interfoveolar space [3, 7, 10]. Treatment of HP can rapidly cause disappearance of the organism and reversion of the abnormalities. However, if the treatment fails to eradicate the organism totally, the improvement will be temporary [12].

Thus the goal of anti-HP treatment is a stable eradication of the bacteria. Single therapy with bismuth salts or antibiotics (amoxicillin, metronidazole, tinidazole, tetracycline, and others) was found to be poorly effective [5, 8, 9]. A combination of these drugs can achieve a better rate of eradication: the best results have been obtained with a triple therapy using bismuth salts and antibiotics for oral treatment of 2–4 weeks [1, 2].

This combined treatment has the disadvantage of a high rate of side effects so that recently the efforts have been devoted to reducing the treatment period. The aim of this study was to evaluate the effect of a short-term triple therapy on HP-associated gastritis.

Materials and Methods

Subjects

Twenty consecutive patients affected by HP-positive non-ulcer dyspepsia were treated for 10 days with colloidal bismuth subcitrate (CBS 120 mg q.i.d.), tinidazole (500 mg b.i.d.), and amoxicillin (1 g b.i.d.). The infection was detected by the histological evaluation of four antral biopsies taken during endoscopy and by a urease test. Six weeks after therapy the patients were submitted to a new endoscopy and the previously cited evaluations were performed.

Patologia Medica B, Ospedale San Vito, Strada San Vito 34, 10133 Torino, Italy

J.M. Pajares et al. (Eds.)
H. pylori and Gastroduodenal Pathology

Histological Evaluation

Hematoxylin and eosin-stained biopsies were examined and the diagnosis of the type of gastritis was made according to Correa's classification [7].

- Superficial gastritis (SG): characterized by a band-like infiltrate of lymphocytes and plasma cells occupying the superficial portion of the gastric mucosa, mostly at the level of the gastric pits and the gland necks.
- Diffuse antral gastritis (DAG): characterized by a dense infiltrate of lymphocytes and plasma cells occupying the full thickness of the antral mucosae. The infiltrate expands the lamina propria and separates the gastric glands giving a false impression of gland loss or athrophy. Lymphoid follicles may be prominent.
- Reflux gastritis (RG): characterized by stromal edema, widening and tortuosity of the foveolar cells, vascular congestion, paucity of inflammatory infiltrate, and occasional smooth muscle fibers in the lamina propria. The hyperplastic foveolar changes may be marked.
- Diffuse corporal atrophic gastritis (DCAG): there is a diffuse loss (atrophy) of the oxyntic glands (corpus and fundus).
- Cluster atrophic gastritis (CAG): independent foci of atrophy accompanied by an inflammatory mononuclear infiltrate which decreases in intensity as the atrophy progresses.

Any of these lesions may have a series of changes which implicate acute injury ("active gastritis"). The grade of activity was made according to the grade of polymorphonuclear infiltrate, depletion of cytoplasmic mucus in foveolar cells, and a mild degree of architectural distorsion of the foveolar portion of the mucosa.

The main morphological changes of inflammation, atrophy, activity were considered for the classification of intensity of gastritis (mild, moderate, severe).

A simple stain (Giemsa stain without differentiation) was used for the detection of HP. The density of the organisms was graded as follows: sporadic microorganism (+), spotty (+ +), and disseminated layer (+ + +).

Statistical Analysis

The results were elaborated statistically as follows: (a) chi-square test to determine if the therapy causes a significant reduction of intensity and activity of gastritis ($p < 0.05$); (b) Spearman test to evaluate whether there is a relationship between HP positivity and histological damage (activity and intensity).

Results

All patients completed the treatment; no significant side effects were reported. At second endoscopy, made 6 weeks after therapy, 14 patients out of 20 (70%) were negative to the urease test for detection of HP infection. This result was

Table 1. Histological pattern of treated patients

		Basal			After therapy		
Patient No.	Sex	HP	Activity	Intensity	HP	Activity	Intensity
1	F	+ + +	Sev	Sev	+	–	Sev
2	F	+ +	Sev	Mod	+	Mild	Mild
3	M	+ + +	Sev	Mod	+ +	Mod	Sev
4	F	+ + +	Sev	Mod	+ +	Mod	Sev
5	M	+ + +	Sev	Mod	+ + +	Mod	Mod
6	F	+ +	Sev	Mild	+ +	Sev	Mod
7	M	+ +	Sev	Mild	–	–	Mild
8	M	+ + +	Mild	Mild	–	–	Mild
9	M	+ + +	Mod	Mod	–	–	Mild
10	M	+	–	Mod	–	–	Mod
11	F	+ + +	Mod	Mild	–	–	Mild
12	F	+ + +	Sev	Mod	–	–	Mild
13	M	+ + +	Mild	Mod	–	–	Mod
14	M	+	Mild	Mild	–	Mild	Mild
15	F	+ + +	Sev	Mod	–	Mild	Mod
16	M	+ + +	Mod	Mod	–	Mild	Mild
17	M	+ + +	Mild	Mild	–	Mild	Mild
18	M	+	Sev	Mod	–	Mild	Sev
19	F	+	Mod	Mod	–	Mod	Mod
20	M	+ +	Mod	Sev	–	Mod	Mild

Sev, severe; Mod, moderate.

confirmed by histological assessement. Table 1 reports the histological pattern of each patient at basal time and after treatment. It was graded under consideration of the activity of gastritis, the intensity of the lesions and the presence of HP.

It appears clear that HP eradication is associated with a decrease in activity of gastritis: in 50% of eradicated cases we observed a complete disappearance of polymorph infiltration ($p < 0.005$, chi-square test) (Fig. 1). The spearman test, used to correlate HP positivity with the activity of gastritis, was highly significant ($p < 0.01$). The therapy did not obtain a significant improvement in intensity of gastritis.

Discussion

Triple therapy for 10 days seems to be effective in the treatment of HP-associated gastritis. Before therapy 60% of the patients presented a layer distribution of HP on the antral mucosa, 20% a spotty distribution, and in 20% of patients HP was present sporadically. One month after treatment HP was eradicated in 70% of patients while a layer of bacteria was observed in only 5% of cases.

The inflammatory process was characterized by lymphocyte and polymorph infiltration of the superficial lamina propria in the earliest stage of gastritis,

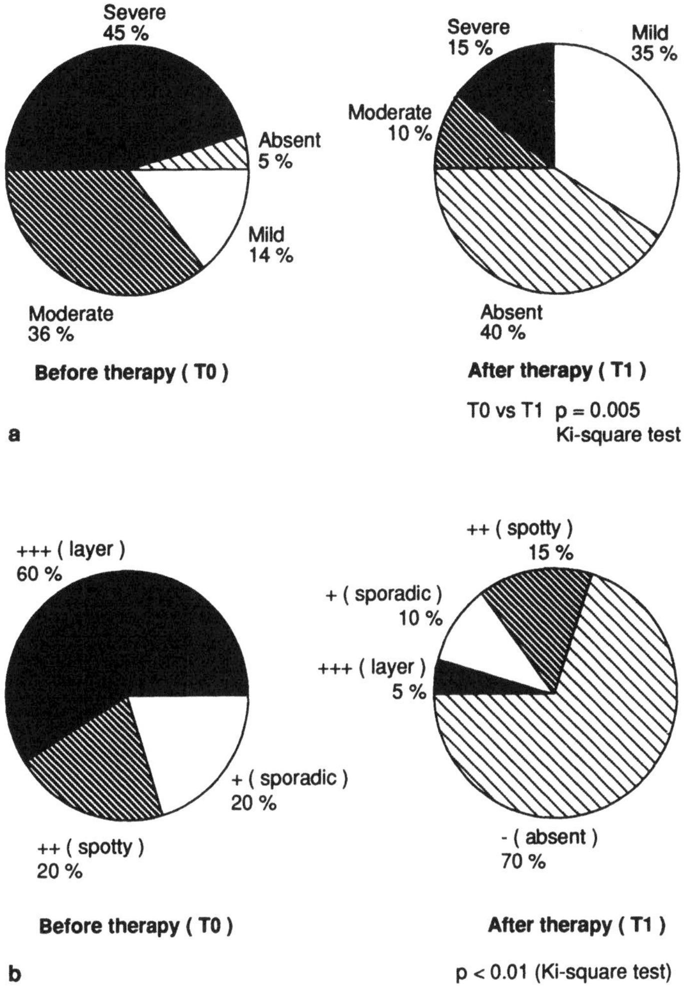

Fig. 1a, b. Histological outcome after therapy. **a** Grade of polymorph infiltration; **b** HP positivity (see text)

while in mild and severe cases infiltration diffused towards the mucosal surface. However, in the patients in whom there was only a mild inflammatory cell infiltration, a mid-zonal infiltrate with spearing of the superficial mucosa could be observed.

In terms of polymorph infiltration, this has been borne out by morphometric methods which have confirmed maximal activity centred on the pit–isthmus region [6]. In attempting to link polymorph response to HP colonization, this represents something of a paradox, since the number of organisms declines steadily from the surface to the foveolar neck. Nevertheless (Table 1), we found an overall correlation, before and after therapy, between the density of HP colonization and polymorph infiltration.

The disappearance of polymorphs, when it occurred, was closely associated with HP eradication. This is in agreement with the observations of Rauws [11] and Glupczynski [5].

The intensity of gastritis was not significantly affected by the reduction or disappearance of HP. It has to be considered that many cases presented CAG, characterized by some histological features such as fibrosis and gland loss which are irreversible lesions. Nevertheless, a regression could be conceivable in earlier stages of gastritis, such as SG and DAG, characterized by lymphocytic infiltration without atrophy. This feature could probably be demonstrated with a longer follow up.

In conclusion, short-term triple therapy is effective for eradicating HP from gastric mucosa and for reducing activity of gastritis even if the intensity of the inflammatory process is unaffected by the disappearance of bacteria 1 month after treatment.

Further study is needed to define the possibility of regression of lesions in earlier stages of gastritis. At the same time, a longer follow up may demonstrate if, even in more severe lesions, a decrease of the intensity can be obtained.

References

1. Borody T, Cole P, Noonan S, Eaves ER, Hansky J (1988) Long term Campylobacter pylori recurrence post-eradication. Gastroenterology 94:43 (abstract)
2. Borsch G Mai U, Opferkuch W (1988) Oral triple therapy may effectively eradicate Campylobacter pylori in man: a pilot study. Gastroenterology 94:44 (abstract)
3. Burette A, Glupczynski Y, Jona C (1986) Signification de la presence du Campylobacter pyloridis dans l'antre gastrique. Acta Gastroenterol Belg 49:70–84
4. Correa P (1988) Chronic Gastritis: a clinical pathological classification. Am J Gastroenterol 83:504–509
5. Glupczynski Y, Berette A, Labbe M, Deprez C, De Reuck M, Deltenre M (1988) Campylobacter pylori associated gastritis: a double-blind placebo-controlled trial with amoxicillin. Am J Gastroenterol 83:365–372
6. Hopwood D (1988) A histometric analysis of gastric biopsies from patients treated with Gastritex: a new drug active against acute or chronic gastritis. J Pathol 54:86A
7. Johnston BJ, Reed PI, Ali MH (1986) Campylobacter-like organism in duodenal and antral endoscopic biopsies: relation to inflammation. Gut 27:1132–1137
8. Marshall BJ, Hislop J, Glancy R, Armstrong J (1984) Histological improvement of active chronic gastritis in patients treated with De-Nol. Aust NZ J Med 14 [Suppl 4]:907
9. Mc Nulty CAM, Gearty JC, Grump B, Davis M, Donovan IA, Melikian V, Lister DN, Wise R (1986) Campylobacter pyloridis and associated gastritis: investigator blind, placebo controlled trial of bismuth salicylate erytromycin ethilsuccinate. Br Med L 293:645–649
10. Misiewicz JJ, Tytgat GNJ, Goodwin CS, Price AB, Sipponen P, Strickland RC, Cheli R (1990) The Sydney System: a new classification of gastritis. Working Party reports, pp 1–10
11. Rauws EAJ, Langenberg W, Houthopf HJ, Zanen HC, Tytgat GNJ (1988) Campylobacter pyloridis associated chronic antral gastritis. Gastroenterology 94:33–49
12. Tytgat GMJ, Rauws E, Langenberg W (1986) the role of Colloidal Bismuth salts in gastric ulcer and gastritis. Scand J Gastroenterol 21 [Suppl 122] :22

Helicobacter pylori in Subtotal Gastrectomies

D. Boixeda[1], A.L. San Roman[1], C. Martin de Argila[1], R. Cantón[2], C. Redondo[3], F. Hernandez Ranz[1], and L. De Rafael[2]

Introduction

Surgery has long been the cornerstone in the management of peptic ulcer disease (PUD) refractory to medical treatment. Although the number of elective interventions has substantially decreased after the marketing of H2 receptor antagonists, the surgical approach maintains its full validity in the management of complicated PUD.

The association between PUD, specially duodenal ulcer, and *Helicobacter pylori* infection is well established. However, the exact contribution of this microorganism to the development of the lesion is still an object of controversy. Different studies have stated that the clearance of *H. pylori* reduces the risk of a recurring ulcer [1, 2].

It has been shown in in vitro experiments that the growth rate of *H. pylori* is inhibited by the addition of a broth of bile salts [3]. The existence of bile reflux after gastric surgery is frequently observed and depends on the surgical technic.

Even though the aforementioned facts should have encouraged more interest, the presence of *H. pylori* in the operated stomach and any possible influence of previous gastric surgery on bacterial colonization has received little interest, and the references in the literature are few.

Our proposal was to study the prevalence of *H. pylori* infection in a group of patients submitted in the past to partial gastrectomy for the management of PUD.

Patients, Methods, and Materials

We studied 22 patients submitted in the past to partial gastrectomy with gastroenteroanastomosis for the treatment of PUD. Six patients had a Billroth I, and the remaining 16 a Billroth II reconstruction. The mean time from surgery was 24.5 ± 9.9 years (57.1% over 25 years). Mean age was 62.5 ± 7 years. Twenty patients were male.

In every case, at least three mucosal samples were taken from each of the following sites: intestinal side of the anastomosis, gastric side of the anastomosis,

[1]Department of Gastroenterology, Hospital "Ramon y Cajal," 28034 Madrid, Spain
[2]Department of Microbiology, Hospital "Ramon y Cajal," 28034 Madrid, Spain
[3]Department of Pathology, Hospital "Ramon y Cajal," 28034 Madrid, Spain

J.M. Pajares et al. (Eds.)
H. pylori and Gastroduodenal Pathology

and gastric fundus. In Billroth I patients, additional samples were obtained from the gastric body. All samples were cultivated on sheep blood (5%) Columbia agar plates, supplemented with polymyxin B (2 mg/ml), trimethoprim (10 mg/l) and vancomycin (15 mg/l), incubated for 7 days at 37°C in an atmosphere containing 5% oxygen, 10% carbon dioxide, and 85% nitrogen. A Gram's staining was also done. The positivity of any of the microbiological methods was considered to be diagnostic of the presence of *H. pylori*.

The remaining samples were processed for histological study following the classical routine (paraffin inclusion and hematoxylin and eosin staining).

Results

The presence of *H. pylori* was detected in 14/22 patients (63.6%). It was never isolated from the samples taken at the intestinal side of the anastomosis. In 13 cases (59%) it was present in the gastric side and in 11 cases (50%) in the gastric body and/or fundus (not significant). There were no significant differences regarding the type of gastroenteroanastomosis (Billroth I 4/6, i.e., 66.7%, and Billroth II 10/16, i.e., 62.5%). The clinical symptoms at the time of the first consultation (nonspecific dyspepsia), the endoscopical diagnosis (remnant gastritis), and the histological diagnosis (chronic gastritis) were uniform. The patient's age did not influence the recovery rate of *H. pylori*.

A correlation between the presence of the microorganism and the time elapsed from surgery could nevertheless be shown: the subgroup submitted to surgery 25 or more years ago had an incidence of *H. pylori* detection of 83.3%, while in those operated upon less than 25 years before, the bacteria could be detected in only 33.3% ($p < 0.01$). In no case submitted to surgery less than 3 years before the sampling could the microorganism be isolated.

Discussion

For many years, surgical management of PUD has been a fundamental resource in the treatment of PUD whenever a failure of medical measures was apparent.

A high prevalence of *H. pylori* detection in patients with partial gastric resections and peptic ulcer has been repeatedly demonstrated [4–6]. However, the prevalence of colonization in similar patients without peptic ulcer, as well as any influence of surgery on this prevalence, have scarcely been studied.

Our results show an overall prevalence of 63.6%, a rate consistent with previously published results (53%–76%) [7–9]. Much lower rates have occasionally been communicated [10]. A possible explanation for this fact is that *H. pylori* colonization rates may increase with the time elapsed from surgery, as our own results seem to show.

In patients submitted to partial gastrectomy, a narrow correlation between the absence of *H. pylori* recovery and the following facts has been demonstrated by O'Connor et al. [11]: high indexes of bile reflux, hypochlorhydria (pH $\geqslant$ 4), and high gastric bile salt concentration ($\geqslant$ 1 mmol/l). Offerhaus et al. [6]

studied a group of patients with previously demonstrated PUD and *H. pylori* infection submitted to different surgical approaches for the treatment of their disease: all patients in whom a Roux-en-Y reconstruction (a procedure virtually eliminating bile reflux) had been chosen failed to clear the microorganism, while it was still present after surgery in 55% of patients submitted to a Billroth II gastroenteroanastomosis. This seems to support the potential of bile reflux in inhibiting the growth of *H. pylori.*

Steer [12] reported on patients submitted to selective or truncal vagotomy for the treatment of PUD: they showed a statistically significant decrease in the total number of *Helicobacter*-like bacteria following surgery. A possible explanation could be a theoretical competitive growth of other bacteria on the operated stomach. These microorganisms would represent the normal colonizing flora of oropharynx and the large intestine.

A frequent discussed topic is a possible role for *H. pylori* in gastric carcinogenesis. The fact that recovery rates rise with the time elapsed from surgery attracted our attention: it is known that the development of gastric remnant neoplasms follows a similar tendency. However, these data must be cautiously considered, given that the number of patients studied is rather low. Larger studies would be needed to ascertain this observation.

References

1. Coghlan JC, Gilligan D, Humphries H et al. (1987) Campylobacter pylori and recurrence of duodenal ulcer. A 12-month follow-up study. Lancet 2:1109–1111
2. Marshall BJ, Goodwin CS, Warren JR et al. (1988) Prospective double-blind trial of duodenal ulcer relapse after eradication of Campylobacter pylori. Lancet 2:1437–1441
3. Tompkins DS, West AP (1987) Campylobacter pylori, acid and bile. J Clin Pathol 40:1387
4. O'Connor HJ, Dixon MF, Wyatt JI, Axon ATR, Ward DC, Deward EP (1986). Effect of duodenal ulcer surgery and enterogastric reflux on Campylobacter pyloridis. Lancet 2:1176–1181
5. Rauws EAJ, Langenberg W, Houthoff HJ, Zonen HC, Tytgat GNJ (1988) Campylobacter pyloridis associated chronic active antral gastritis; a prospective study of its prevalence and the effects of antibacterial and anti ulcer treatment. Gastroenterology 94:33–40
6. Offerhaus GJA, Reu PNMA, Jansen JBMJ, Joosten HJM, Lambers CBHW (1989) Prospective comparative study of the influence of postoperative bile reflux on gastric mucosaal histology and Campyolobacter pylori infection. Gut 30:1552–1557
7. Janisch HD, Witzel L, Niedobitek F, Klein M (1989) Campylobacter pylori in the operated stomach and its relation to symptoms in those patients. In: Megraud F, Lamouliatte H (eds) Proceedings of the 1st workshop on gastroduodenal pathology and Campylobacter pylori. Excerpta Medica, Amsterdam, p 151
8. Loffeld R, Loffeld B, Arends J, Flendrig J, Van Spreeuwel J (1988). The prevalence of Campylobacter associated gastritis in post gastrectomy patients. Has Campylobacter still a role after gastric resection? In: Megraud F, Lamouliatte H (eds) Proceedings of the 1st workshop on gastroduodenal pathology and Campylobacter pylori. Excerpta Medica, Amsterdam, p 152
9. Herz R, Lombardi E, Gregor V, Stolte M (1989) Contribution of Campylobacter pylori infection to chronic gastritis in the operated stomach. Klin Wochenschr 67 (Suppl 18): 30
10. Mackenroth T, Arnholdt H, Burmester E, Herhan J, Loehrs U, Otte M (1989) Gastric or jejunal ulcer and Campylobacter pylori in operated patients. Klin Wochenschr 67 (Suppl 18): 43
11. O'Connor HJ, Wyatt JI, Dixon MF, Axon ATR (1986). Campylobacter-like organisms and reflux gastritis. J Clin Pathol 39:531–534
12. Steer H (1984). Mucosa-related bacteria in the stomach. Lancet 2:528

Helicobacter pylori and Duodenal Ulcer

C. O'Morain and R. Collins

The causation of duodenal ulcer is thought to be an imbalance between aggressive and host resistance factors. Up to now the most important aggressive factor was acid, and there is a dictum stating that if there is no acid there is no ulcer [1]. However, if one analyses this claim it is clear that only 50% of patients who have duodenal ulcers are acid hypersecretors as determined by various stimulatory tests, therefore acid is not the only factor to cause duodenal ulcer [2].

In the 1970s the discovery of H2 antagonist receptors was a significant contribution in the treatment of duodenal ulcer. These agents effectively reduce acid secretion. It was first suggested that the dose of cimetidine should be taken five times a day. Since then there has been progress in that suppressing nocturnal acid secretion is important and that, by administering a single dose at night, is effective in healing ulcer. There are now several pharmaceutical companies making H2 antagonists. There are few advantages of one over the other. Claims are made for some in that they do not interact with other drugs [3], however, this is not usually a major problem in clinical practice.

H2 antagonists can achieve 96% healing of duodenal ulcer if given for a period of 8 weeks [4]. However, once the ulcer is healed it is likely to recur again. up to 80% of patients relapse within 1 year [5]. The strategy for treating patients who frequently relapse is to continue to treat with H2 antagonist at a maintenance dose, usually half the healing dose, or prescribe other drugs such as mucosal protective agents or recommend surgery.

However, these recommendations preceded the discovery of *Helicobacter pylori*, a Gram-negative spiral organism which resides deep in the gastric pits protected from the acidic environment of the stomach by the gastric mucus layer. It was first described in association with chronic active gastritis [6]. However, in the literature there are several reports suggesting urease-producing organisms in patients operated on for peptic ulcer disease [7].

H. pylori—Pathogen or Commensal

Whether *H. pylori* is a pathogen or a commensal has been widely debated. In support of a pathogenic role is that it adheres to cell culture in vitro [8]. Several

Department of Gastroenterology, Meath / Adelaide Hospital and Trinity College, Dublin 8, Ireland

J.M. Pajares et al. (Eds.)
H. pylori and Gastroduodenal Pathology

enteropathogenic organisms have this property. This enables it to bind closely to the gastric epithelium through receptor sites and prevents its dislodgment by peristaltic waves of the gastrointestinal tract, and close adherence can increase local concentration of toxic enzymes. It also elicits an antibody response both locally and systemically [9], and secretes a variety of cytotoxic enzymes including urease, mucinase and phospholipases.

Transmission of *H. pylori*

Studies from both humans and animals show that it can be transmitted [10, 11]. The incidence of *H. pylori* infection increases with age and is more frequent in underdeveloped countries and more common among the poor than the rich. This points to a person-to-person transmission.

We carried out a study of relatives of patients who had *H. pylori* in association with duodenal ulcer and relatives of controls who had a normal endoscopy and in whom culture of the antral biopsy did not grow *H. pylori*. A total of 90% of the relatives of patients with a duodenal ulcer associated with *H. pylori* had a raised antibody level, whereas only 10% of the controls had a high antibody level [12]. *H. pylori* was found in most patients with duodenal ulcer, and those patients in whom it was not found were invariably taking non-steroidal anti-inflammatory drugs [13].

Different Strains of *H. pylori*

Helicobacter pylori is found in 95% of patients with chronic gastritis, 70% of patients with gastric ulcer and 90% of patients with duodenal ulcer. The wide variety of gastroduodenal diseases would point to the supposition that there are different strains associated with the various conditions. In order to see if this is the case we looked at virulence factors. *H. pylori* secretes a wide range of cytotoxic enzymes including urease, oxidase, mucinase and phospholipase. Some of these enzymes are used in its identification. Of particular interest to us is phospholipase [14]. Phospholipases act on lysolecithin, an integral part of the cell membrane, and the product of phospholipase activity on lecithin itself is cytotoxic. There are a number of phospholipases and they are named according to the site where they cleave. Patients who have a duodenal ulcer have more phosopholipase C activity than patients with gastritis. This suggests that the strain associated with duodenal ulcer is more virulent.

H. pylori and Duodenal Ulcer

There is a strong association between *H. pylori* and type B gastritis and peptic ulceration. Colloidal bismuth subcitrate is associated with a lower relapse of duodenal ulcer on account of its antibacterial properties, but it is difficult to prove that *H. pylori* infection causes peptic ulcer disease. We have contributed three important studies to this debate: (a) we found that bismuth, but not cimetidine, improves associated gastritis and suppresses *H. pylori* in duodenal ulcer [15] ; (b) colloidal bismuth subcitrate heals *H. pylori*-positive ulcers but not *H. pylori*-negative ulcers [16] ; (c) eradication of *H. pylori* lowers the relapse rate of duodenal ulcer [17].

We have extended our studies and found the overall recurrence rate in 50 patients in whom we successfully eradicated *H. pylori* to be 10%. This compares more than favourably with the use of H2 receptor antagonist in maintenance treatment. Duodenal ulcers only occurred in patients who were reinfected with *H. pylori*. In a follow up of patients, 18 were reinfected, ten had gastritis and five had duodenal ulcer. This suggests the sequence of events to be *H. pylori* reinfection leading to gastritis and then to a duodenal ulcer.

H. pylori and Pathogenesis of Duodenal Ulcer

The link between *H. pylori* and duodenal ulcer may appear tenuous since *H. pylori* is found only in gastric epithelium. However, gastric metaplasia in the duodenum is found in at least 70% of patients with duodenal ulcer. We also found that *H. pylori* from the antrum can invade the gastric metaplastic tissue [18]. In patients with duodenal ulcer who have been successfully treated with metronidazole and colloidal bismuth subcitrate, gastrin levels fall following successful eradication of the bacteria. *H. pylori* can cause local gastritis and, by increasing local pH, this will result in increased gastrin [19]. Gastrin in turn stimulates acid hypersecretion. A low pH in the duodenum is injurious to the duodenal mucosa and gastric metaplasia occurs in response to the hyperacidity. Gastric metaplasia can be infected by bacteria releasing cytotoxic enzymes and resulting in duodenal ulcer.

Helicobacter pylori therefore plays a central role in the aetiology of duodenal ulcer. The old dictum of "no acid, no ulcer" needs to be revised in that if there is no *H. pylori*, no hypersecretion of acid can occur, and no ulcer will develop.

Colloidal Bismuth Subcitrate

Colloidal bismuth subcitrate has been used for some considerable time in the treatment of peptic ulcer disease. It enables healing by enhancing prostaglandin production, improving the mucus layer and increasing alkaline production [20]. It also has an effect on *H. pylori* in vitro [21]. Previous studies have shown its effect to be superior compared to H2 antagonists in that patients have a lower relapse rate [22–28]. In a study from our unit [26], we randomised patients to one of two treatment regimens: colloidal bismuth subcitrate or cimetidine. The patients were endoscoped at the beginning and at the end of treatment to see if their ulcers had healed and if *H. pylori* was present. The presence of *H. pylori* was documented by histology, Gram stain and culture. We followed up the patients and endoscoped them if they became symptomatic or at the end point of 1 year. We found that patients who remained *H. pylori* positive after their ulcer had healed had an 80% relapse rate, whereas only 10% relapsed of those who had the bacteria eradicated. Patients who were *H. pylori* negative and relapsed had been reinfected with *H. pylori*. Since then there has been worldwide confirmation of this observation. There are eight independent studies with over 500 patients with duodenal ulcer in whom *H. pylori* had been eradicated with a very low relapse rate, and of those who relapsed all were reinfected with *H. pylori* [29–36].

Colloidal bismuth subcitrate only eradicates *H. pylori* in 30% of cases. In our initial studies we took biopsies from the patients immediately after cessation of treatment. It is now generally accepted that 4 weeks should elapse before biopsies are taken and before eradication can be claimed because some drugs have a bacteriostatic effect. We have found that colloidal bismuth subcitrate at a dose of one tablet four times a day is superior to two tablets twice a day [37].

Antibiotic Therapy

Helicobacter pylori is sensitive to a wide range of antibiotics in vitro, and it is logical that antibiotics should be used in treatment regimens. The most successful in eradication is with metronidazole. The ideal antibiotic is one that would be secreted into the stomach. Metronidazole given intravenously is secreted into the gastric juice, whereas ampicillin is not [38]. Metronidazole is not dependent on pH for its activity, which would make it an ideal choice for the treatment of *H. pylori*. A combination of metronidazole and colloidal bismuth subcitrate achieves an eradication rate of up to 70% [39]. However, in some patients *H. pylori* is resistant or develops resistance to metronidazole. Triple therapy with a second antibiotic improves eradication rates, but side effects are common including some potentially serious ones such as *Clostridium difficile*-induced diarrhoea [40].

Our treatment of choice at the moment is triple therapy with colloidal bismuth subcitrate, one tablet four times a day, along with tetracycline 500 mg t.d.s. for 1 week, and metronidazole 400 mg t.d.s for 1 week. With this regimen we get an eradication rate of up to 90% [17]. We feel that the eradication rates would be improved if the regimen were more simple which would improve patient's compliance.

Omeprazole and *H. pylori*

Omeprazole is an innovative drug in that it renders the patients completely achlorhydric by inhibiting the proton pump. We have shown that omeprazole heals duodenal ulcers resistant to H2 antagonists and the associated gastritis also improved [41]. We found that *H. pylori* was suppressed while on treatment.

However, 4 weeks after discontinuing treatment all the patients had *H. pylori* back again. The temporary suppression may be due to achlorhydria as the bacteria do not have to secrete urease to keep their micro-environment alkaline. We have shown that by combining omeprazole with antibiotics, metronidazole and tetracycline, the *H. pylori* can be eradicated in 60% of patients. We are currently studying the efficacy of this regimen long term in preventing relapse.

References

1. Schwartz K (1910) Ueber penetrierende Magen-und Jejunalgeschwüre. Bruns Beitr Klin Chir 67:96–128
2. Tytgat GNT(1987) Colloidal bismuth subcitrate in peptic ulcer—a review. Digestion 37 (Suppl 2):31–41
3. Grant SM, Langtry HD, Brogden RN (1989) Ranitidine. An updated review of its pharmacodynamic and pharmaco-kinetic properties and therapeutic use in peptic ulcer disease and other allied diseases. Drugs 37/6:801–870
4. Arvanitakis C, Nikopoulos A, Theoharidis A et al. (1990). A comparative clinical trial of duodenal ulcer healing with two regimens of cimetidine: 800mg once nightly and 400mg twice daily. J Int Med Res 18/5:430
5. Bardhan K, Cole DS, Hawkins BW, Franks CR (1982). Does treatment with cimetidine extended beyond initial healing of duodenal ulcer reduce the subsequent relapse rate? Br Med J 284:621–623
6. Warren JR, Marshall BJ (1983) Unidentified curved bacilli on gastric epithelium in active chronic gastritis. Lancet i:1273–1275
7. Fitzgerald O, Murphy P (1950) Studies on the physiological chemistry and clinical significance of urease and urea with special reference to the stomach. Ir J Med Sci 292:99–153
8. Goodwin CS, Armstrong JA, Marshall BJ (1986) Campylobacter pyloridis, gastritis and peptic ulceration. J Clin Pathol 39:353–365
9. Rathbone BJ, Wyatt JI, Worsley BW et al. (1986) Systemic and local antibody responses to gastric Campylobacter pyloridis in non-ulcer dyspepsia. Gut 27:642–647
10. Ramsay EJ, Carey KV, Peterson WL et al. (1979) Epidemic gastritis with hypochlorhydria. Gastroenterology 76:1449–1457
11. Morris A, Nicholson G, Lloyd G, Haines D et al. (1986) Seroepidemiology of Campylobacter pyloridis. N Z Med J 89:657–659
12. Collins R, Patchett S, Keane C, Drumm B, O'Morain C (1990). Reinfection with Helicobacter pylori due to intra-familial clustering of the organism. Enferm Digest:0–15 (abstr)
13. O'Riordan TG, Tobin A, O'Morain C (1991) Helicobacter pylori: infection in elderly dyspeptic patients. Age Ageing 20:189–192
14. Daw MA, Cotter L, Healy M, O'Moore R et al. (1990) Phospholipases and cytoxic activity of Helicobacter pylori. Enferm Digest [Suppl I] 78(25):39
15. Dooley CP, McKenna D, Humphreys H, Bourke S et al. (1988) Histological gastritis in duodenal ulcer: relationship to Campylobacter pylori and effect of therapy. Am J Gastroenterol 83/3:278–281
16. Humphreys H, Bourke S, Dooley C et al. (1988) Effect of treatment on Campylobacter pylori in peptic disease. A randomised prospective trial. Gut 29:279–283
17. Patchett S, Beattie S, Lean E, Keane C, O'Morain C (1991) The role of Heliocobacter pylori eradication on the natural history of duodenal ulcer disease. Aliment Pharmacol Ther
18. Tobin A, Leen E, Gilligan D, O'Morain C (1989) Duodenal Campylobacter pylori and gastric metaplasia in duodenal ulcernation. In: Megraud F, Lamouliatte H (eds) Gastroduodenal pathology and campylobacter pylori. Excerpta Medicà, Amsterdam
19. Levi S, Haddad G, Ghosh P et al. (1989) Campylobacter pylori and duodenal ulcers: the gastrin link. Lancet i:1167–1168

20. Konturck SJ, Bilski J, Kiviecien N et al. (1987) De-Nol stimulates gastric and duodenal alkaline secretion through a prostaglandin dependent mechanism. Gut 28:1557–1563
21. Goodwin C, Blake P, Glincow E (1986) The minimum inhibitory and bactericidal concentrations of antibiotics and anti-ulcer agents against Campylobacter pyloridis. J Antimicrob Chemother 17:309–314
22. Smith AC, Price AB, Borriello P, Levi AJ (1988) A comparison of ranitidine and tripotassium dicitrato bismuthate (TBD) in relapse rates of duodenal ulcer. The role of Campylobacter pylori (CP). Gastroenterology 94/5, 2 A431
23. Martin DF, Hollanders D, May SJ, Ravenscroft MM, Tweedle DEF, Miller JP(1981) Difference in relapse rates of duodenal ulcer after healing with cimetidine or tripotassium dicitrato bismuthate. Lancet i:7–10
24. Hamilton I, O'Connor HJ, Wood NC, Bradbury I, Axon ATR (1986) Healing and recurrence of duodenal ulcer after treatment with tripossium dicitrato bismuthate (TDB) tablets or cimetidine. Gut 27:106–110
25. Lee FI, Samlof IM, Hardman M (1985) Comparison of tripotassium dicitrato bismuthate tablets with ranitidine in healing and relapse of duodenal ulcers. Lancet i:1299–1301
26. Coghlan JG, Gilligan D, Humphreys H, McKenna D, Dooley C, Sweeney E, Keane C, O'Morain C (1987) Campylobacter pylori and recurrence of duodenal ulcers-12 months follow-up study. Lancet ii:1109–1111
27. Kang JY, Piper DW (1982) Cimetidine and colloidal bismuth subcitrate in the treatment of chronic duodenal ulcer, comparison of initial healing and recurrence after healing. Digestion 23:73–79
28. Shreeve DR, Klass HJ, Jones PE (1983) Comparison of cimetidine and tripossium dicitrato bismuthate in healing and relapse of duodenal ulcers. Digestion 28:96–101
29. Smith AC, Peice AB, Borriello P, Levi AJ (1988) A comparison of ranitidine and tripotassium dicitrato bismuthate (TDB) in relapse rates of duodenal ulcer. The role of Campylobacter pylori (CP). Gastroenterology 94/5, 2:A431
30. Goodwin CS, Marshall BJ, Blincow ED et al. (1988) Prevention of Nitroimidazole resistance in Campylobacter pyloridis by co-administration of colloidal bismuth subcitrate: clinical and in vitro studies. J Clin Pathol 41:207–210
31. Marshall BJ, Warren JR, Blincow ED et al. (1988) Prospective double-blind trial of duodenal ulcer relapse after eradication of Campylobacter pylori. Lancet ii:1437–1441
32. Borody TJ, Cole P, Noonan S et al. (1989) Recurrence of duodenal ulcer and Campylobacter pylori infection after eradication. Med J Aust 151/8:431–435
33. Rauws EAJ, Tytgat GNJ (1990) Cure of duodenal ulcer associated with eradication of Helicobacter pylori. Lancet 335/8700:1233–1235.
34. Blum AL, Armstrong D, Damman H et al. (1990) The effect of Helicobacter pylori on the healing and relapse of duodenal ulceration. Gastroenterology 98/5:A22
35. George LL, Borody TJ, Andrews P et al. (1990) Cure of duodenal ulcer after eradication of Helicobacter pylori. Med J Aust 153/3:145–149
36. Patchett S, Beattie S, Leen E et al. (1990) The role of Helicobacter pylori eradication in the natural history of duodenal ulcer disease.Enferm Digest:264 (abstr)
37. Coghlan J, Hutchinson L, Gilligan D, McKenna D et al. (1990) Dosage of colloidal bismuth subcitrate in duodenal ulcer healing and clearance of Campylobacter pylori. Aliment Pharmacol Ther 4:49–54
38. Hollingsworth JA, Goldie J, Silletti Li Y, Richardson H, Hunt RH (1987) Gastric secretion of antibiotics used for Campylobacter pyloridis. Gut 28:Ai409
39. O'Riordan T, Mathai E, Tobin A, McKenna D et al. (1990) Adjuvant antibiotic therapy in duodenal ulcers treated with colloidal bismuth subcitrate. Gut 31:999–1002
40. Borody T, Cole P, Noonan S, Morgan A, Ossip G et al. (1988) Long-term Campylobacter pylori recurrence post eradication. Gastroenterology 94/15, 2:A43
41. Daw MA, Deegan P, Leen E, O'Morain C (1991) Effect of short term treatment with omeprazole on Helicobacter pylori and associated gastritis in persistent duodenal ulcer. Aliment Pharmacol. 5:435–439

Helicobacter pylori and the Oesophagus

N. Ectors[1], K. Geboes[1], J. Janssens[2], V. Desmet[1] and G. Vantrappen[2]

Introduction

The occurrence and pathogenic properties of *Helicobacter pylori* have been studied extensively in the stomach [1–7] and in the duodenum [8–11]. Despite the prevalent nature of acid–related disorders in the oesophagus, the role of *Helicobacter pylori* in this localisation has received little or no attention and therefore remains to a large extent unknown.

Helicobacter pylori has been reported to be present in Barrett's oesophagus [12–19]. Of Paull and Yardley's [12] patients with a Barrett's oesophagus and an *H. pylori*-positive gastric biopsy, 40% ($n = 4/10$) showed *H. pylori* in the Barrett's mucosa. Of the patients of R. Talley et al. [13] with a Barrett's oesophagus and an *H. pylori*-positive gastric biopsy, 68% ($n = 13/19$) showed *H. pylori* at the lower oesophageal sphincter and 54% ($n = 7/13$) were colonized at a distance of 2 and/or 4 cm. Of the unselected patients of N. Talley et al. [14] with Barrett's oesophagus, 52% ($n = 12/23$) had *H. pylori* in the oesophagus. Only 15% ($n = 3/20$), 29% ($n = 2/7$) and 23% ($n = 19/82$) of the patients of, respectively, Hazell et al. [15], Walker et al. [18] and Stuart et al. [19] showed the presence of *H. pylori* under the same circumstances. Despite some variation in the figures, there is unanimity in the finding of *H. pylori* in a vast proportion of patients with Barrett's oesophagus. In contrast, there is an obvious divergence in the finding of associated inflammatory changes. All four of Paull and Yardley's patients with *H. pylori* in their Barrett's oesophagus showed acute inflammation of this mucosa [12]. The presence of *H. pylori* was associated with only a mild inflammatory reaction in the patients of R. Talley et al. [13]. Whether *H. pylori* was found or not, the scores for acute and chronic inflammation were similar in the patients of N. Talley et al. [14]. The frequency of ulceration was similar in *H. pylori*-positive or -negative biopsies in Paull and Yardley's patients [12]. Similarly, patients with a Barrett's oesophagus complicated by stricture and/or ulcer (24.4%) were no more likely to be *H. pylori* positive than uncomplicated cases (21.6%) in the group of Stuart et al. [19]. However, all three patients of Hazell et al. [15] (100%) demonstrated active

[1] Laboratory for Histo- and Cytochemistry, Department of Pathology II, U.Z.St. Rafaël, Catholic University of Leuven, Minderbroedersstraat 12, 3000 Leuven, Belgium
[2] Department of Medical Research, University Hospital, Catholic University of Leuven, Minderbroedersstraat 12, 3000 Leuven, Belgium

J.M. Pajares et al. (Eds.)
H. pylori and Gastroduodenal Pathology

ulceration, this was in contrast to only 29% of patients in whom *H. pylori* was absent.

Helicobacter pylori has only been documented in ectopic gastric mucosa of the upper oesophagus in our multicentric study; the results have been discussed at length [20].

In a study on reflux oesophagitis, *H. pylori* was isolated on oesophageal cultures and reported to be present on squamous as well as columnar epithelium [21]. The organisms were found on ulcerated squamous mucosa. *H. pylori* has also been cultured from six squamous epithelium-lined oesophageal biopsies by Walker et al. [18] but they could not be demonstrated morphologically at this site by light microscopy or electron microscopy. In two of these six cases the oesophagus was the only site of isolation.

Materials and Methods

Barrett's Oesophagus

We reviewed endoscopic biopsies from 49 patients with a Barrett's oesophagus. These patients were subdivided according to the presence ($n = 26$) or absence ($n = 23$) of ulceration in the columnar-lined oesophagus. The composition of the mucosa was analysed and described as junctional type, body type and specialized intestinal type mucosa. We assessed the presence and localisation of *H. pylori* together with the presence and nature of an inflammatory reaction in the mucosa. The parameters related to *H. pylori* and the inflammatory reaction were also analysed in gastric biopsies taken during the same endoscopic session ($n = 7$) or taken at another occasion ($n = 24$). The analyses were performed in a semi-quantitative way. The results were evaluated by means of the Fisher exact test. The biopsy material was Bouin's or formalin fixed, paraffin embedded and stained with haematoxylin and eosin and cresylviolet.

Oesophageal Heterotopic Gastric Mucosa

A total of 56 cases of heterotopic gastric mucosa in the upper oesophagus were reviewed in a multicentric study. Gastric biopsies taken during the same endoscopic session were available in 18 cases. The procedure was entirely comparable with the one described for the Barrett's oesophagus cases.

Normal Oesophagus

We examined endoscopic oesophageal biopsies obtained in the upper ($n = 20$), middle ($n = 20$) and distal ($n = 20$) oesophagus of various adult patients. The selection criteria for these biopsies were the presence of squamous epithelium-lined mucosa with submucosal oesophageal glands. Gastric biopsies were available in all cases. Again, a similar procedure was applied.

Results

Barrett's Oesophagus with Ulcer

The Barrett's oesophagus was lined with junctional-type mucosa in 23 cases, specialised intestinal-type mucosa in 20 cases and contained body-type glands in three cases (Table 1). *H. pylori* were present in 17 cases out of 26 (65%). They were mainly found on the mucosal surface as well as in the crypts ($n = 13$) and were numerous in nine cases. In these *H. pylori*-positive biopsies an associated inflammatory process was found in five cases and a lymphoid hyperplasia in four cases.

Gastric biopsies were available in ten cases, half of them were taken during the same endoscopic session. All gastric biopsies were *H. pylori* positive, and nine of them showed an associated inflammatory process, with activity in four cases. Of these *H. pylori*-positive gastric biopsies 70% were associated with an *H. pylori*-positive Barrett's oesophagus with ulcer.

Barrett's Oesophagus Without Ulcer

The lining mucosa was of junctional type in all cases together with specialized intestinal type mucosa in 20 cases and body type glands in two cases (Table 2). *H. pylori* were present in seven cases out of 23 (30%). Again, they were mainly found on the mucosal surface as well as in the crypts ($n = 5$). They were, however, scant in the majority of cases ($n = 5$). Only three biopsies showed an associated inflammatory process.

Table 1. Results—Barrett's oesophagus with ulcer ($n = 26$)

	(n)	(%)
Type of mucosa		
Junctional-type mucosa	23	89
Body-type glands	3	12
Intestinal-type mucosa	20	77
Presence of *H. pylori*	17	65
Associated inflammation	9	35
Gastric Biopsies	10	
Presence of *H. pylori*	10	100
Associated inflammation	9	90

Table 2. Results—Barrett's oesophagus without ulcer ($n = 23$)

	(*n*)	(%)
Type of mucosa		
Junctional-type mucosa	23	100
Body-type glands	2	9
Intestinal-type mucosa	20	87
Presence of *H. pylori*	7	30
Associated inflammation	3	13
Gastric biopsies	14	
Presence of *H. pylori*	14	100
Associated inflammation	6	43

Gastric biopsies were available in 14 cases, but only two were taken during the same endoscopic session. Again, all gastric biopsies were *H. pylori* positive, but with an associated inflammatory reaction in only six cases. This inflammatory reaction appeared active in four cases. Five of these *H. pylori*-positive gastric biopsies were associated with an *H. pylori*-positive Barrett's oesophagus without ulcer.

Oesophageal Heterotopic Gastric Mucosa

The heterotopic gastric mucosa consisted of antral type mucosa in six cases while the other 50 cases contained body-type glands (Table 3). A chronic inflammatory process was found in 23 cases (41%). Activity was only present in three cases. *H. pylori* were observed in three of the 56 cases (5.3%) (Fig. 1). All three cases contained body-type glands. The heterotopic mucosa was normal in one case and showed an active chronic inflammatory process in the remaining

Table 3. Results—oesophageal heterotopic gastric mucosa ($n = 56$)

	(*n*)	(%)
Type of mucosa		
Antral-type mucosa	6	11
Body-type glands	50	89
Associated inflammation	23	41
Presence of *H. pylori*	3	5
Associated inflammation	2	67
Gastric biopsies	18	
Associated inflammation	5	28
Presence of *H.: pylori*	5	28

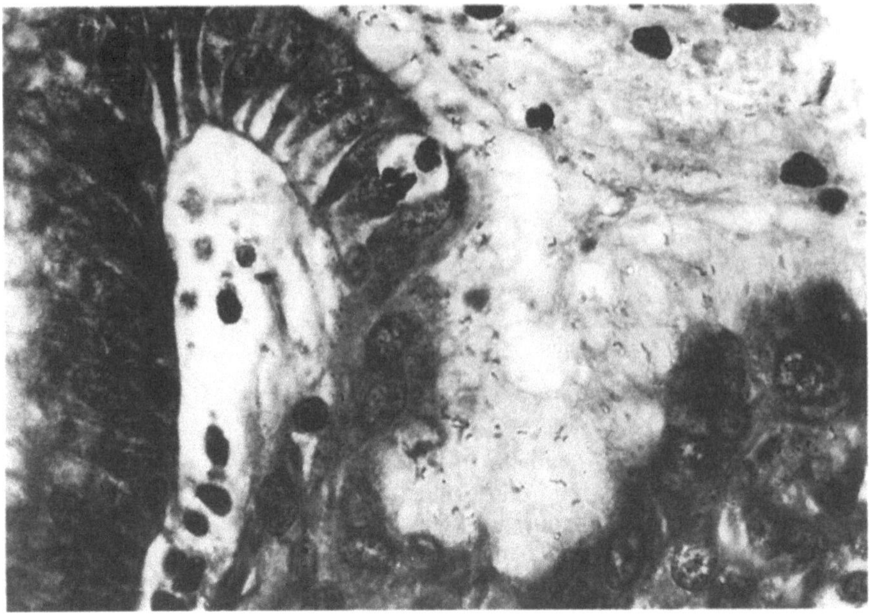

Fig. 1. Numerous *H. pylori* were seen in the mucus overlying the surface of the ectopic columnar epithelium. The epithelium as well as the lamina propria contained a number of inflammatory cells as can be seen in this figure (cresyl violet)

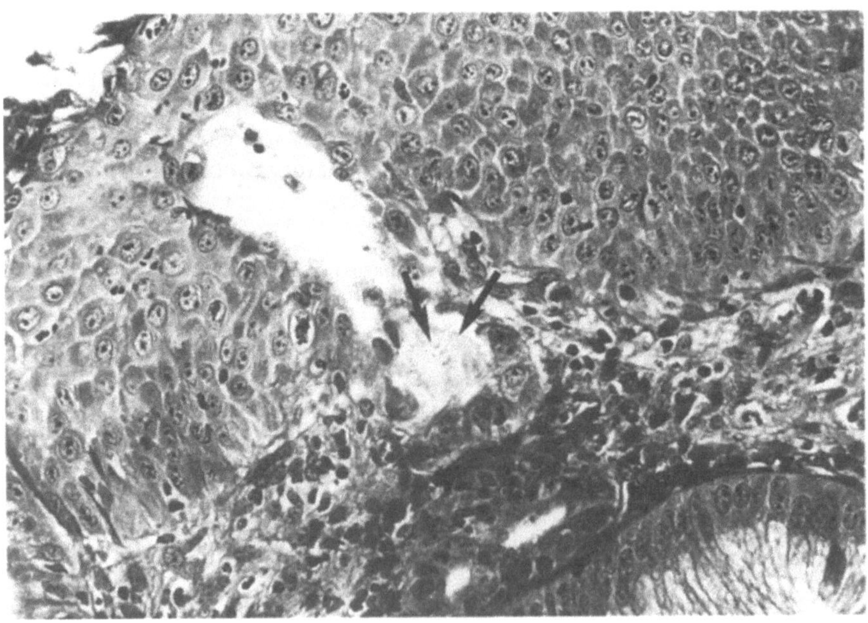

Fig. 2. *H. pylori* were found to be present in the lumen of the oesophageal submucosal glands. In this figure they can be seen in the lumen of the crypt of an oesophageal submucosal gland connecting to the surface through the squamous epithelium (cresyl violet)

Table 4. Results—presence of *H. pylori*

	Total (*n*)	*H. pylori* positive	
		(*n*)	(%)
Barrett's oesophagus with ulcer	26	17	65
Barrett's oesophagus without ulcer	23	7	30
Oesophageal heterotopic gastric mucosa	56	3	5
Normal oesophagus			
Upper	20	4	20
Middle	20	1	5
Distal	20	6	30

two cases. During the same endoscopic session gastric biopsies were taken in 18 patients. Five of these biopsies showed an active chronic gastritis with *H. pylori*. Two of them were associated with *H. pylori*-positive oesophageal heterotopic gastric mucosa.

Normal Oesophagus

In the 20 biopsies from the upper, middle and distal oesophagus the submucosal oesophageal glands contained *H. pylori* in four, one and six biopsies, respectively (Fig. 2; Table 4). These figures correspond to 20%, 5% and 30%. Gastric biopsies were available from all these cases. All the *H. pylori*-positive oesophageal biopsies were associated with *H. pylori*-positive gastric biopsies. No associated inflammatory reaction was found near the submucosal glands.

Discussion

Since the rediscovery of *H. pylori* [1], attention has been focussed on its presence and role in the pathogenesis of gastric diseases [2–7]. Nowadays, *H. pylori* is strongly implicated in the aetiology of non-immune chronic gastritis and peptic ulcer disease [1–11]. Morphological observations demonstrated the close relationship of *H. pylori* to gastric mucin cells. Heterotopic and metaplastic gastric epithelium can be found in all portions of the alimentary tract. By analogy *H. pylori* was looked for on gastric mucin cells in other, heterotopic or metaplastic, localisations. *H. pylori* is present in areas of gastric metaplasia in the duodenum and has been implicated in the aetiopathogenesis of duodenal ulcers [8–11]. *H. pylori* has also been reported to colonise heterotopic gastric mucosa in the rectum [22]. The search for *H. pylori* in gastric mucosa in Meckel's diverticulum was successful too [23]. Although the search for *H. pylori* in Barrett's oesophagus, as a complication of an acid-related disorder, seems obvious, it has received a somewhat limited interest [12–19].

Barrett's Oesophagus

First we focussed our attention on Barrett's oesophagus. *H. pylori* is present in Barrett's oesophagus with an incidence of 40% and 68% in patients with *H. pylori* in the stomach [12, 13]. Independently of the gastric *H. pylori* status, the incidence varies from 15% to 52% [14, 15, 18, 19]. However, as far as the relationship with inflammation and ulceration is concerned, the literature findings are controversial. *H. pylori* has [12, 13] or has not been [14] associated with active and/or severe inflammation and has [15] or has not been [12, 19] associated with ulceration.

Our results show an overall incidence of 49%. The presence of ulceration, however, appeared to be a discriminating factor. Barrett's oesophagus with or without ulceration was associated with *H. pylori* in 65% and 30% of cases, respectively. This difference is statistically significant ($p = 0.015$). Hazell et al. [15] did not statistically analyse their results obtained in a rather limited group [15]. However, their results are also significant ($p = 0.049$). It may be argued that the difference in associated inflammation (35% versus 13%) may be related, in a number of cases, to the ulceration rather than directly to the presence or absence of *H. pylori.* However, although in all our cases of Barrett's oesophagus gastric *H. pylori* were present, it appeared that this presence was related to an inflammatory reaction in the stomach in 90% and 43%, respectively, in the Barrett's oesophagus patients with or without ulceration. These results are also statistically significant ($p = 0.024$). These results lead us to suggest an aetiological role for *H. pylori* in the ulceration of Barrett's oesophagus. Considering the findings related to the associated gastritis, this role may eventually be linked to a variable pathogenic potential of different bacterial strains. This possibility has already been put forward for the gastric mucosa [24]. The importance of *H. pylori* in Barrett's oesophagus may be corroborated by the clinical experience that reflux oesophagitis improves under colloidal subcitrate and cimetidine therapy [21].

Oesophageal Heterotopic Gastric Mucosa

We assessed the presence of *H. pylori* in oesophageal heterotopic gastric mucosa. *H. pylori* were observed in three out of 56 (5.3%) heterotopic gastric mucosa cases. All three cases were of body-type mucosa. The mucosa appeared normal in one case but showed an active chronic inflammation in the remaining two cases. This finding contrasts with the absolute absence of *H. pylori* in Barrett's oesophagus above 4 cm proximally of the lower oesophageal sphincter as reported by R. Talley et al. [12], but should not be unexpected since *H. pylori* can be found as far away from the stomach as in the rectum [22]. The finding of *H. pylori* in heterotopic gastric mucosa in the upper oesophagus could be explained by the existence of gastro-oesophageal reflux as well as by the location of the ectopy on the infection route. Although initially thought to be of a congenital nature, oesophageal heterotopic gastric mucosa could share a common aetiopathogenetic link with Barrett's oesophagus [25]. Besides, *H. pylori* has never been cultured from the oropharynx [18]. The prior hypothesis may therefore be more likely.

Normal Oesophagus

The results obtained by Walker et al. [18] and Borkent and Beker [21] prompted us to look for *H. pylori* in squamous oesophageal epithelium. Walker et al. cultured *H. pylori* from oesophageal biopsies in 27% ($n = 8/30$) of patients without being able to identify *H. pylori* in the oesophagus by light microscopy or electron microscopy. Only two of these patients had gastric-type mucosa in their oesophagus. Borkent and Beker isolated *H. pylori* on oesophageal cultures in 45% ($n = 9/20$) of their patients with severe reflux oesophagitis. We report the presence of *H. pylori* by light microscopical examination in submucosal oesophageal glands. The incidence varied between upper (20%), middle (5%) and the distal third (30%) of the oesophagus and may be related to the distribution of the oesophageal glands [26]. The micro-organisms were seen in the mucus present within the glands. No associated inflammatory changes were found. These findings probably explain the results obtained by Walker et al. and Borkent and Berker. Submucosal oesophageal glands are not often present in endoscopic biopsies, and, if present, attention is not likely to be caught if no pathological changes are present. The absence of inflammation is probably due to the absence of acid, confirming the importance of both *H. pylori* and acid in the aetiopathogenesis of peptic lesions [11]. Like the fundus, the oesophagus could function as a reservoir and be the source of re-infection in a treated patient. All our positive patients also showed *H. pylori* in their gastric biopsies. However, Walker et al. [18] found two cases where the only site of isolation appeared to be the oesophagus. From therapeutic point of view, the most important difference in the fundus or the oesophagus functioning as a reservoir would be the lack of direct/close contact between *H. pylori* and the drugs. From clinical experience it seems this contact is important for therapeutic success.

These findings should caution against the idea that, if patients do not have gastric *H. pylori*, they will not have *H. pylori* in other sites such as the (Barrett's) oesophagus.

References

1. Marshall BJ, Warren JR (1984) Unidentified curved bacilli in the stomach of patients with gastritis and peptic ulceration. Lancet i: 1311–1315
2. Rathbone BJ, Wyatt JI, Heatley RV (1986) Campylobacter pyloridis—a new factor in peptic ulcer disease? Gut 27: 636–641

3. Rathbone BJ, Wyatt JI, Worsley BW, Shires SE, Trejdosiewicz LK, Heathley RV, Losowsky MS (1986) Systemic and local antibody responses to gastric Campylobacter pyloridis in non-ulcer dyspepsia. Gut 27:642–647
4. Blaser MJ (1987) Gastric Campylobacter-like organisms, gastritis and peptic ulcer disease. Gastroenterology 93:371–383
5. Wyatt JI, Dixon MF (1988) Chronic gastritis—a pathogenetic approach. J Pathol 154:113–124
6. Yardley JH, Paull G (1988) Campylobacter pylori: a newly recognized infectious agent in the gastrointestinal tract. Am J Surg Pathol 12:89–99
7. Rauws EAJ, Langenberg W, Houthoff HJ, Zanen HC, Tytgat GNJ (1988) Campylobacter pyloridis-associated chronic active antral gastritis. A prospective study of its prevalence and the effects of antibacterial and antiulcer therapy. Gastroenterology 94:33–40
8. Johnston BJ, Reed PI, Ali MH (1986) Campylobacter-like organisms in duodenal and antral endoscopic biopsies : relationship to inflammation. Gut 27:1132–1137
9. Wyatt JI, Rathbone BJ, Dixon MF, Heatley RV (1987) Campylobacter pyloridis and acid induced gastric metaplasia in the pathogenesis of duodenitis. J Clin Pathol 40:841–848
10. Andersen LP, Holck S, Povlsen CO Elsborg L, Justesen T (1987) Campylobacter pyloridis in peptic ulcer diease: I. Gastric and duodenal infection caused by C. pyloridis: histopathologic and microbiologic findings. Scand J Gastroenterol 22:219–224
11. Goodwin CS (1988) Duodenal ulcer, Campylobacter pylori, and the "leaking roof" concept. Lancet ii:1467–1469
12. Paull G, Yardley JH (1988) Gastric and esophageal Campylobacter pylori in patients with Barrett's esophagus. Gastroenterology 95:216–218
13. Talley R, Weinstein WM, Marin-Sorensen M, Schneidman D, Reedy TJ, van Deventer G (1988) Campylobacter pylori colonization of Barrett's esophagus. Gastroenterology 94:A454
14. Talley NJ, Cameron AJ, Shorter RG, Zinsmeister AR, Phillips SF (1988) Campylobacter pylori and Barrett's esophagus. Mayo Clin Proc 63:1176–1180
15. Hazell SL, Carrick J, Lee A (1988) Campylobacter pylori can infect the oesophagus when gastric tissue is present. Gastroenterology 94:A178
16. Graham DY (1988) Campylobacter pylori and Barrett's esophagus. Mayo Clin Proc 63:1258–1260 (editorial)
17. Kalogeropoulos NK, Whitehead R (1988) Campylobacter-like organisms and Candida in peptic ulcers and similar lesions of the upper gastrointestinal tract: a study of 247 cases. J Clin Pathol 41:1093–1098
18. Walker SJ, Birch PJ, Stewart M, Stoddart CJ, Hart CA, Day DW (1989) Patterns of colonisation of Campylobacter pylori in the oesophagus, stomach and duodenum. Gut 30:1334–1338
19. Stuart RC, Henihan R, McKenna J, Nolan N, Gorey TF, McEntee G, O'Morain C, Hennessy TPJ (1989) Campylobacter pylori and Barett's mucosa: an association of prognostic significance. Gut: A729
20. Fléjou JF, Potet F, Molas G, Bogomoletz WV, Nasca S, Rigaud C, Feydy P, Ectors N, Geboes K (1990) Campylobacter-like organisms in heterotopic gastric mucosa of the upper oesophagus. J Clin Pathol 43:961–968
21. Borkent MV, Beker JA (1988) Treatment of ulcerative reflux oesophagitis with colloidal subcitrate in combination with cimetidine. Gut 29:385–389
22. Dye KR, Marshall BJ, Frierson HF, Pambianco DJ, McCallum RW (1990) Campylobacter pylori colonizing heterotopic gastric tissue in the rectum. Am J Clin Pathol 93:144–147
23. De Cothi GA, Newbold KM, O'Conner HJ (1989) Campylobacter-like organisms and heterotopic gastric mucosa in Meckel's diverticula. J Clin Pathol 42:132–134
24. Burnie J, Lee W, McNulty C, Dent J (1988) Virulence of campylobacter strains and degree of gastritis. Lancet i:302
25. Bogomoletz WV, Geboes K, Feydy P, Nasca S, Ectors N, Rigaud C (1988) Mucin histochemistry of heterotopic gastric mucosa of the upper oesophagus in adults: possible pathogenic implications. Hum Pathol 19:1301–1306
26. Bargmann W (1967) Histologie and mikroscopische Anatomie des Menschen, 6th edn. Thieme, Stuttgart

Gastritis: A Short Appraisal of Classification and the Sydney System

A.B. Price

The confusion surrounding the classification of gastritis conjures up that in the biblical story of the Tower of Babel. There are many classifications in use [1–11], but there is limited common language. For example, type B gastritis according to Wyatt and Dixon [10] is aetiological, referring to a bacterial gastritis, but Yardley's type B gastritis [9] is morphological and indeed specifically excludes a bacterial pathogenesis, in particular *Helicobacter pylori*. Indeed, for clinicians and pathologists alike, especially those without a detailed knowledge of the field, the situation is nearly as confusing as the classification of lymphomas. To try and solve that dilemma an expert group was set up and produced a working formulation: a simple system which acted as a matrix on which the various classifications could be interrelated. The Working Party on Gastritis, set up by the 9th World Congress of Gastroenterology in Sydney, produced the Sydney System for the Classification of Gastritis [12]. This was prompted by the contradictions between existing classifications and the acute need for widely acceptable terminology following the deluge of new data, the result of the discovery of *H. pylori* [13] as one of the main causes of chronic gastritis.

In this brief critique it is worth stressing that the Sydney System does not incorporate a new discovery but is merely an attempt to rationalize the current classifications. The System has an endoscopic division and a histological division, and both will need to be tested in the field of everyday gastroenterology. Here only a brief resumé of the histological limb is appropriate and a short appraisal of its role in the context of the other classifications and the natural history of gastritis.

The gold standard for any classification must be one based on aetiology. The discovery that *H. pylori* may be responsible for over 80% of cases of chronic gastritis has now made this a feasible goal. To this end, the Sydney System incorporates aetiology along with two other limbs, topography and morphology, into a flexible formula for classification (Fig. 1). Most of the previous classifications have been based on only one of these limbs. The terminology of the System remains purely descriptive. Alphabet letters such as A, B, C to designate patterns [4–6, 10, 11] are abandoned because of past confusion, as are functional terms, e.g. hypersecretory [7]. Only three basic forms of gastritis are recognized: acute, chronic and special forms. Integral to the System is the separate assessment of antral and corpus pathology (two biopsies from each

Department of Pathology, Northwick Park Hospital and CRC, Harrow, Middlesex, UK

J.M. Pajares et al. (Eds.)
H. pylori and Gastroduodenal Pathology

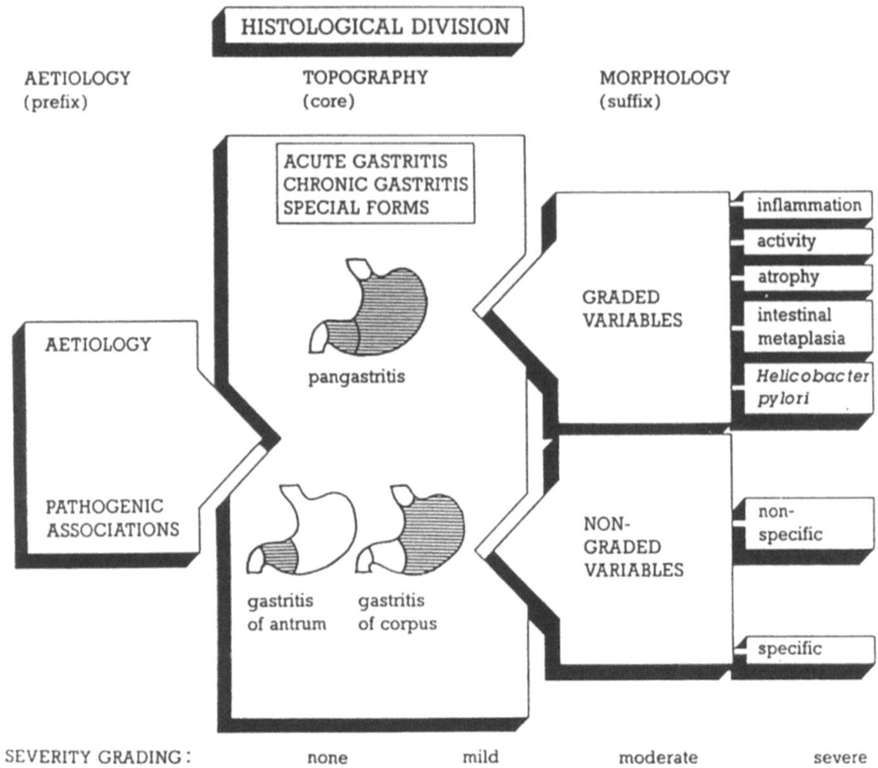

Fig. 1. Histological division of the Sydney System [12]

compartment are recommended) recognizing that each compartment has a differing role in the natural history of gastritis, determining the development of duodenal ulcer, gastric ulcer and even gastric cancer [14, 15].

In more detail, in the morphological limb, inflammation (chronic inflammatory cells), activity (neutrophil polymorph component), atrophy, metaplasia and numbers of *H. pylori* are graded on a simple scale of mild, moderate or severe (0, 1, 2, 3). These five attributes, at the present time, best correlate with outcome and allow scope for quantitation of the progression or response to therapy. Other morphological features, "non-specific" (Fig. 1), are merely documented as appropriate, not graded. The category of "specific" (Fig. 1) refers to granulomas, significant numbers of eosinophils [16], lymphocytes [17] or similar distinctive features that suggest a specific pathogenesis, even though the aetiology may not yet be known. The System recommends any one of the simple special stains for *H. pylori* to be a routine requirement but not yet the detailed breakdown of the three patterns of intestinal metaplasia.

After completion of the morphological limb the distribution of disease is encapsulated in the topography as gastritis of the corpus, gastritis of the antrum or pangastritis, the latter qualified, if necessary, by an indication of the predominant compartment involved. Finally, the aetiology, when known, is integrated into the conclusion according to a formula placing it as the prefix to the topography and morphology as suffix. Two examples of the completed classification might be (a) *Helicobacter*-associated pangastritis, antrum predominant; (b) drug-induced acute antral gastritis with erosions. The drug-induced example illustrates that clinical data as well as pathology are required to complete the classification in some cases.

The System with this combination of aetiology, topography and morphology has several advantages. It has built-in flexibility and it takes account of current knowledge on aetiology. It employs simple morphological terms that ought to be within the capability of general pathologists. It uses a simple grading scale permitting comparisons between biopsies and between investigators. Abandoning the alphabetical codes and any commitment to a particular theory of pathogenesis eliminates semantic confusion yet still provides an easily understood reference matrix for workers too wedded to older classifications to change.

A purist might criticize the Sydney System for not being strictly a classification if defined as a way of grouping nosological entities. It is more a hybrid between a classification and a method of reporting. However, it overcomes the ambiguities amongst the existing classifications, most of which also fall outside the above definition. The main controversies are over chronic gastritis, for acute gastritis, although a clinical problem, is short-lived and rarely seen in biopsy work. However, the acute phase of *H. pylori* gastritis is becoming increasingly recognized [18]. Special forms of gastritis, such as granulomatous gastritis, are seldom problems of classification.

Regardless of classification, certain topographical patterns of chronic gastritis are well recognized in peptic ulcer disease [19]. The chronic gastritis associated with duodenal ulceration is predominantly antral, that of gastric ulceration usually a pangastritis. *H. pylori*-associated gastritis is invariably associated with chronic antral gastritis, though it is often more extensive, especially initially, whilst autoimmune gastritis is a chronic gastritis of the corpus and in a small proportion of cases may extend to include the antrum. Most of the classifications in use recognize these different topographical patterns. The differences lie in how each incorporates what is known of the natural history of gastritis. To some extent these differences reflect the Atlantic divide [20]. American authors, namely Correa [8] and more recently Yardley [9], view *H. pylori*-associated gastritis as a distinct entity that may or may not occur in conjunction with other distinct patterns of gastritis characterized by atrophy and metaplasia. By contrast, mainly based on the longitudinal studies from Finland, gastroenterologists and pathologists in Europe, whatever their terminology, favour a progressive spectrum of change in chronic gastritis. This is likely to be initiated by *H. pylori*, with atrophy and metaplasia then developing in a certain proportion in response to an as yet unidentified set of genetic or environmental factors. The effect of these factors on the pathology of the corpus, antrum or both, in particular the development of atrophy and metaplasia, determine gastric function and hence the risk of peptic ulcer and probably gastric cancer [21–24].

Although the discovery of *H. pylori* was a giant step, providing an aetiology for the large majority of cases of chronic gastritis, it is only a tip-toe towards unravelling the factors that determine its progression. These are now the key issues, for on them rests how peptic ulcer and gastric cancer evolve from chronic gastritis and ultimately whether *H. pylori* requires eradication on a large scale.

This dynamic concept of chronic gastritis is better suited to the Sydney System than other less flexible classifications. In this respect, the malleable nature of the histological division of the Sydney System, combining aetiology, topography and morphology, provides an unambiguous way to document gastric inflammatory disease unfettered by previous concepts.

It would be unrealistic to think it might not need some future modifications. Paramount is the need for rigorous tests of reproducibility between observers and investigators. Other limitations also exist. For example, in chronic gastritis when *H. pylori* is not seen, nor any other obvious aetiology or special form, the unsatisfactory term "idiopathic" still has to be invoked. At present, in routine diagnostic work, one cannot ascertain in such cases if *H. pylori* initiated the inflammation, as seems likely from some serological studies [25], nor can one account for the persistence of inflammation in the absence of the organism. Like "idiopathic", the term "pangastritis" is not ideal, implying diffuse abnormality. In the Sydney System it is used to convey involvement of both gastric compartments by what is probably a patchy process. Finally, the term "reactive gastritis", proposed to replace the terms "reflux gastritis" and "type C gastritis" [26, 27] so as to reflect the histology and accommodate a wider range of aetiologies, may not be strictly appropriate, though adequate for the time being. However, most of the above limitations are semantic and the System does offer workers in the field a clear-cut method of reporting, classification and comparative research. New advances may be integrated into any of the limbs, as may eventual modifications in terminology.

Only use and experience, reflected in publications over the next few years, will show whether this classification finds widespread acceptance.

References

1. Rauws EAJ, Langenberg W, Houthoff HJ, Zanan HC, Tytgat GNJ (1988) Campylobacter pyloridis-associated chronic active antral gastritis. Gastroenterology 94:33–40
2. Whitehead R, Truelove SC, Gear MWL (1972) The histological diagnosis of chronic gastritis in fibreoptic gastroscope biopsy specimens. J Clin Pathol 25:1–11
3. Cheli R, Giacosa A (1983) Chronic atrophic gastritis and gastric mucosal atrophy:one and the same. Gastrointest Endosc 29:23–25
4. Strickland RG, Mackay IR (1973) A reappraisal of the nature and significance of chronic atrophic gastritis. Dig Dis 18:426–440
5. Glass GB, Pitchumoni CS (1975) Atrophic gastritis. Hum Pathol 6:219–250
6. Kekki M, Siurala M, Varis K, Sipponen P., Sistonen P, Nevanlinna RH (1987) Classification principles and genetics of chronic gastritis. Scand J Gastroenterol 22 [Suppl]:1–28
7. Correa P (1980) The epidemiology and pathogenesis of chronic gastritis: three etiologic entities. Front Gastrointest Res 6:98–108
8. Correa P (1988) Chronic gastritis: a clinico-pathological classification. Am. J Gastroenterol 83:504–509
9. Yardley JH (1990) Pathology of chronic gastritis and duodenitis. In: Goldman H, Appelman HD (eds) Gastrointestinal pathology. Williams and Wilkins, Baltimore, pp 69–143

10. Wyatt JI, Dixon MF (1988) Chronic gastritis: a pathogenic approach. J Pathol 154:113–124
11. Stolte M, Heilmann KL (1989) New classification of gastritis. Leber Magen Darm 19:220–226
12. Price AB (1991) The Sydney System: histological division. J Gastroenterol Hepatol 6:209–222
13. Marshall BJ, Warren JR (1984) Unidentified curved bacilli in the stomach of patients with gastritis and peptic ulceration. Lancet i:1311–1315
14. Sipponen P, Seppala K, Aarynen M., Helske T, Kettunen P (1989) Chronic gastritis and gastroduodenal ulcer: a case control study on risk of co-existing duodenal or gastric ulcer in patients with gastritis. Gut 30:922–929
15. Sipponen P, Kekki M, Haapokoski J, Ihamaki T, Siurala M (1985) Gastric cancer risk in chronic atrophic gastritis: statistical calculations of cross sectional data. Int J Cancer 35:173–177
16. Johnstone JM, Morson BC (1978) Eosinophilic gastroenteritis. Histopathology 2:235–246
17. Haot J, Hamuchi, Walley L, Mainguet P (1988) Lymphocytic gastritis: a newly described entity. A retrospective endoscopic and histological study. Gut 29:1258–1264
18. Sobala GM, Crabtree JE, Dixon MF, Schorah CJ, Taylor JD, Rathbone BJ, Heatley RV, Axon ATR (1991) Acute Helicobacter pylori infection: clinical features, local and systemic immune response, gastric mucosal histology and gastric juice ascorbic acid concentrations. Gut 32:1425–28
19. Dixon MF (1991) Helicobacter pylori and peptic ulceration: histological aspects. J Gastroenterol Hepatol 6:125–130
20. Correa P, Yardley JH (1992) Grading and classification of chronic gastritis: one American response to the Sydney System. Gastroenterology 102:355–359
21. Foreman D, Sitas F, Newell, Stacey AR, Boreham J, Peto R, Campbell TC, Li J, Chen J (y1990) Geographic association of H. pylori antibody prevalence and gastric cancer mortality in rural china. Int J Cancer 46:608–611
22. Wyatt J (1991) Gastritis and its relation to gastric carcinogenesis. Sem Diagn Pathol 8:137–148
23. Parsonnet J, Friedman GD, Vandersteen DP, Chang Y, Vogelman J, Orentreich N, Sibley RK (1991) Helicobacter pylori infection and the risk of gastric carcinoma. N Engl J Med 325:1127–1131
24. O'Connor HJ (1992) Helicobacter pylori and gastric cancer: a review and hypothesis. Eur J Gastroenterol Hepatol 4:103–109
25. Karnes WE, Samloff IM, Siurala M, Kekki M, Sipponen P, Kim SWR, Walsh JH (1991) Positive serum antibody and negative tissue staining for Helicobacter pylori in subjects with atrophic body gastritis. Gastroenterology 101:167–174
26. Sobala GM, King RFG, Axon ATR, Dixon MF (1990) Reflux gastritis in the intact stomach. J Clin Pathol 43:303–306
27. Dixon MF, O'Connor HJ, Axon ATR, King RFJG, Johnston D (1986) Reflux gastritis: distinct histopathological entity? J Clin Pathol 39:524–530

V. Immunology

Stress Proteins and Local Immune Response in *Helicobacter pylori*-Associated Chronic Gastritis

L. Engstrand

Introduction

Interest in gastric immunity has increased since the rediscovery of *Helicobacter pylori*. The induction of local immune response elicited by these bacteria has focused the interest on stomach-associated immunology (this part of the gut was almost neglected for a long time due to its considered lack of immunocompetence). Today, histologic evaluation of gastritis requires an immunologic basis.

There still remain significant gaps in our knowledge of the pathogenic mechanisms of *H. pylori*. It is likely that the disease mechanisms are both complex and multiple. Colonization factors as well as disease-causing factors have been proposed as virulence factors. The local immune response to *H. pylori* is well established, and there are characteristic immunohistologic patterns. We and others [3, 19] have previously shown that there is an increased number of T lymphocytes in the epithelium and an induced expression of class II antigens on the epithelial cells. Moreover, lymphocyte aggregates mainly comprising CD4-positive T cell subsets as well as lymphoid follicles with a predominance of B cells and plasmocytosis with a dominance of IgA secreting plasma cells are strongly associated with *H. pylori*-associated chronic gastritis and indicate that the bacteria may initiate local immune response.

Since we found an increased number of intraepithelial T cells in gastric biopsy specimens from patients with *H. pylori*-associated chronic gastritis of the antrum [3], we posed the question whether these T cells express the γ/δ receptor. The close adhesion of *H. pylori* to the epithelial cells [1] and the eventuality of invading organisms (Evans, personal communication) [1, 4] might stress these cells as well as the environment in the stomach possibly stressing the bacteria itself. Therefore we decided to investigate whether *H. pylori* and the gastric epithelium express heat shock proteins.

Stress Proteins

Heat shock proteins (HSP) or stress proteins are a highly conserved group of proteins found in prokaryotic and eukaryotic cells. To cells exposed to stress like

Department of Clinical Microbiology, University Hospital, Akademiska sjukhuset, 751 85 Uppsala, Sweden

J.M. Pajares et al. (Eds.)
H. pylori and Gastroduodenal Pathology

heat, inflammation, and anoxia [16] these HSP become the major transcriptional products. Induction of HSP is rapid and intense [17] but HSP can also be expressed at lower levels in normal cells indicating a physiologic role in maintaining normal cell functions [20]. HSP are the major antigens of many pathogens [14]. Increased synthesis of HSP in these pathogens may be caused by stress imposed by the host [14]. HSP 65 kDa is a major antigen of *Mycobacteria leprae, Mycobacteria tuberculosis* and other species of mycobacteria [5]. A homologue of the 65-kDa HSP of *Mycobacteria bovis* that shares 65% sequence homology on the protein level has been identified in humans [12].

γ/δ T Cells: Frontiers in the Immune System

Data imply that T cells with the γ/δ receptor are specialized to recognize the mycobacterial 65-kDa HSP. In humans, γ/δ T cells account for 1%–5% of peripheral blood T cells and they are distributed mainly in the lymphoid organs [6]. γ/δ T cells are capable of producing cytokines including interleukin 2 (IL2), IL3, IL4, and interferon-γ (IFN-γ) [2, 22]. We have only just begun to understand the physiologic function of γ/δ T cells, and direct demonstration of their role in the immune response is lacking. However, distinct γ/δ T cell subsets may have different roles in immunity. γ/δ T cells localized in epithelial layers form a first barrier between host and pathogen. An increased number of intraepithelial γ/δ T cells have been demonstrated in the intestine of patients with celiac disease [7]. They may be directed against intestinal microorganisms or specialized for the elimination of transformed, infected, or otherwise stressed autologous cells [11]. Perhaps, some γ/δ T cells may have a dual function, both eliminating stressed autologous cells, as well as possessing immunity to bacteria. One hypothesis is that γ/δ T cells are involved in autoimmunity elicited by bacterial infections and cross-react to autologous HSP, which may be expressed at high levels by stressed cells at infection sites [14]. During the last year several reports have shown that T cells with the γ/δ receptor are activated by mycobacterial antigens [8, 10]. HSP 65 kDa-reactive γ/δ T cells have been isolated from peripheral blood after stimulation with purified protein derivative (PPD) [8] and from the synovial fluid of a patient with rheumatoid arthritis [10].

Association of *H. pylori*, 65-kD HSP and γ/δ T Cells

A total of 31 patients were examined by endoscopy because of symptoms in the upper gastrointestinal tract. Two biopsy specimens were taken in the prepyloric area and processed as described elsewere [3]. In 23 patients gastritis was diagnosed histologically, and growth of *H. pylori* was demonstrated by culture. Eight patients showed no histologic gastritis and were *H. pylori* negative by culture. For the immunohistochemical staining we used a mouse monoclonal antibody (MAb) ML30, recognizing the 65-kDa HSP of mycobacteria, a homologous stress protein to the 58–64-kDa protein family encoded by the chromosomal groE1 gene of *Escherichia coli* [9]. This MAb, originally raised to *M. leprae*, reacts with the human homologue of groEl and shows a widespread staining of human tissues [5, 13]. Immunoperoxidase staining using ML30 demonstrated positive staining of groEl in the epithelial cells in all *H. pylori*-positive biopsy specimens (Fig. 1). Two different patterns of epithelial staining

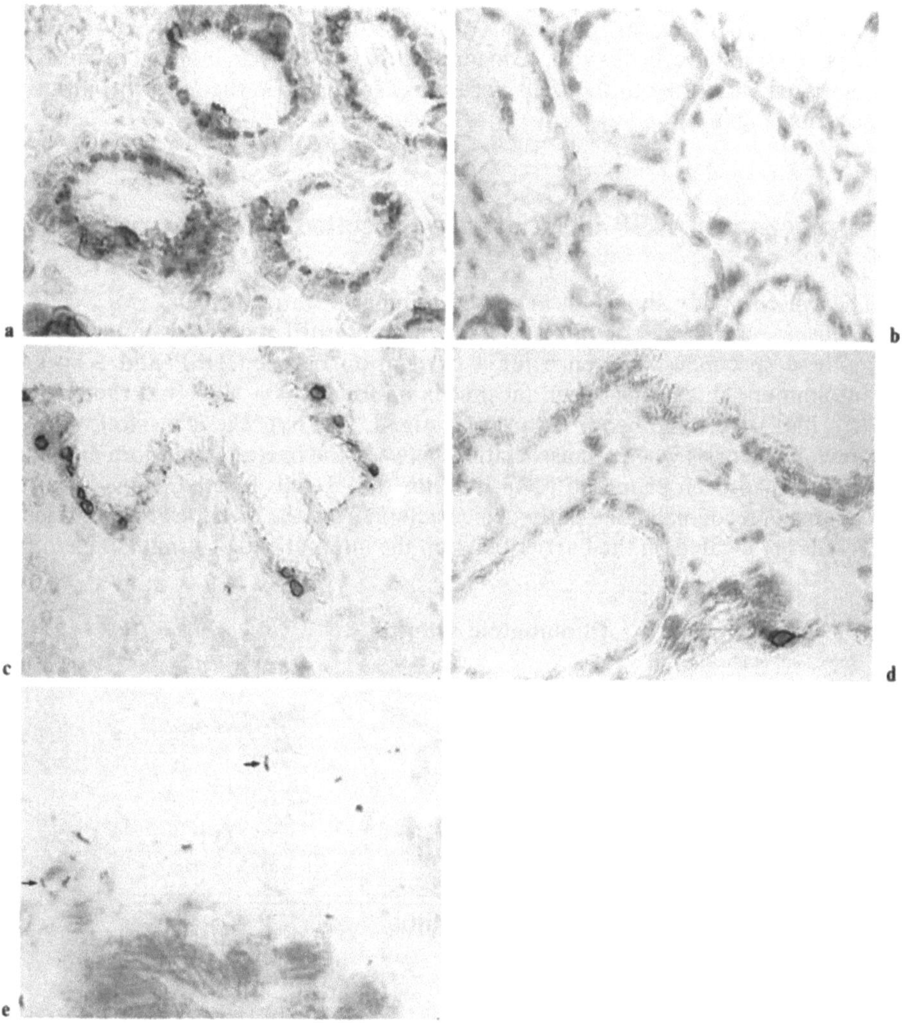

Fig. 1a-e. Immunoperoxidase staining of frozen sections of gastric biopsy specimens from a patient with gastritis (**a, c, e**) and a patient with normal antral mucosa (**b, d**) . Note presence of the groE1 stress protein homologue in epithelial cells (**a**) and an increase of intraepithelial γ/δ T cells (**c**) . *H. Pylori* are indicated by *arrows* (**e**) . In **b** there is no staining of the epithelial cells, and in **d** only single γ/δ T cells are present. Staining was performed as previously described(10). In **a, b** and **e** staining was with the ML30 antibody. In **c** and **d** staining was with a γ/δ T cell-specific antibody. The sections are counterstained with Mayer hematoxylin

was observed: (a) intense staining of groEl located to the area of the Golgi apparatus (Fig. 1a); and (b) more widespread in the cytoplasm with a granular pattern. In the *H. pylori*-negative biopsy specimens staining with ML30 was never observed in the epithelial cells (Fig. 1b). *H. pylori* located in the gastric pits above the epithelial cells were also stained by ML30 (Fig. 1e)

A MAb recognizing a determinant on the δ chain of the human γ/δ T cell antigen receptor was used. In the patients with *H. pylori*-associated chronic gastritis we found an increase of γ/δ T cells preferentially located within the epithelia rather than in the lamina propria; 10–200 cells per biopsy specimen determined in five fields with a magnification of 400 per biopsy specimen as compared with none to five cells per biopsy specimen in the *H. pylori*-negative patients (Fig. 1).

Significance of HSP in *H. pylori*-Associated Chronic Gastritis

The present study shows an increased number of intraepithelial γ/δ T cells in *H. pylori*-positive gastric biopsy specimens and groEl expressing epithelial cells in these specimens. Evidence for a correlation between HSP and a specific autoimmune disease has been most striking for 65-kDa HSP and rheumatoid arthritis. A working theory is shown in Fig. 2. This hypothesis is a follow up to previous theories where an association between the bacterial gut flora and joint diseases had been proposed [18]. It seems that T cells located in the joint are capable of recognizing the pathogenic bacteria from the intestine. Intraepithelial T cells are located on the barrier between the internal and external milieu in the

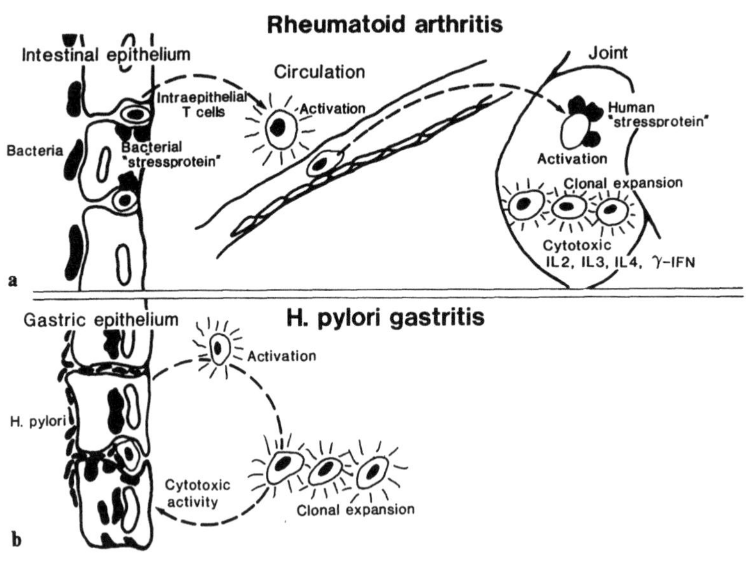

Fig. 2a, b. A working hypothesis for a correlation between HSP and rheumatoid arthritis (**a**) and for a correlation between HSP and *H. pylori*-associated chronic gastritis (**b**) (see text)

gut, thus a kind of frontier in the immune system. Following the challenge by pathogenic bacteria in the intestine, the T cells pass a "clonal expansion" and migrate to the joint. These activated T cells may find their way to the joint either by homing receptors or by recognizing autologous HSP expressed by synovial cells [13]. Cross-reactivity with autologous HSP in the joint may then cause destruction of the synovial cells.

The idea that gastric intraepithelial γ/δ T cells are directed against HSP-synthesizing *H. pylori* and cross-react with autologous HSP in the epithelial cells is attractive. Especially since in vitro as well as in vivo studies describe intracellular uptake of *H. pylori* [1, 4]. Pathogenic mycobacteria live and reproduce inside macrophages, and it can be assumed that the stress imposed by the activated macrophages induces HSP synthesis in their intracellular parasites [15]. If *H. pylori* invades the gastric epithelial cells, they may induce HSP synthesis in these cells, which will then result in a strong primary immune response by intraepithelial γ/δ T cells specialized to recognize this antigen. Whether *H. pylori* is stressed by the host, i.e., the environment in the stomach such as acid, or by the activated epithelial cells remains to be elucidated. Furthermore, entrance of *H. pylori* into the cytoplasm of epithelial cells may explain why the organism can avoid antimicrobial therapy. This may also explain recurrence of infection post treatment.

Our findings support the hypothesis [11] that γ/δ T cells may play a role in the primary host defense. Further studies are required to investigate whether these T cells in *H. pylori*-associated chronic gastritis are specialized to recognize stress proteins expressed both by the bacteria and by the gastric epithelial cells.

References

1. Bode G, Malfertheiner P, Ditschuneit H (1988) Pathogenic implications of ultrastructural findings in Campylobacter pylori related gastroduodenal disease. Scand J Gastroenterol 23 (suppl 142): 25–29
2. Cron RQ, Koning F, Maloy WL et al. (1988) Peripheral murine CD3 +, CD4 −, CD8 − T lymphocytes express novel T cell receptor γ/δ structures. J Immunol 141: 1074–1082
3. Engstrand L, Scheynius A, Påhlson C et al. (1989) Association of Campylobacter pylori with induced expression of class II transplantation antigens on gastric epithelial cells. Infect Immun 57: 827–832
4. Evans DG, Evans DJ Jr, Graham DY (1992) Adherence and internalization of Helicobacter pylori by HEP-2 cells. Gastroenterology 102: 1557–1567
5. Evans DJ, Norton P, Ivanyi J. (1990) Distribution in tissue sections of the human groEl stress protein homologue. APMIS 98: 437–441
6. Groh V, Porcelli S, Fabbi M et al. (1989) Human lymphocytes bearing T cell receptor γ/δ are phenotypically diverse and evenly distributed throughout the lymphoid system. J Exp Med 169: 1277–1294
7. Halstensen TS, Scott H, Brandtzaeg P (1989) Intraepithelial T cells of the TcRγ/δ^+CD8$^-$ and Vδ1/Jδ1$^+$ phenotypes are increased in coeliac disease. Scand J Immunol 30: 665–672
8. Haregewoin A, Soman G, Hom RC et al. (1989) Human γ/δ^+ T cells respond to mycobacterial heat-shock protein. Nature 340: 309–312
9. Hendrix RW, Tsui L (1978) Role of the host in virus assembly: cloning of the Escherichia coli groEl gene and identification of its protein product. Proc Natl Acad Sci USA 75: 136–139

10. Holoshitz J, Koning F, Coligan JE et al. (1989) Isolation of CD4 – CD8 – mycobacteria-reactive T lymphocyte clones from rheumatoid arthritis synovial fluid. Nature 339:226–229
11. Janeway CA (1988) Frontiers of the immune system. Nature 333:804–806
12. Jindal S, Dudani AK, Singh B et al. (1989) Primary structure of a human mitochondrial protein homologous to the bacterial and plant chaperonins and to the 65-kilodalton mycobacterial antigen. Mol Cell Biol 9:2279–2283
13. Karlsson-Parra A, Söderström K, Ferm M et al. (1990) Presence of human 65 kD heat shock protein (hsp) in inflamed joints and subcutaneous nodules of RA patients. Scand J Immunol 31:283–288
14. Kaufmann SHE (1990) Heat shock proteins and the immune response. Immunol Today 11:129–136
15. Koga T, Wand-Württenberger A, DeBruyn J et al. (1989) T cells against a bacterial heat shock protein recognize stressed macrophages. Science 245:112–114
16. Lindquist S (1986) The heat shock response. Annu Rev Biochem 55:1151–1191
17. McMullin TW, Hallberg RL (1987) A normal mitochondrial protein is sectively synthesized and accumulated during heat shock in Tetrahymnea thermophila. Mol Cell Biol 7:4414–4423
18. Olhagen B (1980) Postinfective or reactive arthritis. Scand J Rheumatol 9:193–202
19. Papadimitriou CS, Ioachim-Velogianni EE, Tsianos ER et al. (1988) Epithelial HLA-DR expression and lymphocyte subsets in gastric mucosa in type B gastritis. Virchows Archiv [A] 413:197–204
20. Pelham HR (1989) Heat shock and the sorting of luminal ER proteins. EMBO J 8:3171–3176
21. Sternberger LA (1979) Immunocytochemistry. Wiley, New York
22. Tentori L, Pardoll DM, Zuniga JC et al. (1988) Proliferation and production of IL-2 and B cell stimulatory factor 1/IL-4 in early fetal thymocytes by activation through Thy-1 and CD3. J Immunol 140:1089–1094

Local and Systemic Antibody Responses During *Helicobacter pylori* Infections

A.R. Stacey, P.R. Hawtin, and D.G. Newell

Introduction

The majority of patients with a *Helicobacter pylori* infection mount a systemic antibody response which is easily detectable by serological assays such as the enzyme-linked immunosorbent assay (ELISA) [1]. Positive results in bacterial agglutination and complement fixation tests [2] suggest that these antibodies possess some anti-bacterial activity. However, the infection persists suggesting an underlying ineffectiveness in the ability of the systemic antibody response to eradicate the infection.

There are several plausible explanations for this apparent ineffectiveness [3]. The possibility exists that no active immune response occurs at the site of infection, as a consequence of this the gastric mucosal surface could be an immunologically privileged site. This is not unlikely as little is known about the immunology of the stomach. However, the presence of the substantial infiltrate of immunocompetent cells in areas of *H. pylori*-associated gastritis would appear to negate this possibility. Although the antigenic specificity of the infiltrating T cells is, as yet, undetermined, local antibody responses have been detected [3]. However, in contrast to the systemic antibodies, the major class of antibodies produced by gastric mucosal biopsies and lymph node cell cultures is IgA, although some IgG antibodies of the IgG1, IgG2 and IgG4 subclasses are present. Previous preliminary studies have shown that these local antibodies are directed against a variety of protein fractions of *H. pylori* including the urease and the flagella-containing fractions. However, the molecular basis of the antigenic specificity of these local antibodies has not been investigated. Nevertheless, it seems much more likely that the organism has developed an array of mechanisms by which it is able to avoid or nullify the effects of a locally produced immune response [3].

In this study the antigenic specificity and bactericidal activity of antibodies against *H. pylori* produced close to, or at, the site of infection were investigated following the tissue culture of gastric mucosal biopsies and lymph node biopsies. Additionally, antibodies produced by lymph node cell suspensions following transformation by the Epstein-Barr virus were studied.

The molecular specificity of the antibodies against *H. pylori*, in serum and gastric mucosal biopsy tissue culture supernatants, for *H. pylori* surface proteins was investigated by Western blotting techniques. Such antibodies gave complex

PHLS Centre for Applied Microbiology and Research, Porton Down, Salisbury, SP4 0JG, UK

J.M. Pajares et al. (Eds.)
H. pylori and Gastroduodenal Pathology

patterns of reactivity which were compared with respect to origin and isotype for a group of five patients (Fig. 1). Positive bacterial culture and Gram stain were used as the criteria for *H. pylori* positivity in these patients.

In general the gastric mucosal IgA antibodies (tracks d) gave patterns of reactivity which were reflected by the matching serum IgG antibodies. Although patient-to-patient variation was observed, several polypeptides were consistently immunoblotted, including the urease polypeptides of 28 kDa and 61 kDa, and the 56-kDa and 120-kDa polypeptides. The most frequent reaction was observed with a 54-kDa polypeptide which has been identified as being flagella associated. However, in all cases the serum IgG antibodies demonstrated a broader spectrum of reactivity. This trend was also observed when the IgG antibodies from the serum and the gastric mucosal were compared. Moreover, serum IgA antibodies also reacted with a larger number of polypeptides than the gastric mucosal IgA antibodies. However, in two cases the gastric mucosal IgA antibodies reacted with some polypeptides more strongly than the serum IgA antibodies.

One patient, who was negative for *H. pylori* by both culture and Gram stain, had no anti-*H. pylori* serum antibody response as detected by ELISA, although some weak staining was observed by Western blotting. No local antibody response was detected in this patient by either technique.

The antibodies secreted by lymph node cells, transformed by Epstein-Barr virus in order to increase antibody production, reacted with a 54-kDa polypeptide present in a whole cell-sonicated antigen and also in 17/18 fast protein

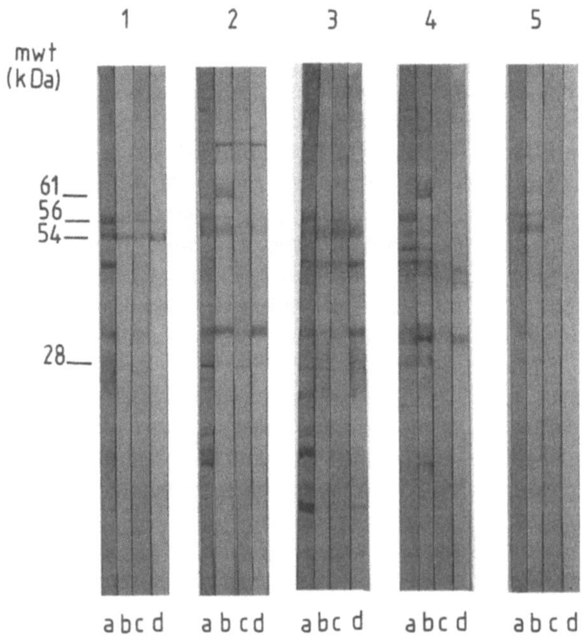

Fig. 1. The patterns of antigenic reactivity for five patients, as detected by Western blotting, with serum IgG (*track a*) and IgA antibodies (*track b*) were compared with gastric mucosal IgG (*track c*) and IgA antibodies (*track d*)

liquid chromatography fractions [3]. This polypeptide has now been identified as the flagellin of *H. pylori*.

Bactericidal Activity of Antibodies Produced in the Local Anti-*H. pylori* Antibody Response

Preliminary studies have been undertaken to establish the biological activity of the locally produced antibodies against *H. pylori*. To this end, the bactericidal activity of these antibodies was studied by incubating *H. pylori* with varying dilutions of mucosal biopsy tissue culture supernatants or matching serum samples for periods of up to 60 min.

Incubation of *H. pylori* with a dilution of serum of 5% or more resulted in the killing of organism; however, there was no difference in the bactericidal activity of sera from patients who were positive or negative for *H. pylori*. Conversely, increasing concentrations of biopsy supernatant did not effect the survival of *H. pylori* after 30 min of incubation, irrespective of *H. pylori* status (Fig. 2). The addition of 2.5% normal human serum as a source of complement had no effect on *H. pylori* survival.

Although bactericidal activity occurred following 60 min of incubation with a 20% concentration of biopsy supernatant containing 2.5% complement (Fig. 3), reduced *H. pylori* survival was noted with supernatants from patients who were positive or negative for *H. pylori*, suggesting that the effect reflected non-specific bactericidal activity of normal serum.

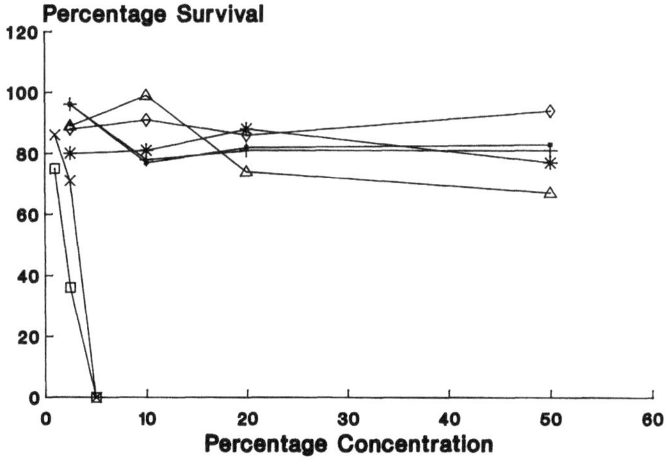

Fig. 2. The survival of *H. pylori* following incubation with increasing concentrations of gastric mucosal biopsy supernatants from *H. pylori*-positive (+) and -negative patients (-■-) was compared with the effect of incubation with matching *H. Pylori*-positive (-×-) and -negative sera (-□-). The surival of *H. pylori* in fresh tissue culture medium (✱) was used as a control. The effect of complement was investigated by the addition of 2.5% human serum to antibody-positive (-△-) and antibody -negative (-◇-) biopsy supernatants

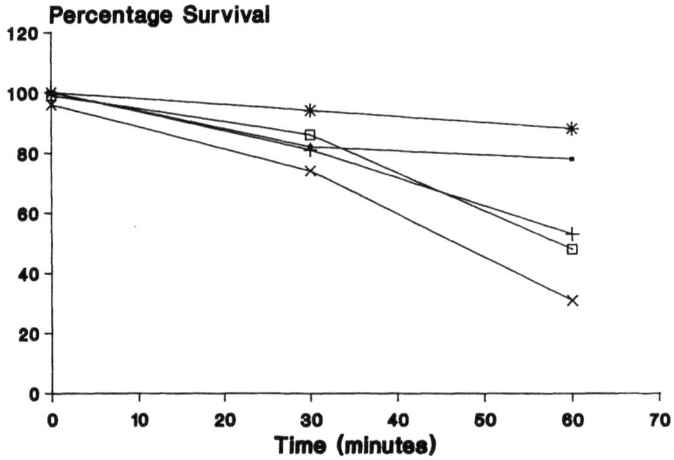

Fig. 3. The survival of *H. Pylori* during incubation with dilutions of antibody-positive (+) and antibody-negative (–■–) biopsy supernatants. Normal human serum (2.5%) was also added to the positive (–×–) and negative (–□–) biopsy supernatants. The survival of the organism during incubation with fresh tissue culture medium (✱) was used as a control

Conclusions

This study has further demonstrated the specificity of the serum anti-*H. pylori* antibody response for *H. pylori* antigens separated by sodium dodecyl sulphate–polyacrylamide gel electrophoresis (SDS-PAGE). These results confirm previous studies using ELISA of fractionated antigens [3] that the patterns of antigenic reactivity demonstrated by the serum antibodies are mirrored by the antibodies produced at, or close to, the site of infection. Notably antibodies from both sources react with the major *H. pylori* surface antigens including urease and the flagella. In general, Western blotting demonstrated that the serum IgG and IgA antibodies had a broader specificity than those antibodies of the same isotypes produced by the gastric mucosal biopsies. This may be due, at least in part, to the production of low concentrations of antibodies, of all specificities, at the gastric mucosal surface which is reflected in the biopsy culture supernatants. However, following amplification of the mucosal antibody response in secondary lymphoid tissue, such as the spleen and lymph nodes, increased levels of all of these antibodies in the serum could give an apparently wider pattern of reactivity.

The presence of the specific antibody response at the gastric mucosa strongly suggests that the ineffectiveness of the immune response during *H. pylori* infections is not due to immunoprivilege of, or immuno-incompetence at, the site of infection. Nevertheless, very little is known about the biological activity of these antibodies.

Although in vitro *H. pylori* was susceptible to the complement-mediated killing of normal and human immune serum, we have been unable to demonstrate any bactericidal activity, specific or otherwise, with the antibodies produced in the local immune response. This may be due to the fact that the majority of the antibodies present were of the IgA class, which is known to be an inefficient activator of complement. The primary immunological role of this class of antibody is to prevent the adherence of pathogens to mucosal surfaces. The ability of these antibodies to prevent the adherence of *H. pylori* to the gastric epithelium needs to be investigated.

References

1. Newell DG, Rathbone BJ (1989) The serodiagnosis of Campylobacter pylori infection—a review. Serodiagn Immunother 3:1–6
2. Jones DM, Lessels AM, Eldridge J (1984) Campylobacter-like organisms on the gastric mucosa: culture, histology and serological studies. J Clin Pathol 37:1002–1006
3. Stacey AR, Hawtin PR, Newell DG (1990) Local immune responses to Helicobacter pylori infections. In: Malfertheiner P, Ditschuneit H (eds) Helicobacter pylori, gastritis and peptic ulcer. Springer, Berlin Heidelberg New York, pp 162–166

Local Immunoglobulin A Subclass Alteration in the Gastric Mucosa of *Helicobacter pylori*-Infected Patients

M.M.C. van der Est[1], R.A. Veenendaal[1], A.S. Peña[2], I. Kuiper[1], and C.B.H.W. Lamers[1]

Introduction

The serum antibody response in patients with *Helicobacter pylori*-associated antral gastritis consists mainly of the immunoglobulin G (IgG) and IgA immunoglobulin classes [1, 2]. IgA represents 15%–20% of the human serum immunoglobulin pool. It is the predominant immunoglobulin in the mucosa of the gastrointestinal tract where it prevents the attachment of microorganisms to the epithelial surface [3]. Another characteristic of IgA immunoglobulins is their inability to activate the classical complement system. This is probably a natural system to prevent local tissue damage which could be caused by a phlogistic IgG or IgE responses. The IgA response probably avoids a T cell-mediated hypersensitivity against non-pathogenic antigens [4].

Studies of IgA subclasses in various secretions and in sera have shown that in serum approximately 90% is of the IgA1 subclass. In secretions only 40%–50% of IgA is of the IgA1 subclass, whereas 60%–50% is of the IgA2 subclass [5, 6]. One reason for this difference in the systemic and local IgA subclass immune response can be found in the structure of the IgA1 and IgA2 immunoglobulins. IgA2 is relatively resistant to proteolytic enzymes produced by several microorganisms whereas IgA1 is not [7, 8].

We have found that the systemic human specific IgA response against *H. pylori* consists mainly of the IgA1 subclass [9]. In this study we amplify these results by investigating the specific local IgA subclasses against *H. pylori* in the gastric mucosa. IgA1 and IgA2 immunoglobulins in biopsy specimens and sera were assayed using specific monoclonal antibodies to these immunoglobulin subclasses with a modified enzyme-linked immunosorbent assay (ELISA) technique. Additionally the total IgA subclass in biopsy specimens was assayed by an ELISA technique.

Patients and Materials

Twelve patients were studied. All patients were referred for upper gastrointestinal endoscopy because of the presence of symptoms compatible with non-ulcer dyspepsia. Antral biopsy specimens were taken during endoscopy for culture,

[1] Departments' of Gastroenterology, Leiden University Hospital, Leiden, and
[2] Free University Hospital, P.O. Box 7057, 1007 MB Amsterdam, The Netherlands

J.M. Pajares et al. (Eds.)
H. pylori and Gastroduodenal Pathology

histology, and measurement of the local antibody response. Serum of each patient was obtained for the determination of specific anti-*H. pylori* antibodies on the day of the endoscopy. All patients had a *H. pylori* infection demonstrated by culture and/or histology. Three of the 12 patients had a *H. pylori* infection according to the criteria above without a significant systemic IgA antibody response to these bacteria.

Biopsy specimens taken during endoscopy for serological tests were weighed and homogenized in 300 μl phosphate-buffered saline containing 0.005% Tween 20 (PSBT). Finally, the homogenates were diluted with PBST to a concentration of 1 mg/100 μl before they were sonicated. The protein concentration of the biopsy specimens was quantitated by the method of Lowry et al. [10].

ELISA for Specific Local and Systemic IgA Antibodies Against *H. pylori*

An antigen suspension was obtained by sonicating a pool of six *H. pylori* strains for 6 min on a Branson sonifier (Stage 4, 30000 cycles per second). The ELISA technique for IgA has been previously described [11]. Briefly, each well of a flat-bottomed polystyrene microtiter plate (Dynatech Laboratories, UK; M129A) was coated with 100 μl antigen solution (1.0 μg/100 μl carbonate buffer pH 9.6) overnight at 20°C. The plates were washed with PBST, whereafter sera diluted 1:200 in PBST, or biopsy specimen homogenates diluted 1:100 in PBST were added to each well. After incubation and washing, goat IgG antihuman-IgA peroxidase (GαHIgA/PO; Pasteur Institutes, Paris, France; code No. 75041 diluted 1:2500 in PBST), specific for the heavy chain of the human IgA, was added to each well. The plates were then incubated, washed, and 100 μl substrate (2,2-azinobis-3-ethylbenzthiazoline sulfonic acid, ABTS; Sigma A-1888 containing 0.005% H_2O_2) for serum; *o*-phenylenediamine (OPD) for biopsy specimens) was added with a multichannel pipet. The incubation time was 30 min for each well, the reaction being stopped with 50 μl 0.01% NaN_3 in 0.1 *M* citric acid. The optical density (OD) was read at 405 nm on the Titertek Multiscan (Flow Laboratories, Irvine, Scotland, UK) plate reader.

ELISA for Local and Systemic Specific IgA1 and IgA2 Subclasses Against *H. pylori*

The previously described ELISA technique was modified to be able to measure the low IgA subclass content expected in sera and biopsy specimens. This ELISA was made more sensitive than the technique described above by using monoclonal antibodies against the subclasses and by changing the colored substrate for the serum ELISA from ABTS to OPD. The optimal OD for this substrate is read at 492 nm. The ODs of the blanks and the control reference sera were used to correct the absorbance.

The ELISA plates for the IgA1 and IgA2 assays were coated as previously described [11]. The plates were incubated with sera diluted 1:20 or biopsy specimen homogenates diluted 1:20 in PBST for 1.5 h at 20°C and washed three times with PBST. Next, the plates were incubated with 100 μl mouse antihuman IgA1, IgA2 monoclonal antibodies (MαHIgA1/MαHIgA2), respectively (Nordic, Tilburg, The Netherlands; code no. 3498 and 3460 diluted 1:2500 and 1:1500, respectively) for 1.5 h at 20°C and washed three times with PBST. As a conjugate 100 μl diluted sheep antimouse IgG (whole molecule) peroxidase (SαMIgG/PO) was added (Sigma, St. Louis, MO 63178, USA; code no. A-6782 diluted 1:5000). After incubation for 1.5 h at 20°C and washing, 100 μl substrate (0.40 mg/ml OPD; Sigma, St. Louis, MO 63178, USA; code no. P-1526; containing 0.4 μl 30% H_2O_2) was added. Care was taken to make sure that the reaction time was the same for all the wells. After 30 min the reaction was stopped by adding 50 μl 2.5 *M* H_2SO_4 to each well. On a Titertek Multiscan the OD was read at 492 nm.

ELISA for Total Local and Systemic IgA, IgA1, and IgA2

The ELISA technique for measuring total IgA, IgA1, and IgA2 in sera and biopsy specimens differed from the ELISA technique for specific IgA, IgA1, and IgA2 against *H. pylori* described above in the antigen suspension used. The plates to measure the total IgA, IgA1, and IgA2 antibody response were coated with an antigen suspension consisting of 400 ng/ml rabbit immunoglobulins to human IgA (α-chains; Dakopatts A 262).

Results

In the ELISAs for biopsy specimen homogenates, the OD value of each biopsy specimen homogenate was corrected for the protein concentration of the homogenate. The protein concentration was set at 10%. The range of the OD values of the ELISAs for specific local IgA against *H. pylori* was between 0.58 and 3.01. The range of OD values of the IgA1-specific local anti-*H. pylori* ELISA was between 0.15 and 1.35, whereas for IgA2 this range was 0.00–0.23. In Fig. 1 the distribution of local specific IgA subclass responses to *H. pylori* are shown. This figure shows the percentage contribution of the IgA subclasses to the total IgA OD value. Therefore the sum of the OD values of both the IgA1 and IgA2 ELISAs was set at 100%. Next, according to this percentage distribution, the contribution of the IgA subclasses to the OD value of the specific IgA ELISA was determined. The low IgA2 values indicated a low IgA2 local response against *H. pylori.*

Total IgA1 and IgA2 in Gastric Antral Mucosa

The OD values of the ELISAs for the IgA1 and IgA2 content in the homogenates were in a range of 1.05–2.33 and 0.15–0.80, respectively. Table 1 shows the relative distribution of specific anti-*H. pylori* and total IgA1 and IgA2 subclass antibodies in biopsy specimens expressed as percentages of either specific anti-*H. pylori* or total IgA subclass antibody response of each patient.

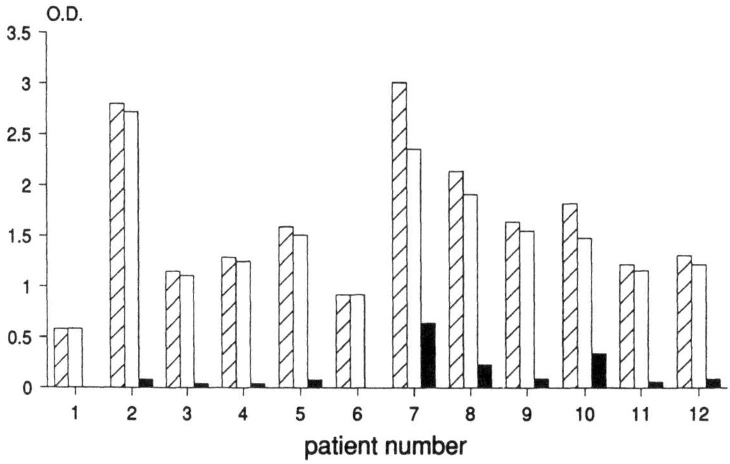

Fig. 1. Distribution of local specific IgA subclass responses to *H. pylori* in biopsy specimens expressed as OD values. The sum of the OD values of both the local IgA1 and IgA2 ELISAs was set at 100%. Next, according to this percentage distribution, the contribution of the IgA subclasses to the OD value of the specific IgA ELISA was determined. *Hatched columns*, IgA; *open columns*, IgA1; *solid columns*, IgA2

Table 1. Relative distribution of specific anti-*H.pylori* and total IgA1 and IgA2 subclass antibodies in biopsy specimens expressed as percentages. For both ELISAs (total IgA subclasses and specific anti-*H. pylori* IgA subclasses) the sum of the IgA1 and IgA2 OD values of each patient was set at 100%

Patient number	anti-*H.pylori* local IgA1 %	anti-*H.pylori* local IgA2 %	Total local IgA1 %	Total local IgA2 %
1	100	0	89.66	10.34
2	97.12	2.88	80.84	19.16
3	96.23	3.77	81.71	18.29
4	97.22	2.78	84	16
5	94.87	5.13	84.92	15.08
6	100	0	81.88	18.12
7	78.90	21.10	74.44	25.56
8	89.32	10.68	78.53	21.47
9	94.59	5.41	85.54	14.46
10	81.48	18.52	81.53	18.47
11	95.08	4.92	83.05	16.95
12	93.18	6.82	83.59	16.41

Total IgA1 and IgA2 in the Serum

The OD values of the ELISA technique for systemic specific IgA antibodies to *H. pylori* were around 0.50 (range: 0.26–1.17). The ELISA OD values for the systemic specific anti-*H. pylori* IgA1 subclasses ranged from 0.36 to 1.00.

Specific IgA2 antibodies against *H. pylori* could hardly be detected in the sera, with OD values lower than 0.05. In Fig. 2 the distribution of these systemic specific IgA subclass responses to *H. pylori* are shown. The calculation has been done as described above for the local IgA subclass.

Next, we examined the systemic total IgA subclass distribution. The OD values of the total IgA1 and IgA2 ELISAs were close to 1.30 (range: 1.23–1.37) and 0.10 (range: 0.03–0.24), respectively. Table 2 shows the relative distribution

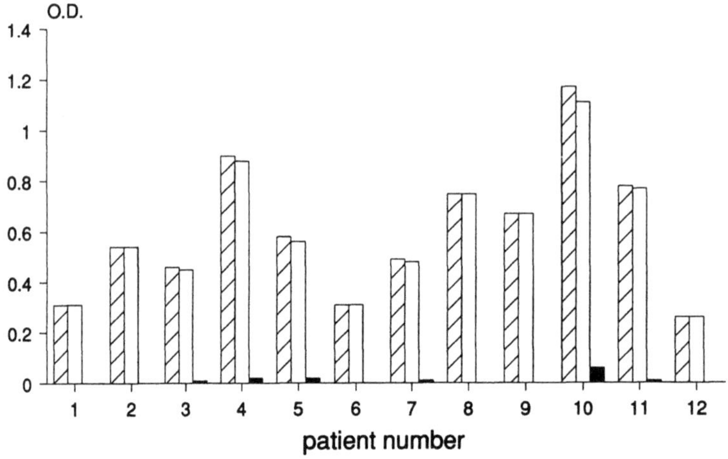

Fig. 2. Distribution of systemic specific IgA subclass responses to *H. pylori* expressed as OD values. The sum of the OD values of both the systemic IgA1 and IgA2 ELISAs was set at 100%. Next, according to this percentage distribution, the contribution of the IgA subclasses to the OD value of the specific IgA ELISA was determined. *Hatched columns*, IgA; *open columns*, IgA1; *solid columns*, IgA2

Table 2. Relative distribution of the systemic specific anti- *H.pylori* and total IgA1 and IgA2 subclass antibodies expressed as percentages. For both ELISAs (total IgA subclasses and specific anti- *H.pylori* IgA subclasses) the sum of the IgA1 and IgA2 OD values of each patient was set at 100%

Patient number	Anti-*H.pylori* serum IgA1 (%)	Anti-*H.pylori* serum IgA2 (%)	Total serum IgA1 (%)	Total serum IgA2 (%)
1	100	0	94.62	5.38
2	100	0	93.57	6.43
3	98.33	1.67	94.93	5.07
4	97.67	2.33	84.21	15.79
5	96.01	3.39	89.21	10.79
6	100	0	91.55	8.45
7	98.53	1.47	97.74	2.26
8	100	0	94.20	5.80
9	100	0	89.80	10.20
10	95.24	4.76	87.82	12.18
11	99	0	86.09	13.91
12	100	0	90.71	9.29

of specific anti-*H. pylori* and total IgA1 and IgA2 subclass antibodies in sera expressed as percentages of either specific anti-*H. pylori* or total IgA subclass antibody response of each patient.

Discussion

We have previously shown [9] that the systemic specific IgA antibody response against *H. pylori* consists mainly of the IgA1 subclass. This distribution of IgA subclass corresponds to other studies in serum were it has been found [5, 6] that the systemic IgA subclass response consists mainly of the IgA1 subclass, IgA1 represents only 40%–50% of the IgA response in secretions. Therefore we expected that the specific anti-*H. pylori* IgA subclass distribution in the gastric mucosa would involve the IgA2 subclass to a greater extent than in the serum. As shown in the results, the local specific IgA response to these bacteria is also of the IgA1 subclass. This finding supports the pathogenetic role *H. pylori* plays in producing inflammation in the gastric mucosa. It appears that, when inflammation occurs, signals are given to the bone marrow to stimulate the production of IgA1. It is possible that this response is the normal gastrointestinal response to inflammation since similar local IgA subclass alterations have been observed in ulcerative colitis and Crohn's disease of the colon both by paired immunofluorescence staining in colonic specimens [12] and spontaneous secretion of IgA by intestinal mononuclear cells isolated from both diseases [13]. Studies to determine the specific local IgG subclass in *H. pylori*-infected patients are indicated.

In summary, the specific IgA response in serum and at the level of the gastric mucosa against *H. pylori* is mainly of the IgA1 subclass.

References

1. Newel DG, Rathbone BJ (1989) The serodiagnosis of Campylobacter pylori infections. Serodiagn Immunother 3:1–6
2. Vaira D, Holton J, Cairns SR, Faizon M, Polydoran A, Dowsett JF, Salmon RR (1988) Antibody titre to Campylobacter pylori after treatment for gastritis. Br Med J 297:397
3. Bienenstock J, Befus AD (1980) Mucosal immunology. Immunology 41:249–270
4. Brandtzaeg P, Halstensen TS, Kett K, Krajci P, Kvale D, Rognum TO, Scott H, Sollid LM (1989) Immunobiology and immunopathology of human gut mucosa: humoral immunity and intraepithelial lymphocytes. Gastroenterology 97:1562–1584
5. Hanson LA, Brandtzaeg P (1980) In: Stiehm ER, Fulginiti VA Immunologic disorders in infants and children 2nd edn. Saunders, Philadelphia, P 137
6. Grey HM, Abel CA, Yount WJ, Kunkel HG (1968) A subclass of human γAglobulins (γA2) which lacks the disulfide bonds linking heavy and light chains. J Exp Med 128:1223–1236
7. Kilian M, Thomsen B, Peterson TE, Bleeg HS (1983) Occurrence and nature of bacterial IgA proteases. Ann NY Acad Sci 409 : 612–624
8. Plaut AG, Heller I, Gilbert J, Rule A (1976) Microbial IgA protease. Characterization and assay of a novel alimentary tract enzyme. Gastroenterology 70:A-69/927
9. van der Est MMC, Veenendaal RA, Peña AS, van Duijn W, Kuiper I, Lamers CBHW (1990) ELISA analysis of IgA subclass antibodies to Helicobacter pylori. Elevated serum IgA1 antibodies in Helicobacter pylori infected patients. J Clin Nutr Gastroenterol 5:185–190

10. Lowry OH, Rosebrough NJ, Farr AL, Randall RJ (1951) Protein measurement with the Folin phenol reagent. J Biol Chem 193:265–275
11. Peña AS, Endtz Hph, Offerhaus GJA, Hoogenboom-Verdegaal A, van Duijn W, de Vargas N, den Hartog G, Kreuning J, van der Reyden J, Mouton RP, Lamers CBHW (1989) Value of serology (ELISA and immunoblotting) for the diagnosis of Campylobacter pylori infection. Digestion 44:131–141
12. MacDermott RP, Nash GS, Bertovich MJ, Mohrman RF, Kodner IJ, Delacroix DL, Vaerman J-P (1986) Altered patterns of secretion of monomeric IgA and IgA subclass 1 by intestinal mononuclear cells in inflammatory bowel disease. Gastroenterology 91:379–385
13. Kett K, Brantzaeg P (1987) Local IgA subclass alterations in ulcerative colitis and Crohn's disease of the colon. Gut 28:1013–1021

Antibodies Against *Helicobacter pylori* as an Indicator of Antrum Gastritis

S.-G. Xu[1], R.B. Zotz[1], G. von Recklinghausen[2], K. Donhuijsen[3], and H. Goebell[1]

Introduction

The Gram-negative spiral bacterium *Helicobacter pylori* (HP) is found in the human stomach and duodenum. It is highly associated with type B antral gastritis and with peptic ulcer disease. HP has been detected in 64%–95% of patients with active chronic gastritis and in 75%–100% of patients with duodenal ulcer disease [1–5]. An etiologic role of HP in the pathogenesis of type B gastritis is indicated by human volunteer studies, animal models, and therapeutic trials with antimicrobial agents [1]. The detection of specific IgG antibodies against HP has been shown to be a valuable method for the diagnosis of HP infection and associated diseases [6–9]. We determined the efficacy of immunoglobulin G (IgG) antibodies against HP, detected by the enzyme-linked immunosorbent assay (ELISA), as an indicator for histologically proven antrum gastritis and antral HP colonization.

Patients and Methods

A total of 154 patients (96 men, 58 women; aged 17–77, mean age 50.4 years) attending the endoscopy unit of the Medical Clinic of Essen University Hospital were randomly examined. At least one antrum biopsy was taken for histology, HP culture plus Gram-stained smear, and for urease testing (BUT). Antrum gastritis was evaluated by the presence of inflammatory cells (polymorphonuclear leukocytes, lymphocytes, plasma cells) infiltrating the antral mucosa (hematoxylin and eosin stain). For the HP culture the biopsy was rolled on two nonselective and two selective agar media followed by microaerophilic incubation for 2–7 days [10]. Isolates were identified as HP by positive urease, catalase, and oxidase reactions. A Gram-stained smear was made from the same biopsy. Biopsy urease was tested with 1 ml unbuffered 1% w/v urea solution, pH 4.5, with phenol red as the indicator. Color change was monitored after 20 min, 3 h, and 24 h, respectively. In all patients serum was tested for IgG

[1] Division of Gastroenterology, University of Essen, 4300 Essen 1, FRG
[2] Department of Microbiology, University of Essen, 4300 Essen 1, FRG
[3] Department of Pathology, University of Essen, 4300 Essen 1, FRG
Corresponding Author: Shi-Guang Xu, Medizinische Klinik Abt. für Gastroenterologie Universitätsklinikum Essen

J.M. Pajares et al. (Eds.)
H. pylori and Gastroduodenal Pathology

antibodies by an ELISA as described elsewhere [9]. In brief, microtiter plates were coated with an acid-glycine-extracted antigen of HP ATCC 43504 [11]. Patient sera were diluted 1:500 with phosphate-buffered saline (PBS) and incubated for 90 min at 37°C. The specific binding of IgG antibodies was visualized using affinity-purified anti-human-IgG peroxidase conjugate from the gout (Kierkegaard and Perry, Asbach, FRG), dilution 1:40000, and 2,2-azinobis-3-ethylbenzthiazoline sulfonate (ABTS) as the substrate. Optical density (OD) was measured at 405 nm after a 30-min substrate incubation at room temperature. Sera were evaluated as positive if OD exceeded more than one third of a positive calibrator serum, e.g., cut off at OD 0.400 with positive control at OD 1.200.

Results

In 154 patients HP status and presence of gastritis were determined. A total of 118 patients (77%) showed histologic signs of gastritis, in 108 patients (70%) IgG antibodies against HP were detected, and HP was cultured from 88 (57%) patients. The results of the different methods of HP detection in relation to antral inflammation are shown in Table 1.

In 136 patients (88%) the presence or absence of antral inflammation was predicted correctly by the IgG antibody test. With respect to antrum gastritis, the test has a positive predictive value (PPV) of 96%. The negative predictive value (NPV) is 70%, because in 14 of 46 IgG negative patients gastric antral inflammation was present (Table 2).

Eight patients with a negative IgG test result and without detectable HP colonization had histologic signs of gastritis. In three of these patients, inflammatory bowel disease was the underlying condition (two with Crohn's disease,

Table 1. Antrum gastritis, IgG antibodies against HP, and biopsy detection of HP in 154 patients

		Inflammatory cells in antrum biopsy					
		Present		Absent		Total	
		(*n*)	(%)	(*n*)	(%)	(*n*)	(%)
Total		118	77	36	23	154	100
IgG antibody	+ ve	104	68	4	3	108	70
ELISA	– ve	14	9	32	21	46	30
Culture	+ ve	86	56	2	1	88	57
	– ve	25	16	34	22	59	38
	ND*	7	5	0		7	5
Gram	+ ve	65	42	4	3	69	45
stain	– ve	47	31	32	21	79	51
	ND*	6	4	0		6	4
Biopsy	+ ve	95	62	3	2	98	64
urease	– ve	22	14	32	21	54	35
test	ND*	1	0.5	1	0.5	2	1

+ ve, positive; – ve, negative; ND, not done

Table 2. Performance criteria of the HP IgG antibody test to detect antrum gastritis and HP colonization (culture, Gram stain and/or biopsy urease test)

	Antrum gastritis		HP colonization	
	(%)	(*n*)	(%)	(*n*)
Sensitivity	88	104/118	93	101/109
Specificity	89	32/36	84	38/45
Positive predictive value	96	104/108	94	101/108
Negative predictive value	70	32/46	83	38/46
Efficiency	88	136/154	90	139/154

one with colitis ulcerosa), one had a leiomyoma of the stomach, and one had non-Hodgkin's lymphoma with stomach involvement. In three patients the cause of antral inflammation remained unknown.

With respect to HP colonization (i.e., culture, Gram stain, BUT, at least one test positive), the IgG antibody test performs with an efficiency of 90%, a PPV of 94%, and a NPV of 83% (Table 2).

Discussion

A high correlation between nonspecific active chronic gastritis and HP colonization of the inflamed gastric mucosa has been demonstrated in previous studies (for a review see Blaser [1]). In the collective presented here, again 73% of patients with antral gastritis (86/118) were infected with HP as shown by cultural detection. However, the sensitivity of the HP culture and other biopsy-based detection methods is limited because of the patchy distribution and variable cell count of HP on the mucosal surface [12]. Therefore indirect detection of HP infection by measuring the systemic immune response induced by HP may be more reliable and efficient than biopsy detection [1].

An antibody detection method based on a partially purified antigen like the acid-glycine extract used in our study provides a high correlation with biopsy-proven HP infection. The value of our test system to predict HP colonization as detected from antral biopsy is over 90%, which is confirmed by other authors [7, 13]. Therefore a high correlation between HP antibody detection and gastritis can be assumed. Our data show that the presence or absence of antrum gastritis is correctly predicted in 136 of 154 patients (88%). In fact, only four of 108 patients with positive IgG antibody test results had no inflammation cells in the antral biopsy specimen evaluated parallel. Thus, in the present study the value of a positive IgG antibody test to predict antral gastritis is over 95%.

In contrast, the value of a negative test result to predict the absence of antral inflammation is markedly reduced. The NPV of the serotest depends on the proportion of patients with gastritis of another etiology. In our study, in five of eight IgG antibody-negative and biopsy test-negative patients with gastric antral inflammation the underlying causes for gastritis could be identified as inflammatory bowel disease and tumor. In three gastritis patients the etiology remained unclear. If these patients with gastritis etiology other than HP are excluded, the NPV rises to 84%.

The data indicate that the IgG antibody test can undoubtedly act as a tool for the prediction of HP-associated gastritis. It is much easier to perform and less stressful to the patient than invasive diagnostic methods. However, as stated before [14], a clear management plan for the patient with a positive or negative test result is required.

References

1. Blaser MJ (1990) Helicobacter pylori and the pathogenesis of gastroduodenal inflammation. J Infect Dis 161:626–633
2. Graham DY (1989) Campylobacter pylori and peptic ulcer disease. Gastroenterology 96:615–625
3. Dooley CP, Cohen H, Fitzgibbons PL, Bauer M, Appleman MD, Perez-Perez G, Blaser MJ (1989) Prevalence of Helicobacter pylori infection and histologic gastritis in asymptomatic persons. N Engl J Med 321:1562–1566
4. McKinlay AW, Upadhyay R, Gemmell CG, Russell RI (1990) Helicobacter pylori: bridging the credibility gap. Gut 31:940–945
5. Von Recklinghausen G, Kehler U, Breuer N, Ansorg R (1989) Assessment of a biopsy urease test for Campylobacter pylori. In: Mégraud F, Lamouliatte H (eds) Gastroduodenal pathology and Campylobacter pylori. Elsevier, Amsterdam, pp 61–63
6. Newell DG, Rathbone BJ (1989) The serodiagnosis of Campylo-bacter pylori infection. Serodiagn Immunother 3:1–6
7. Von Wulffen (1990) Methods of studying the immune response. In: Malfertheiner P, Ditschuneit H (eds) Helicobacter pylori, gastritis and peptic ulcer. Springer, Berlin Heidelberg New York, pp 137–140
8. Wyatt JI, Rathbone BJ (1989) The role of serology in the diagnosis of Campylobacter pylori infection. Scand J Gastroenterol 24 [Suppl 160] :27–34
9. Von Recklinghausen G, Zotz RB, Ansorg R (1989) Wertigkeit des qualitativen Nachweises von Serum-IgG-Antikörpern gegen Campylobacter pylori als Indikator der Magenbesiedlung mit C. pylori. Lab Med 13:421–424
10. Ansorg R, von Recklinghausen G, Pomarius R, Schmid EN (1991) Evaluation of techniques for isolation, subcultivation, and preservation of Helicobacter pylori. J Clin Microbiol 29:51–53
11. Newell DG (1987) Identification of the outer membrane proteins of Campylobacter pyloridis and antigenic cross-reactivity between C. pyloridis and C. jejuni. J Gen Microbiol 133:163–170
12. Morris A, Maher K, Thomesen M, Miller G, Nicholson G, Tasman-Jones C (1988) Distribution of Campylobacter pylori in the human stomach obtained at postmortem. Scand J Gastroenterol 23:257–264
13. Hirschl AM, Rotter ML (1990) Serodiagnosis of Helicobacter pylori infections: suitability of various antigen preparations. In: Malfertheiner P, Ditschuneit H (eds) Helicobacter pylori, gastritis and peptic ulcer. Springer, Berlin Heidelberg New York, pp 141–146
14. Sobala GM (1990) Possible clinical uses of serology of Helicobacter pylori. In: Malfertheiner P, Ditschuneit H (eds) Helicobacter pylori, gastritis and peptic ulcer. Springer, Berlin Heidelberg New York, pp 147–153

Antibodies to Vacuolating Toxin of *Helicobacter pylori* in Dyspeptic Patients

N. Figura[1], M. Bugnoli[2], P. Guglielmetti[1], R.A. Musmanno[3], M. Russi[3], S. Quaranta[4], A. Barberi[1] A. Rossolini[1], and P.F. Bayeli[5]

Introduction

Certain *Helicobacter pylori* (HP) strains are capable of inducing a cytopathic effect in mammalian cells in vitro consisting in the formation of intracytoplasmic vacuoles [1, 2]. In vitro cells exposed to broth culture filtrate (BCF) of vacuolating HP (VHP) strains die more rapidly than do cells exposed to BCF of non-VHP ones and to uninoculated broth. In addition, VHP causes a strong reduction of the proliferation index of Epstein-Bass Virus (EBV) -transformed B lymphocytes [2].

The substance or the substances responsible for the vacuolating phenomenon is/are thought to be proteinaceous in nature as they are heat labile, ammonium sulfate precipitable, and protease sensitive [1]. Concentrated BCF (CBCF) of VHP strains contains some proteins which are not present [2] or are present less frequently [3] in nonvacuolating CBCF. One of these proteins is approximately 130 kDa (128 kDa according to Cover et al. [3]), shared by all cytotoxic strains tested, and reacts with nearly 100% of serum samples from dyspeptic patients infected by cytotoxic HP strains [2]. Thus this protein can be considered as a vacuolating activity marker.

In a study of ours, HP strains isolated from patients with peptic ulcers, especially with duodenal ulcers, produced vacuolating toxin more frequently than did isolates from patients without ulcers (66.6% of strains, versus 30.1% of strains [4, 5]). Thus cytotoxigenicity could contribute to the development of peptic ulceration.

The aims of this study were to determine the prevalence of serum antibodies to 128-kDa vacuolating activity-associated protein (VP) in 116 dyspeptic patients (DP) who underwent diagnostic endoscopy and in 40 children without gastric symptoms to see whether there were prevalence/sex and prevalence/age correlations, and to verify whether DP infected by HP and with anti-VP antibodies had more serious endoscopic and histological features and more intense dyspepsia symptoms than did infected DP without anti-VP antibodies.

[1] Instituto di Malattie Infettive, Universita di Siena, Via P.A. Mattioli 10, 53 100 Siena, Italy
[2] Sclavo Research Center, 53 100 Siena, Italy
[3] Dipartimento di Biologie Moleclare, Sez. di Microbiologie, Universita di Siena, 53 100 Siena, Italy
[4] Ospedale di Grosseto, Div. Medica II, Grosseto, Italy 53 100 Siena, Italy
[5] Istituto di Patologia Speciale Medica, Universita di Siena, 53 100 Siena, Italy

J.M. Pajares et al. (Eds.)
H. pylori and Gastroduodenal Pathology

Materials and Methods

Vacuolating *H. pylori* strain G-32 was cultured in brucella broth (Difco, DID, Milan) containing 5% fetal calf serum, 1% Vitox (Oxoid Italiana S.p.A., Milan), 10 mg/1 vancomycin, 10 mg/1 trimethoprim, 5000 U/1 polymyxin B, and 5 mg/1 amphotericin B.

The broth culture was incubated in a microaerobic environment at 37°C at 150 oscillations per minute for 72 h. The culture was centrifuged at 4°C and filtered through a 0.22-μm-pore filter. Proteins were precipitated with 50% saturated ammonium sulfate at 4°C for 24 h, then were suspended with the minimal amount of PBS pH 7.4 at which no precipitate was observed. CBCF was then dialyzed with cellulose tubing which retained proteins with a molecular weight greater than 12000 versus PBS pH 7.4 for 48 h and stored at − 70°C. CBCF contained 4.4 mg protein per milliliter (versus 0.07 mg protein per milliliter of unconcentrated broth culture) and induced vacuolization of ⩾ 50% intestine 407 cells in vitro up to 1:160 dilution.

CBCF proteins were subjected to electrophoresis in sodium dodecylsulfate polyacrylamide gel by using 3.5% stacking gel and 6% separation gel (each gel line contained 15 μg proteins) and transferred to nitrocellulose. Protein-free sites were satured with 50 m*M* Tris-HCl pH 7.4 supplemented with 0.15 *M* sodium chloride, 3% defatted milk and 0.1% triton-X 100 (TMT) at room temperature for 30 min. Sheets were then dipped in serum samples diluted 1:100 in TMT and incubated on a shaker at room temperature overnight. After washing with TMT, antigen-coated sheets were shaken at room temperature for 90 min with alkaline phosphatase-labeled anti-human immunoglobulin G (IgG; Cappel, Cooper Biomedical, Malvern, PA, USA) diluted 1:2500 in TMT. Strips were then washed in Tris-saline-triton, and rinsed in Tris-saline; the reaction was determined with Alkaline Phosphatase Conjugate Substrate Kit (Bio-Rad, Richmond, CA, USA).

We examined serum samples from 20 children aged 2–5 years, from 20 children aged 6–10 years without gastric symptoms, and from 116 adult DP who underwent diagnostic gastroduodenal endoscopy and who were not receiving treatment with drugs potentially active against HP. Children were not examined endoscopically. Five biopsies, taken from the gastric antrum and/or the edge of ulcers, were obtained from each DP. One biopsy was examined histologically with a hematoxylin and eosin stain. The severity of gastritis was assessed by the degree of round and polymorphonuclear cell infiltrates and the presence of erosions. The other biopsies were cultured on Columbia agar with 7% horse blood, 5 mg/1 amphotericin B, and the Skirrow mixture of antibiotics, smeared for microscopic examination after staining with Gram stain and acridine orange, and tested for rapid urease activity.

Suspected colonies were identified as HP as previously reported [5].

Results and Discussion

One of the 20 asymptomatic children (5%) aged 2–5 years and three of the 20 children (15%) aged 6–10 years had IgG antibodies to VP. These four children also had antibodies to a glycine extract of HP detected by an enzyme-linked immunosorbent assay. Out of 116 adult patients with dyspepsia, 89 (76.7%) had antibodies to VP (Fig. 1). No prevalence/sex and prevalence/age correlations were found: 55 out of 71 male (77.4%) and 34 of 45 female patients (75.5%) had antibodies to VP; 12 patients out of 19 (63.1%) under 40 had antibodies versus 77 patients out of 97 (79.3%) $\geqslant 40$ (the most marked difference in prevalence of antibodies to VP according to age was in these two groups of patients). Out of 89 patients with and 27 patients without antibodies to VP, 75 (84.2%) and ten (37.1%), respectively were infected by HP on the basis of culture, microscopic examination, and rapid urease activity of gastric biopsies. A significantly higher prevalence of endoscopic diagnosis of duodenal (15 patients) or combined ulcers (two patients) was found in patients with antibodies to VP (Table 1). The prevalence of the other endoscopic diagnoses did not differ significantly in the two groups of patients. The possession of antibodies to the toxin was not associated with the severity of histological gastritis (Table 2) and was not associated with the prevalence of the major symptoms of dyspepsia in the same 75 infected patients (Table 3).

Vacuolization is an in vitro cellular response to noxious substances and a noxious environment. If we add urea, for instance 10 m*M*, to a cellular medium or if we grow a strain in the presence of urea, even noncytotoxic strains can induce vacuoles. However, cells in culture exposed to BCF and supernatants of HP whole cell suspensions show an evident vacuolization only at levels of urea exceeding by far those contained in media for cells [6]. We added fetal calf

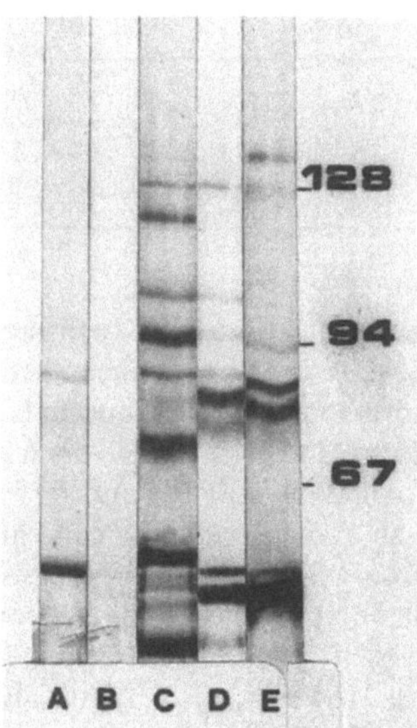

Fig. 1. Immunoblots of vacuolating CBCF with five serum samples. Patient A was infected by nonvacuolating HP; patient B was not infected; patients C, D, and E were infected by vacuolating HP strains (numbers represent standard molecular masses in kilodaltons)

Table 1. Endoscopic diagnosis in 89 patients with antibodies and 27 patients without antibodies to vacuolating cytotoxin-associated proteins

Endoscopic diagnosis	Patients with antibodies		Patients without antibodies	
	(*n*)	(%)	(*n*)	(%)
Gastric ulcer	9	10.1	3	11.3
Duodenal or combined ulcer	17	19.1*	0	0*
Stoma ulcer	0	0	1	3.7
Erosive gastritis	18	20.2	6	22.2
Absence of lesions	45	50.5	17	62.9
Total	89	100	27	100

*$p < 0.05$ (chi-square test, 1 df)

Table 2. Histological diagnosis in 75 HP-infected patients with antibodies and ten infected patients without antibodies to vacuolating cytotoxin-associated proteins

Degree of gastritis	Patients with antibodies		Patients without antibodies	
	(*n*)	(%)	(*n*)	(%)
Mild	4	5.3	2	20
Moderate	17	22.6	3	30
Severe	24	32.0	2	20
Erosive	29	38.6	3	30
Atrophic	1	1.3	0	0

Table 3. Dyspepsia symptoms in 75 HP-infected patients with antibodies and ten infected patients without antibodies to cytotoxin-associated proteins

Symptom	Patients			
	With antibodies		Without antibodies	
	(*n*)	(%)	(*n*)	(%)
Pain	48	64.0	8	80
Heartburn	34	45.3	5	50
Nausea	16	21.3	4	40
Vomiting	16	21.3	5	50

serum (which contains 6 m*M* urea [6] at 5% concentration both to brucella broth and to a medium for cells cultured in vitro. At 0.3 m*M* final concentration of urea, no vacuolating effect could be seen with ultrasonicates and supernatants of cytotoxic HP suspensions (personal observation), and with BCF which did not contain cytotoxin-associated proteins (approximately 130, 95, and 80 kDa [2], Besides HP, other ureolytic bacteria, such as *Proteus vulgaris*, can cause vacuolization in the presence of urea (personal observation). In all these cases it is the urease enzyme which is involved in the vacuolating effect. The vacuolization can, in fact, be induced by ammonium even at low concentrations (we observed vacuoles from 10 m*M* ammonium on). However, also nonureolytic bacteria such as *Vibrio cholerae* and *Escherichia coli* can induce vacuolization in intestine 407 cells (personal observation). Thus there are other substances different from urease which can cause vacuole formation.

As far as HP strains are concerned, (a) there is a very close association between the ability to induce vacuoles (without the addition of urea to cellular medium) and the presence of an approximately 130-kDa protein in CBCF [2]; (b) this protein is absent in noncytotoxic BCF [2]; (c) levels of urease in cytotoxic and noncytotoxic BCF are similar (personal observation); (d) glycine extract, which mostly contains urease subunits, of one cytotoxic strain does not cause vacuolization (personal observation); (e) the addition of urea to cellular culture medium does not increase the level of toxicity of a cytotoxic HP BCF; (f) a urease-deficient strain of HP was derived by chemical mutagenesis from a urease-producing cytotoxin organism, both strains showed vacuolating activity at the same titer [7].

For all these reasons, we think that HP strains whose BCF cause vacuolization without adding urea to the cell medium differ from those which do not cause vacuolization under the same conditions due to the production of a substance which is also produced in vivo (since it stimulates antibody response) and which has many of the toxins' characteristics.

The fact that only the four children infected by HP also had antibodies to VP could indicate that VHP strains circulate more intensely than the nonvacuolating ones or that they are more virulent. This hypothesis could be verified by testing a higher number of serum samples from children. Histological degrees of gastritis and the frequency of dyspepsia symptoms do not seem related to the possession of antibodies to VP or to the frequency of infection with vacuolating HP strains.

In conclusion, the finding that all patients with duodenal ulcer, besides being infected by HP had antibodies to VP suggests that the production of the toxin by HP strains could contribute to the development of duodenal ulcerative lesions. Duodenal ulceration should be suspected in patients with serum antibodies to the toxin.

Acknowledgement. We gratefully acknowledge Ms. C. Rinina, Istituto di Microbiologia, Universita di Siena, for her skilled technical assistance.

References

1. Leunk RD, Johnson PT, David BC, Kraft WG, Morgan DR (1988) Cytotoxic activity in broth-culture filtrates of Campylobacter pylori. J Med Microbiol 26:93–99
2. Figura N, Bugnoli M, Cusi MG, Pucci AM, Lusini P, Quaranta S, Barberi A, Rossolini A, Di Tommaso AL, De Magistris T, Rappuoli R, Marri L, Musmanno RA, Russi M, Guarna M, Losi M (1990) Pathogenic mechanisms of Helicobacter pylori: production of cytotoxin. In: Malfertheiner P, Ditschuneit H (Eds) Helicobacter pylori, gastritis and peptic ulcer. Springer, Berlin Heidelberg New York, pp 86–95
3. Cover TL, Dooley CP, Blaser MJ (1990) Characterization of and human serologic response to proteins in Helicobacter pylori broth culture supernatants with vacuolizing cytotoxin activity. Infect Immun 58:603–610
4. Figura N, Guglielmetti P, Rossolini A, Cusi MG, Musmanno RA, Russi M, Ibba L, Pucci AM, Losi M, Quaranta S (1989) Does the production of cytotoxin by Campylobacter pylori play a

pathological role in the development of peptic ulcer? In: Mégraud F, Lamouliatte H (Eds) Gastroduodenal pathology and Campylobacter Pylori. Excerpta Medica, Amsterdam, pp 385–389

5. Figura N, Guglielmetti P, Rossolini A, Barberi A, Cusi G, Musmanno RA, Russi M, Quaranta S (1989) Cytotoxin production by Campylobacter pylori strains isolated from patients with peptic ulcers and from patients with chronic gastritis only. J Clin Microbiol 27: 225–226
6. Xu JK, Goodwin CS, Cooper M, Robinson J (1990) Intracellular vacuolization caused by the urease of Helicobacter pylori. J Infect Dis 161: 1302–1304
7. Leunk RD, Ferguson MA, Williams EA, Eaton KA (1990) Adherence and production of cytotoxin by urease-producing and urease-deficient strains of Helicobacter pylori. Rev Esp Enferm Dig 78 (Suppl 1): 54

Cytokines and Mucosal Immune Responses to *Helicobacter pylori*

J.E. Crabtree[1], T.M. Shallcross[1], R.V. Heatley[1], and J.I. Wyatt[2]

Cytokines are well recognised as being important regulators of inflammatory responses [1]. They are synthesised by several different cell types, and individual cytokines have multiple overlapping regulatory immune functions [1] as well as affecting physiological responses [2]. Cytokines are generally short acting and produced locally. The role of cytokines in regulating human intestinal mucosal responses has not been studied in detail, and the cytokine responses of the gastrointestinal mucosa to bacterial colonisation are unclear. Recent studies in animal models suggest that cytokine production at mucosal sites in response to bacterial or endotoxin administration is distinct from systemic events [3, 4]. A knowledge of cytokine production at mucosal sites is therefore important to our understanding of host–bacterial interactions and the immunopathology of chronic infections at mucosal surfaces.

It is now well established that the colonisation of the human gastric mucosa with *Helicobacter pylori* results in long-term chronic inflammation which resolves following successful clearance of the bacteria [5]. The lipopolysaccharide (LPS) of Gram-negative bacteria is a potent stimulator of tumour necrosis factor alpha (TNF-α) [6]. Considerable heterogeneity in *H. pylori* LPS profiles has been demonstrated [7]. Recent studies have shown that interleukin-6 (IL-6) is also produced in response to LPS stimulation, human serum levels being

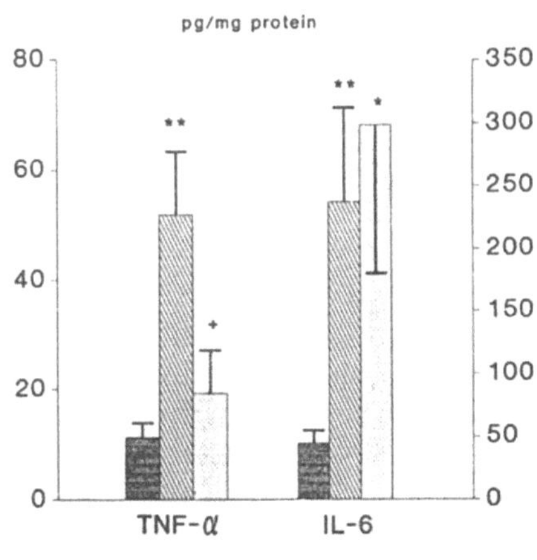

Fig. 1. Concentration (mean + S.D.) of TNF-α and IL-6 in 24-h culture supernatants of antral mucosa of patients with normal histology (*solid columns*) and *H. pylori*-associated gastritis. *Hatched columns*, active gastritis; *dotted columns*, inactive gastritis; $^{*}p < 0.05$; $^{**}p < 0.001$ from normal; $^{+}p < 0.03$ from active gastritis

[1] Department of Medicine, St. James's University Hospital, Leeds LS9 7TF, UK
[2] Department of Pathology, St. James's University Hospital, Leeds LS9 7TF, UK

J.M. Pajares et al. (Eds.)
H. pylori and Gastroduodenal Pathology

raised following endotoxin [8] or TNF-α exposure [9]. Our recent observations have demonstrated that TNF-α [10] and IL-6 (Fig. 1) are produced locally by *H. pylori*-positive antral mucosal biopsies during short-term in vitro culture. Furthermore, TNF-α but not IL-6 secretion profile varied with the activity of the gastritis (Fig. 1), suggesting that in vivo these two cytokines could be important in host–bacterial interactions within the mucosa. In this review, we consider the possible origin of these cytokines and their potential involvement in inducing the mucosal changes observed in chronic gastritis.

Origin of Mucosal Cytokines

Macrophages are considered to be the major source of TNF-α [11], and the local production of TNF-α in *H. pylori*-positive patients with active gastritis is therefore evidence of mucosal macrophage activation. T cells, however, can also synthesise and secrete TNF-α [12]. Increased T cell infiltration into the mucosa is associated with *H. pylori* infection [13], and immunohistological studies have shown increased expression of T cell stimulation markers [14]. Recent studies with in vitro cultured gastric T cells from *H. pylori*-infected subjects [15] have demonstrated TNF-α secretion by cloned gastric T cells (J.E. Crabtree, in preparation). The mucosal TNF-α production in subjects with active gastritis (Fig. 1), therefore, could be derived partly from T cells. Antral lymphoid follicles, found particularly in subjects with active gastritis relative to those with inactive gastritis [16], are a further possible source of local cytokine production. Relatively little is known about mucosal T cell subsets in *H. pylori*-associated gastritis. There is no apparent increase in mucosal T cells expressing the γ/δ chain in *H. pylori*-associated gastritis [17], although there is one reported case of an increase in γ/δ T cells in lymphocytic gastritis [18]. The cellular origin of gastric IL-6 is currently unclear. IL-6 can be produced by lymphoid and non-lymphoid cells including activated macrophages, fibroblasts, T cells and endothelial cells [19].

Effect of Cytokines on B Cells

It has long been established that non-autoimmune gastritis is associated with an increase in mucosal plasma cells and epithelial expression of secretory component, lysozyme and lactoferrin [20, 21]. Recent studies have shown that numerous cytokines can influence humoral responses with different factors acting on quiescent B cells, stimulating B cell proliferation and plasma cell differentiation [22]. IL-6, in particular, will induce terminal differentiation of B cells, and this cytokine was initially termed B cell stimulatory factor [23]. An effect of TNF-α on B cell differentiation has not been unequivocally demonstrated, but it may have a co-stimulatory effect on B cell function and immunoglobulin secretion in conjunction with other factors [24, 25]. Other cytokines have recently been implicated in immunoglobulin class switching [23], but as yet little is known about the action of such factors in the mucosal micro-environment.

The initial observation of *H. pylori* IgA antibodies in gastric juice of subjects with *H. pylori*-associated gastritis [26] suggested that a component of the mucosal plasma cell response represented a local immune response to *H. pylori*. In vitro secretion of *H. pylori*-specific IgA by gastric biopsies of colonised patients [27] and duodenal mucosa of patients with duodenitis [28, 29] confirmed these early observations. More recently, the antigen specificity of the local IgA response to *H. pylori* has been characterised by immunoblotting [30, 31]. Whilst IgA has been shown to have several protective functions at mucosal sites, inhibiting bacterial adherence, preventing antigen uptake and neutralising biologically active antigens [32], its functional role in modifying host–bacterial interactions in *H. pylori*-associated gastritis is presently unclear. IgA-dependent cellular cytotoxicity against enteropathogenic bacteria, however, has been described in mice when cells from gut-associated lymphoid tissue were used as effector cells [33, 34].

Cytokines and Other Mucosal Cells

A feature of chronic gastritis is the increased expression of HLA class II determinants on the epithelium [13, 35] which is closely related to lymphocytic infiltration [13]. Epithelial expression of class II antigens is thought to be cytokine mediated; interferon gamma (IFN-γ) induction of HLA-DR on intestinal epithelial cell lines [36] and fetal intestinal epithelium [37] have both been demonstrated. TNF-α, does not directly induce epithelial HLA-DR expression, but can augment IFN-γ-induced expression [38]. TNF-α, however, up-regulates the expression of secretory component on epithelial cell lines [38]. Additionally, TNF-α stimulates lysozyme production by mononuclear cells [39]. Local production of TNF-α in *H. pylori*-associated gastritis may therefore be partly responsible for the increased lysozyme and secretory component expression observed [20, 21].

Gastric *H. pylori* infection is particularly associated with a neutrophilic response (active gastritis) [40]. Whilst in vitro experiments demonstrate that IL-6 and TNF-α are both produced locally by the antral mucosa of patients with active gastritis and *H. pylori* infection, in inactive gastritis where neutrophils are not found in the epithelium (Fig. 1), it is only IL-6 that is significantly increased. Cytokines have been shown to modify the functional activity of neutrophils [41]. Whilst IL-6 is not chemotactic or chemokinetic for neutrophils, it does increase lysozyme and lactoferrin secretion and the oxidative burst of neutrophils [42, 43]. Interestingly, recent murine studies have demonstrated a similar dissociation between local IL-6 secretion and mucosal polymorphonuclear cell infiltration to sites of bacterial infection [44, 45]. In contrast to IL-6, TNF-α stimulates neutrophil chemotactic factor (interleukin-8) production from endothelial cells [46] and fibroblasts [47]. The observation therefore that TNF-α secretion, but not IL-6, is significantly higher in patients with active gastritis relative to those with inactive gastritis, is in accordance with the known in vitro functional properties of these two cytokines. Several other cytokines are also likely to have direct or synergistic effects on neutrophil activation and function [41] within the mucosa.

Cytokines and Gastric Physiology

Cytokines produced within the gastric mucosa could influence non-specific defence mechanisms such as mucus secretion. In vitro studies have shown that mucus glycoprotein synthesis and secretion are increased in *H. pylori*-associated gastritis [48]. It has recently been demonstrated in the small intestine that T cells have a regulatory role in mucus secretion [49], thus raising the possibility of cytokine involvement not only in regulating the local inflammatory cell infiltrate in patients with *H. pylori*-associated gastritis, but also in influencing other physico-chemical events at the gastric mucosal surface.

Antral *H. pylori* infection is associated with an increase in serum gastrin [50, 51]. It has been proposed that the high urease activity of the bacterium and concomitant local production of ammonia disrupts the normal regulatory mechanisms controlling gastrin release [51]. As inhibition of the urease activity with acetohydroxamic acid fails to modify the post-prandial gastrin response [52], other gastric micro-environmental factors may be important in inducing perturbations in gastric physiology. There is preliminary limited evidence from animal studies that gastrin release may be stimulated by gastric immune responses. Specifically, antigen administration to the stomach was shown to induce gastrin release following pre-immunisation [53]. Additionally, the cytokines IL-2 and IFN-γ have been shown to stimulate gastrin release from perfused canine antrum [54].

TNF-α is known to have many important metabolic effects [55], and the local generation of this cytokine in *H. pylori* infections may be important in modifying gastric physiology. Two recent pharmacological studies have shown that agents which will modify TNF-α production induce changes in plasma gastrin concentrations. Indomethacin, which increases macrophage TNF-α secretion in vitro [56], has been shown to significantly augment post-prandial plasma gastrin [57], an effect not associated with changes in basal acid anti-secretory activity [58]. In contrast, enprostil, a synthetic dehydro-prostaglandin E_2 (PGE_2) analogue, decreases post-prandial gastrin release [57]. PGE_2 suppresses TNF-α release from activated macrophages and is considered an important factor in down-regulation of macrophage activation [59]. Cytokines may also directly affect acid secretion. It is well established that acid secretion is reduced in certain infections, for example, some small intestinal nematode and cestode infections [60]. Infection with *H. pylori* is associated with an initial hypochlorhydria [60] of varying duration. The cytokine IL-1 has recently been shown to inhibit gastric acid secretion in rats [61]. Whether other cytokines similarly have a regulatory effect on acid secretion is unknown.

Conclusions

Cytokines can clearly have many immunoregulatory effects influencing host–bacterial interactions. Studies on cytokines at the site of disease activity are important, and the multiplicity of possible cytokine functions are only just beginning to be unravelled. Further studies are required to delineate which bacterial products of *H. pylori* stimulate mucosal cytokine production and the cellular origin of the cytokines.

References

1. Blackwell FR, Burke F (1989) The cytokine network. Immunol Today 10:299–304
2. Beultler B, Cerami A (1988) Cachectin (tumour necrosis factor) : a macrophage hormone governing cellular metabolism and inflammatory response. Endocr Rev 9:57–66
3. Leist TP, Frei K, Kam-Hansen S, Zinkernagel RM, Fontano A (1988) Tumour necrossis factor alpha in cerebrospinal fluid during bacterial but not viral meningitis. J Exp Med 167:1743–1748
4. Nelson S, Bagby GJ, Bainton BG, Wilson LA, Thompson JJ, Summer WR (1989) Compartmentalization of intraalveolar and systemic lipopolysaccharide-induced tumour necrosis factor and the pulmonary inflammatory response. J Infect Dis 159:189–194
5. Rauws EAJ, Langenberg W, Houtoff HJ et al. (1988) Campylobacter pyloridis-associated chronic active gastritis. A prospective study of its prevalence and the effects of antibacterial and antiulcer treatment. Gastroenterology 94:33–40
6. Beultler B, Cerami A (1988) The biology of cachectin/TNF—a primary mediator of host response. Annu Rev Immunol 7:625–655
7. Perer-Perez GI, Blaser MJ (1987) Conservation and diversity of Camplobacter pyloridis major proteins. Infect Immun 55:1256–1263
8. Fong Y, Moldawer LL, Maranc M et al. (1989) Endotoxemia elicits increased circulating β2-IFN/IL-6 in man. J Immunol 142:2321–2324
9. Jablons DM, Mule JJ, McIntosh JK et al. (1989) IL-6/IFN-β-2 as a circulating hormone. Induction by cytokine administration in humans. J Immunol 142:1542–1547
10. Crabtree JE, Shallcross TM, Wyatt JI, Heatley RV (1990) Tumour necrosis factor alpha secretion by Helicobacter pylori colonised gastric mucosa. Gut 31:A600
11. Le J, Vilcek J (1987) Tumour necrosis factor and interleukin 1: cytokines with multiple overlapping biological activities. Lab Invest 56:588–602
12. Cristmas SE, Meager A, Moore M (1987) Production of interferon and tumour necrosis factor by cloned human natural cytotoxic lymphocytes and T cells. Clin Exp Immunol 69:441–450
13. Engstrand L, Scheynius A, Pahlson C, Grimelius L, Schwan A, Gustavsson S (1989) Association of Campylobacter pylori with induced expression of class II transplantation antigens on gastric epithelial cells. Infect Immun 57:827–832
14. Rathbone BJ, Wyatt JI, Trejdosiewicz LK et al. (1988) Mucosal T cell subset in normal gastric antrum and C. pylori-associated chronic gastritis. Gut 29:A1438
15. Crabtree JE, Rathbone BJ (1990) T cell lines from Helicobacter pylori colonised gastric mucosa. Rev Esp Enferm Dig 78 [Suppl 1] :58A
16. Stolte M, Eidt S (1989) Lymphoid follicles in antral mucosa:immune response to Campylobacter pylori? J Clin Pathol 42:269–1271
17. Trejdosiewicz LK, Calabrese A, Smart CJ et al. (1991) T cell receptor $\gamma/\delta+$ cells of the gastrointestinal mucosa: occurence and V-region gene expression in Helicobacter pylori gastritis, coeliac disease and inflammatory bowel disease. Clin Exp Immunol 84:440–444
18. Kirchner T, Melber A, Fischbach W, Heilmann KL, Muller-Hermelink HK (1989) Immunohistological patterns of the local immune response in Helicobacter pylori gastritis. In : Malfertheiner P, Ditschuneit H (eds) Helicobacter pylori, gastritis and peptic ulcer. Springer, Berlin Heidelberg New York, pp 213–222

19. Kishimoto T (1989) The biology of IL-6. Blood 74:1–10
20. Isaacson P (1982) Immunoperoxidase study of the secretary immunoglobulin system and lysozyme in normal and diseased gastric mucosa. Gut 23:578–588
21. Valnes K, Brandtzaeg P, Elgjo K, Stave R (1984) Specific and nonspecific humoral defense factors in the epithelium of normal and inflamed gastric mucosa. Gastroenterology 86:402–412
22. Callard RE (1989) Cytokine regulation of B-cell growth and differentiation. Br Med Bull 45:371–388
23. Le J, Vilek J (1989) Interleukin 6: a multifunctional cytokine regulating immune reactions and the acute phase protein response. Lab Invest 61:588–602
24. Kehrl JH, Miller A, Fauci AS (1987) Effect of tumour necrosis factor alpha on mitogen-activated human B cells. J Exp Med 166:786–791
25. Jelinek DF, Lipsky PE (1987) Enhancement of human B cell proliferation and differentiation by tumour necrosis factor alpha and interleukin-1. J Immunol 139:2970–2976
26. Rathbone BJ, Wyatt JI, Worsley BW et al. (1986) Systemic and local antibody responses to gastric Campylobacter pyloridis in non-ulcer dyspepsia. Gut 27:642–647
27. Rathbone BJ, Wyatt JI, Tompkins D et al. (1986) In vitro production of Campylobacter pyloridis specific antibodies by gastric mucosal biopsies. Gut 27:A607
28. Crabtree JE, Rathbone BJ, Shallcross TM et al. (1988). Duodenal secretion of Campylobacter pylori specific antibodies in patients with gastritis and duodenitis. Gut 29:A1438
29. Crabtree JE, Rathbone BJ, Heatley RV, Shallcross TM, Wyatt JI, Losowsky MS (1989) Duodenal secretion of Campylobacter pylori-specific antibodies in vitro. In: Megraud F, Lamouliatte H (eds) Gastroduodenal pathology and Campylobacter pylori. Elsevier, New York, pp 341–344
30. Crabtree JE, Taylor JD, Shallcross TM, Rathbone BJ, Heatley RV (1990) Immunoblotting of Helicobacter pylori IgA antibody response in gastroduodenal mucosa. Gut 31:A601
31. Sobala GS, Crabtree JE, Dixon MF et al. (1991) Acute Helicobacter pylori infection: clinical features, local and systemic immune response, gastric mucosal histology and gastric juice ascorbic acid concentrations. Gut 32:1415–1418.
32. Mestecky J, McGhee JR (1987) Immunoglobulin A (IgA) : molecular and cellular interactions involved in IgA biosynthesis and immune response. Adv Immunol 40:153–245
33. Tagliabue A, Boraschi D, Villa L et al. (1984) IgA dependent cell-mediated activity against enteropathogenic bacteria: distribution, specificity, and characterisation of the effector cells. J Immunol 133:988–992
34. Tagliabue A, Nencioni L, Villa L, Keren DF, Lowell GH, Boraschi D (1983) Antibody-dependent cell-mediated antibacterial activity of intestinal lymphocytes with secretory IgA. Nature 306:184–186
35. Papadimitriou CS, Iaoachim-Velogianni, Tsianos EB, Moutsopoulos HM (1988) Epithelial HLA-DR expression and lymphocyte subsets in gastric mucosa in type B chronic gastritis. Virchows Archiv [A] 413:197–204
36. Sollid LM, Gaudernack G, Markussen G, Kvale D, Brandtzaeg P, Thorsby E (1987) Induction of various HLA class II molecules in a human colonic adenocarcinoma cell line. Scand J Immunol 25:175–180
37. MacDonald TT, Weinel A, Spencer J (1988) HLA-DR expression in human fetal intestinal epithelium. Gut 29:1342–1348
38. Kvale D, Lovhaug D, Sollid LM, Brandtzaeq P (1988) Tumour necrosis factor alpha up-regulates expression of secretory component, the epithelial receptor for polymeric Ig. J Immunol 140:3086–3089
39. Lewis CE, McCarthy SP, Lorenzen J, McGee JO'D (1990) Differential effects of LPS, IFN-γ and TNF-α on the secretion of lysozyme by individual human mononuclear phagocytes: relationship to cell maturity. Immunology 69:402–408
40. Bayerdörffer E, Oertel H, Lehn N, Kasper G, Manness GA, Sauerbruch T. Stolte M (1989) Topographic association between active gastritis and Campylobacter pylori colonisation. J Clin Pathol 42:834–839
41. Steinbeck MJ, Roth JA (1989) Neutrophil activation by recombinant cytokines. Rev Infect Dis 11:549–568
42. Kharazmi A, Nielson H, Rechniotzer C, Bendtzen K (1989) Interleukin 6 primes human neutrophils and monocyte burst response. Immunol Lett 21:177–184

43. Borish L, Rosenbaum R, Albury L, Clark S (1989) Activation of neutrophils by recombinant interleukin 6. Cell Immunol 121:280–289
44. Linder H, Engberg I, Van Kooten C, De Man P, Svanborg-Eden C (1990) Effects of anti-inflammatory agents on mucosal inflammation induced by infection with gram-negative bacteria. Infect Immun 58:2056–2060
45. Hedges S, Linder H, De Man P, Swanborg-Eden C (1990) Cyclosporin-dependent, nu-independent mucosal interleukin 6 response to gram-negative bacteria. Scant J Immunol 31:335–343
46. Strieter RM, Kunkel SL, Showell HJ, Marks RM (1988) Monokine-induced gene expression of a human endothelial cell-derived neutrophil chemotactic factor. Biochem Biophys Res Commun 156:1340–1345
47. Strieter RM, Phan SH, Showell HJ et al. (1989) Monokine-induced neutrophil chemotactic factor gene expression in human fibroblasts. J Biol Chem 264:10621–10626
48. Crabtree JE, Rathbone BJ, Wyatt JI, Heatley RV, Losowsky MS (1987) In vitro mucus glycoprotein synthesis and secretion by gastric mucosa colonised with Campylobacter pylori. Gut 28:A1409
49. Crabtree JE, Heatley RV, Trejdosiewicz LK, Losowsky MS (1990) T lymphocyte stimulation of human small intestinal glycoprotein biosynthesis: effects of anti-CD3 antibody on normal and coeliac mucosa. Int Arch Allergy Appl Immunol 93:35–40.
50. Smith JTL, Pounder RE, Nwokolo CU, Lanzon-Miller S, Evans DG, Graham DY, Evans DJ (1990) Inappropriate hypergastrinaemia in asymptomatic healthy subjects infected with Helicobacter pylori. Gut 31:552–525
51. Levi S, Beardshall K, Haddad G, Playford R, Ghosh P, Calam J (1989) Campylobacter pylori and duodenal ulcers: the gastrin link. Lancet i:1167–1168
52. El Nujumi AM, Dorian CA, McColl KEL (1990) Effect of inhibition of Helicobacter pylori urease activity with acetohydroxamic acid on plasma gastrin in subjects with duodenal ulcer. Gut 31:A 1174
53. Teichmann RK, Andress HJ, Liebich H, Seifert J, Brendel W (1984) Possible role of Ia-positive cells in the antrum in gastrin secretion. Eur Surg Res 16:64–65
54. Teichmann RK, Pratschke E, Grab J, Hammer C, Brendel W (1986) Gastrin release by interleukin-2 and gamma interferon in vitro. Can J Physiol Pharmacol 64:62
55. Evans RD, Argiles JM, Williamson DH (1989) Metabolic effects of tumour necrosis factor alpha (cachectin) and interleukin-1. Clin Sci 77:357–364
56. Rook GAW, Taverne J, Leveton C, Steele J (1987) The role of gamma-interefon, vitamin D3 metabolites and tumour necrosis factor in the pathogenesis of tuberculosis. Immunology 62:229–234
57. Lanzon-Miller S, Allison MC, Pounder RE, Ball S, Hamilton MR, Chronos NAF (1988) Enprostil inhibits post-prandial gastrin release: a dose-response study. Aliment Pharmacol Ther 2:317–323
58. Mogard MH, Maxwell V, Reedy TJ, Walsh JH (1987) Gastric acidification inhibits meal-stimulated gastric acid secretion after prostaglandin synthesis inhibition by indomethacin in humans. Gastroenterology 93:63–68
59. Renz H, Gong JH, Schmidt A, Nain M, Gemsa D (1988) Release of tumour necrosis factor alpha from macrophages J Immunol 141:2388–2393
60. Hunt RH (1989) Campylobacter pylori and spontaneous hypochlorhydria. In: Rathbone BJ, Healtley RV (ed) Campylobacter pylori and gastroduodenal disease. Blackwell, Oxford, pp 176–184
61. Uehara A, Okumura T, Sekiya C, Okamura K, Takasugi Y, Namiki M (1989) Interleukin-1 inhibits the secretion of gastric acid in rats: possible involvement of prostaglandins. Biochem Biophys Res Commun 3:1578–1584

Immunological Aspects of *Helicobacter pylori* Infection

A.S. Peña

With the exception of studies on patients with pernicious anaemia, a mouse model of autoimmune gastritis, and experimental studies of gastric tumours, the stomach has not been intensively or systematically studied by immunologists. This is likely to change in the near future. It appears that the major cause of the most common form of antral gastritis is infection with *Helicobacter* pylori. This bacterium may also be involved in the pathogenesis of duodenal ulcer, and infection is perhaps one of the initial events in the transition from gastritis to dysplasia, metaplasia and, eventually, gastrointestinal cancer of the intestinal type. One of the interesting aspects of this infection is that the majority of individuals who acquired the infection cannot get rid of the bacteria, and it is proving extremely difficult to eradicate it. No good explanation of the failure of the local immune response exists in spite of a demonstrable local antibody response. This observation should stimulate further studies in this area of research.

From the presentations at the symposium in Toledo and the papers published in this volume it is quite clear that several groups of investigators have made a good start in investigating this area of gastrointestinal immunity. Shi-Guang et al. from the University of Essen in Germany have confirmed the almost absolute correlation that exists between IgG *H. pylori* antibodies in the serum of the patients and the presence of antral gastritis. Figure et al. from the University of Siena, Italy, have studied the prevalence of serum antibodies directed to the 128-kDa protein which corresponds to the vacuolating activity-associated protein and found that all patients with duodenal ulcer had antibodies to this band, in contrast to patients without duodenal ulcer. This suggests that a high molecular weight protein on the surface of the bacteria may be involved in the pathogenesis of duodenal ulcer. Crabtree and coworkers have reviewed the role of cytokines in the modulation of the immune response against *H. pylori.* An increased production of interleukin-6 and tumour necrosis factor (TNFα) was found in biopsy homogenate supernatants and in culture supernatants of gastric biopsy specimens of patients infected with *H. pylori.* They suggest that TNFα produced locally in the stomach may be responsible for the polymorphonuclear infiltration and, indirectly, for the increased lysozyme and secretory component expression observed in chronic gastritis. Although further studies are necessary to determine the cellular origin of these cytokines it may be that they are important in controlling the extent of the inflammation. An

Free University Hospital, Department of Gastroenterology, P.O. Box 7057, 1007 MB Amsterdam, The Netherlands

J.M. Pajares et al. (Eds.)
H. pylori and Gastroduodenal Pathology

increased number of intraepithelial T cells has been reported in *H. pylori*-associated gastritis. Engstrand from Uppsala, Sweden, has reported that gamma/delta cells appeared to be increased in this condition, and a stress protein detected by a monoclonal antibody directed to the 65-kDa and the 58–64-kDa protein family was found in the epithelial cells of the antrum. This report, if confirmed, will be of special significance in studies of the pathogenesis of gastritis and perhaps in the induction of an autoimmune form of chronic gastritis.

At present it can be concluded that the presence of specific IgG-*H. pylori* antibodies correlates very well with the presence of the bacteria in the stomach. Several enzyme-linked immunosorbent assays (ELISA) are now available. However, it is advisable to validate the performance of these tests with known cases of infected patients before using them in screening of unknown populations. There are still some problems that need to be resolved, apart of the lack of standardization of the different tests available. More data is needed in patients above the age of 70; and in children, the cut off point may be different from that in adults, since it is known that the full IgG response is not reached until the age of 10 years. It is also important to standardize the technique using the same antigen source. If whole cell sonicates are used, it is probably better to include different strains since they may contain different high molecular weight cell-associated proteins. Whenever possible these antigens should be tested by immunoblotting to make sure that all specific bands are present. On the other hand, more studies using single, well-characterized strains with individual patient's serum with this technique are necessary. Studies in this direction may help to clarify why some patients develop a peptic ulcer when *H. pylori* is present.

The importance of selecting an adequate cut-off point in validating different commercial ELISA tests has been remarkably well illustrated by Loffeld et al. in another section of this volume. Contrary to expectations, the local specific response to the bacteria appears to be mainly of the IgA1 subclass instead of IgA2 as reported by Van der Est et al. in gastric biopsy specimens of patients with *H. pylori*-associated gastritis. This finding supports the pathogenetic role of *H. pylori* in producing inflammation of the gastric mucosa. Stacey et al. discuss several possibilities to explain the lack of efficiency of the local gastric humoral immune response in clearing the infection in patients infected with *H. pylori*. The role of specific anti-*H. pylori*-IgA in inhibiting bacterial adherence or uptake is not clear.

With the development of effective treatment regimens, long-term eradication of *H. pylori* is possible in over 90% of patients. There are several excellent tests for the detection of *H. pylori* in untreated patients. After treatment, however, when the number of bacteria in the mucosa is substantially diminished, the number of false-negative test results might increase. Determination of specific IgA and IgG anti-*H. pylori* antibodies by means of an ELISA technique is suitable for screening and follow-up in larger patient populations [1–3]. These serological tests are easy to perform, relatively inexpensive and very acceptable to the patients. It may well be that serological testing will become the preferred method for follow-up after treatment for *H. pylori* infection and will probably replace endoscopy [4].

When reviewing papers in the proceedings of a symposium one cannot be as rigorous as one would be in the case of most scientific publications. The contributions to the proceedings must be provocative and they should stimulate research. From this point of view the contributions here also make clear that *H. pylori* infection of the stomach can be used as a model to study the interaction between host and parasite. Further studies on the systemic and salivary humoral response against the bacteria are needed, as well as studies of genetic factors determining the predisposition to acquire the infection or, perhaps more importantly, the factors determining the lack of the infection, especially in those areas of the world where the infection starts early in life and is widespread.

Patients with chronic active gastritis without duodenal ulcer disease probably have a different predisposition to suffer from upper gastrointestinal disease than duodenal ulcer patients, or they may have been infected by a different strain of *H. pylori*, as discussed in other chapters in this volume. It should be remembered that duodenal ulcer appears to be a heterogeneous group of diseases. The IgG subclass response might depend upon the lipopolysaccharide profile of the *H. pylori* strain and/or virulence factors of the infecting *H. pylori* strain. It is also possible that the genetically determined antibody response of the host plays a role. For example, the IgG subclass response to *H. influenzae* type b polysaccharide vaccine is associated with immunoglobulin allotype G2m [5]. Individuals who are homozygous for the G2m(n) allotype have a higher IgG2 response to pneumococcal antibodies than homozygous G2m(n) allotype negative individuals [6]. These findings indicate that the IgG subclass composition of the immune response is also genetically determined. It is interesting that a selected group of patients with relapsing duodenal ulcer had a significantly ($p < 0.01$) higher IgG2 response than patients with chronic active gastritis [7].

Recent studies [8] have suggested that the gastrin proton pump is the major target recognized by the parietal cell autoantibodies, and certain *H. pylori* profoundly inhibit acid secretion from parietal cells probably involving the gastric proton pump [9]. Studies directed to dissect the autoantigen and the T cell response in patients with the autoimmune form of gastritis may have to include *H. pylori* proteins to help to understand the role which this bacterium may play in precipitating an autoimmune response in the predisposed individual.

More profound knowledge of the bacterium, its antigenic make up and the molecular structure of the adhesins will help us understand the gastric humoral and cellular immune response of the host. Studies on the T cell response may help to clarify the role the bacteria play in the development of chronic gastritis and perhaps in predisposing to the autoimmune form of gastritis.

References

1. Peña AS, Endtz Hph, Offerhaus GJA, Hoogeboom-Verdegaal A, van Duijn W, de Vargas N, den Hartog G, Kreuning J, Lamers CBHW (1989) Value of serology (ELISA and immunoblotting) for the diagnosis of Campylobacter pylori infection. Digestion 44: 131–41
2. Perez-Perez AI, Dwarkin BM, Chodos JE, Blaser MJ (1988) Campylobacter pylori antibodies in humans. Ann Intern Med 109: 11–7

3. Evans DJ, Evans DG, Graham DY, Klein PD (1989) A sensitive and specific serologic test for detection of Campylobacter pylori infection. Gastroenterology 96:1004–1008
4. Veenendaal RA, Peña AS, Meijer JL, Endtz HPh, vd Est MMC, v Duijn W, Eulderink F, Kreuning J, Lamers CBHW (1991) Long term serological surveillance after treatment of Helicobacter pylori infection. Gut 32:1291–1294
5. Granhoff DM, Suarez BK, Pandey JP, Shackelford PG (1988) Genes associated with the G2m(23) immunoglobulin allotype regulate the IgG subclass responses to Heamophilus influenzae type b polysaccharide vaccine. J Infect Dis 1576:1142–1149
6. Sarvas H, Rautonen N, Sipinen S, Mäkelä O (1989) IgG subclasses of pneumococcal antibodies—Effect of allotype G2m(n). Scand J Immunol 29:229–237
7. Bontkes HJ, Veenendaal RA, Peña AS, Goedhard JG, v Duijn W, Meijer JL, Lamers CBHW (1992) Igg subclass response to Helicobacter pylori in patients with chronic active gastritis and duodenal ulcer. Scand J Gastroenterol 27:129–133
8. Burman P, Mårdh S, Norberg L, Karlsson FA. (1989) Parietal cell antibodies in pernicious anemia inhibit H^+, K^+ -Adenoisine triphosphatase, the proton pump of the stomach. Gastroenterology 96:1434–1438
9. Cave DR, Vargas M (1989) Effect of a Campylobacter pylori protein on acid secretion by parietal cells. Lancet ii:187–189

VI. Diagnostic Progress

Usefulness of Two Enzyme-Linked Immunosorbent Assay Systems for the Diagnosis of *Helicobacter pylori*-Associated Gastritis*

R.J.L.F. Loffeld[1], H.L.A. van den Oever[1], and E. Stobberingh[2]

Introduction

Several histological and microbiological detection methods are available for the diagnosis of *Helicobacter pylori*-associated gastritis. However, these require endoscopy with biopsies from the gastric antrum. Therefore they are not feasible for epidemiological studies. To facilitate the diagnosis, non-invasive techniques such as breath tests and serological detection of *H. pylori* antibodies have been developed [1–4]. Home-made serological tests with different antigen preparations are widely employed. Several commercial enzyme-linked immuno sorbent assays (ELISA) have recently become available. A high diagnostic yield, and high sensitivity and specificity values are claimed for these tests. In the present study one such test, the Bio-Rad GAP test, was compared with a home-made ELISA used in our laboratory.

Material and Methods

Serum samples from 72 patients suffering from non-ulcer dyspepsia were used as a reference. The *H. pylori* status was assessed from all patients using the haematoxylin and eosin stain and a modified Giemsa stain; in addition, antral biopsy specimens were cultured. Forty-nine patients showed *H. pylori* in their biopsy material and were regarded as positive in the serological test, the remainder were negative.

A home-made ELISA with a crude sonicate as antigen was carried out as previously described [1]. Serum samples were expressed as positive/negative ratios (P/N) by dividing the optical densities read at 492 nm of the patients sera by the optical density of the pooled negative sera included in every run.

The commercial test, the Bio-Rad GAP test, was obtained from the clinical division of Bio-Rad laboratories, Milan, Italy. The test kit comprises the

* This paper is published in extended form entitled "Usefulness of a new serological test (Bio-Rad) to diagnose *Helicobacter pylori* associated gastritis" in the *Journal of Clinical Microbiology* (1991) 29: 283–286.

[1]Department of Internal Medicine, University Hospital Maastricht, PO Box 5800, 6202 AZ Maastricht, The Netherlands
[2]Department of Microbiology, University Hospital Maastricht, PO Box 5800, 6202 AZ Maastricht, The Netherlands

J.M. Pajares et al. (Eds.)
H. pylori and Gastroduodenal Pathology

reagents including a positive and negative control serum. All procedures were carried out as described by the manufacturers manual. The readings were expressed as P/N ratios by dividing the measured optical density at 450 nm by the optical density of the negative control.

Results

The P/N ratio of the *H. pylori*-positive sera in the home-made ELISA was 2.92 (standard deviation, SD, 1.01), and that of the negative sera was 0.83 (SD 0.58). This difference was statistically significant (chi-square test, $p < 0.0001$). With the receiver-operating curve an optimal cut-off value was determined at a P/N ratio of 1.40. This gives a sensitivity for the test of 91.8%, a specificity of 91.3%, a positive predictive value of 85.7% and a negative predictive value of 84.0%.

The P/N ratios of the *H. pylori*-positive and -negative sera in the Bio-Rad GAP test were 5.19 (SD 1.37) and 2.53 (SD 1.12), respectively. According to the guidelines provided by Bio-Rad, results less than 6.0 are considered to be negative. The specificity at this cut-off point was 100%, the sensitivity was low at 32.7%; the positive and negative predictive values were 100% and 58.9%, respectively.

If the cut-off value of 6.0 was ignored and a receiver-operating curve was made with the 72 patients sera, higher sensitivity and specificity values for the Bio-Rad GAP test were obtained. The optimal cut-off value appeared to be 3.9, resulting in a sensitivity of 87.8%, a specificity of 91.3%, a positive predictive value of 95.6% and a negative predictive value of 77.8%.

Pearson's correlation coefficient for the home-made ELISA and the Bio-Rad GAP test with a cut-off value of 3.9 was 0.88 ($p < 0.0001$); at the cut-off point of 6.0 no correlation was present.

Discussion

For the reliable use of serological tests a known reference population examined with the gold standard usually applied for the diagnosis of the disease should be available. Sera from patients with known *H. pylori* status, determined by the combination of histological and microbiological detection, were used as reference. With a receiver-operating curve it is possible to calculate a cut-off value which will provide high sensitivity and specificity values for the test.

Bio-Rad does not provide data on how the cut-off value was established in their test. In the manufacturer's guide it is stated that control sera were used from patients with or without a history of peptic ulcer disease. No data are provided on how the *H. pylori* status in these patients was assessed. It can be assumed that no endoscopic biopsy was taken. Of course this is not the correct way to establish a proper and reliable control population. Since *H. pylori* is present in 20%–30% of the asymptomatic population, it can be expected that a large number of "control" patients with no history of peptic ulcer disease actually have *H. pylori* infection, resulting in the presence of antibodies. This explains the high cut-off value, the low sensitivity and low negative predictive value, of the Bio-Rad test. In other words, if the Bio-Rad GAP test is used according to the manufacturer's guidelines, almost 66% of the positive reference sera give a negative test result, hence are false negatives. Of course this in unacceptable for a serological test which will be used as a screening method or as a control instrument after treatment.

The manufacturer provides a positive control serum for the quality control of the reagents used and for a semi-quantitative scoring of the test result (not described in this paper). Oddly enough, this positive control serum had a negative test result in our study.

If the cut-off value of the Bio-Rad GAP test is reassessed, the diagnostic yield can be improved considerably and correlates well with the home-made test.

It is important that all commercial diagnostic tests for the diagnosis of *H. pylori* infection are tested with sera of a known reference population before they are introduced into daily practice.

References

1. Loffeld RJLF, Stobberingh E, Flendrig JA, van Spreeuwel JP, Arends JW (1989) Diagnostic value of an immunoassay to detect anti-Campylobacter pylori antibodies in non-ulcer dyspepsia. Lancet i: 1182–1185
2. Hirschl AM, Pletschetter M, Hirschl MH, Berger J, Stanck G, Rotter ML (1988) Comparison of different antigen preparations i an evaluatio of the immune response to Campylobacter pylori. Eur J Clin Microbiol Infect Dis 7: 570–575
3. Perez-Perez GI, Dworkin BM, Chodes JE, Blaser MJ (1988) Campylobacter antibodies in humans. Ann Intern Med 109: 11–17
4. Wyatt JI, Rathbone BJ (1989) The role of serology in the diagnosis of Campylobacter pylori infection. Scand J Gastroenterology 24: 27–34

VII. *Helicobacter pylori* and Non-ulcer Dyspepsia

Helicobacter pylori Infection in Gastroesophageal Reflux Disease

O. Pieramico and P. Malfertheiner

Definition and Relationship Between Gastroesophageal Reflux and Non-ulcer Dyspepsia

Non-ulcer dyspepsia (NUD) is a symptom complex which is very common in patients presenting to the gastroenterologist [1, 2]. Several definitions of non-ulcer dyspepsia have been proposed over the years, but none has been definitively adopted [3– 6]. A recent working party [5] defined NUD as a clinical syndrome including: upper abdominal or retrosternal pain, discomfort, heartburn, nausea, vomiting, or other symptoms considered to be referable to the proximal alimentary tract and lasting more than 4 weeks and for which no focal lesion or systemic disease is responsible.

According to this definition, patients with gastroesophageal reflux symptoms (GERS) might be considered to have NUD. Although including GERS in the definition of NUD may be questioned, we think that patients with retrosternal symptoms as the main symptomatology should be classified under gastroesophageal reflux disease, whereas patients with primary upper abdominal complaints and occasional GERS should be considered as NUD patients. The complexity of the nosological classification of GERS is demonstrated in a study by Klauser et al. [7] showing that heartburn was experienced by 48% of the patients with normal reflux and in 68% of the patients with gastroesophageal reflux disease, as assessed by 24-h pH measurement.

Gastroesophageal Reflux Symptoms in Non-ulcer Dyspepsia and Its Relationship to *Helicobacter pylori* Infection

Heartburn or other GERS frequently occur in the general population and its prevalence ranges between 5% and 20% [8, 9], but the incidence of GERS in patients with NUD [10, 11] is higher than that in the general population. Talley and Piper [10] report GERS in 71 out of 327 patients with NUD (23%), whereas other authors have reported a prevalence of 48% of GERS [11] (Table 1).

Department of Internal Medicine II and Gastroenterology, University of Ulm, R. Koch Str. 8, 7900 Ulm, FRG

J.M. Pajares et al. (Eds.)
H. pylori and Gastroduodenal Pathology

Table 1. GERS in patients with upper abdominal complaints

Reference	Total (*n*)	Patients with GERS (*n*)	Patients with GERS (%)
Edwards et al. 1968	231	88	39
[10]	248	71	29
[11]	660	316	48
Department of Gastroenterology, University of Ulm (1990-1991)	149	75	50

Table 2. GERS in relationship to *H. pylori* infection

	Total		Patients with GERS	
	(*n*)	(%)	(*n*)	(%)
Total Patients	149		75	49%
H. pylori positive	76	51	46	60%*
H. pylori negative	73	49	29	40%

*$p < 0.05$ vs. *H. pylori*-negative

Whether the presence of *Helicobacter pylori*-positive chronic active gastritis is related to GERS and/or abnormal gastroesophageal reflux is still being debated. In a recent prospective clinical study (Table 2), we investigated 331 consecutive patients with upper abdominal complaints. Out of 331 consecutive patients, 149 with upper abdominal complaints could be diagnosed as having NUD according to the above-mentioned inclusion criteria for NUD. In agreement with the available literature [12, 13], 51% of them were *H. pylori* positive (76 out of 149). Out of 76 NUD patients 46 (60%) with *H. pylori* infection had heartburn as a secondary symptom, whereas heartburn was present in only 29 out of 73 (40%) NUD *H. pylori*-negative patients ($p < 0.05$). These data therefore indicate that the prevalence of GERS in patients with NUD and *H. pylori*-positive chronic active gastritis is higher than in *H. pylori*-negative patients with NUD. This supports previous preliminary data [14] which indicated that esophageal symptoms were more frequently present in H. pylori-positive patients than in *H. pylori*-negative patients.

Abnormal Gastroesophageal Reflux Related to *H. pylori*-Positive Chronic Active Gastritis

A disturbed sphincter pressure or wave contraction in the esophagus, an alteration in gastric acid secretion, and a delayed gastric emptying are among the potential mechanisms involved in the pathogenesis of gastroesophageal reflux disease. The complex interplay of secretory and motor changes in the stomach induced by chronic gastritis as well the release of inflammatory mediators might theoretically lead to increased acid reflux. In a previous study [15], chronic gastritis was found to be associated with delayed gastric emptying and increased reflux. If this phenomenon can be confirmed by further studies, one may hypothesize that *H. pylori* infection, by inducing chronic active gastritis, might contribute to increased gastroesophageal reflux. Furthermore, changes in gastrin and gastric acid secretion [16, 17] induced by *H. pylori* infection might be an additional factor favoring reflux. In contrast to this hypothesis, preliminary data [18, 19] have shown that *H. pylori*-positive chronic active gastritis is not associated with reflux esophagitis as assessed by histology. However, inflammatory changes in the distal esophagus are only a small part of the wide spectrum of gastroesophageal reflux disease, and the majority of patients with GERS have no esophagitis.

In order to investigate the relationship between chronic active gastritis and gastroesophageal reflux we performed 24-h esophagus pH measurement in patients with GERS in relationship to the clinical picture and to the *H. pylori* status. For this purpose, *H. pylori* infection was assessed by histological examination of antral biopsies obtained during gastroscopy in 35 consecutive patients who underwent 24-h esophagus pH measurement for heartburn and retrosternal pain. In accordance with a previous report [7], an abnormal gastroesophageal reflux was confirmed by pH measurement in 70% (23 out of 35) of the patients with GERS. When the presence of *H. pylori* infection was considered, six patients out of 12 were *H. pylori* positive and 17 out of the remaining 23 were *H. pylori* negative. The results of the 24-h esophagus pH measurement were not different with regard to *H. pylori*-positive and -negative patients.

Conclusions

The relationship between chronic gastritis and gastroesophageal reflux is still unclear. *H. pylori* infection, by inducing chronic active gastritis, does not appear to be a risk factor for increasing reflux. However, gastroesophageal reflux symptoms occur more frequently in patients with *H. pylori*-positive chronic active gastritis than in *H. pylori*-negative patients. Further studies are needed to clarify the role of altered gastric physiology by *H. pylori* infection on the pathogenesis of gastroesopageal reflux.

References

1. Krag E (1982) Non-ulcer dyspepsia—introduction: epidemiological data. Scand J Gastroenterol 17 [Suppl 79]: 6–8
2. Drossman DA, Grant-Thompson W, Talley NJ et al. (1990) Identification of sub-groups of functional gastrointestinal disorders. Gastroenterol Int 3: 159–172

3. Heatley RV, Rathbone BJ (1987) Dyspepsia: a dilemma for doctors? Lancet 3:779–781
4. Barbara L, Camilleri M, Corinaldesi R et al. (1989) Definition and investigation of Dyspepsia. Dig Dis Sci 34:1272–1276
5. Management of dyspepsia: report of a working party (1988) Lancet 576–579
6. Camilleri M, Thompson DG, Malagelada JR (1986) Functional Dyspepsia. Symptoms and underlying mechanisms. J Clin Gastroenterol 8:424–429
7. Klauser AG, Schindlbeck NE, Müller-Lissner SA (1990) Symptoms in gastrooesophageal reflux disease. Lancet 335: 205–208
8. Heading RC (1989) Epidemiology of oesophageal reflux disease. Scand J Gastroenterol 24 [Suppl 168]:33–37
9. Ruth M, Mansson I, Sandberg N (1991) The prevalence of symptoms suggestive of esophageal disorders. Scand J Gastroenterol 26:73–81
10. Talley NJ, Piper DW (1985) The association between non-ulcer dyspepsia and other gastrointestinal disorders. Gut 20:896–900
11. Johannessen T, Petersen H, Kleveland PM et al. (1990) The predictive value of history in dyspepsia. Scand J Gastroenterol 25:689–697
12. Lambert JR, Dunn K, Borromeo M et al. (1989) Campylobacter pylori—a role in non-ulcer dyspepsia? Scand J Gastroenterol 24 [Suppl 160] :7–13
13. Rokkas T, Pursey C, Uzoechina E et al. (1987) Campylobacter pylori and non-ulcer dyspepsia. Am J Gastroenterol 82:1149–1152
14. Rathbone BJ, Wyatt J, Heatley RV (1988) Symptomatology in C. pylori-positive and negative non-ulcer dyspepsia. Gut 29:1473
15. Fink SM, Barwick KW, DeLuca V et al. (1984) The association of histologic gastritis with gastroesophageal reflux and delayed gastric emptying. J Clin Gastroenterol 6:301–309
16. Smith JTL, Pounder RE, Nwokolo CU et al. (1990) Inappropriate hypergastrinemia in asymptomatic healthy subjects infected with Helicobacter pylori. Gut 31:522–525
17. Malfertheiner P, Pieramico O (1992) Helicobacter pylori. In: Kumar D, Graham D, Gustavsson S (eds) The stomach. Churchill Livingstone, London, p 297
18. Ekstrom P, Unge P, Gnarpe H, Blomqvist C (1989) Is Campylobacter pylori correlated with esophagitis? Klin Wochenschr 67 [Suppl XVIII]:18
19. Weber P, Herz R, Schäfer H, Stolte M (1989) Campylobacter pylori and gastroesophageal reflux disease. Klin Wochenschr 67 [Suppl XVIII]:70

Computerised History Taking for the Assessment of *Helicobacter pylori* in Non-ulcer Dyspepsia: The Quest for the Clinical Syndrome

R.J.L.F. Loffeld and R. Adang

Introduction

Marshall and Warren's original description of *Helicobacter pylori* [1] renewed interest in gastritis research. Until then many investigators thought gastritis to be phenomenon of the stomach brought on by advancing age and without clinical relevance.

Helicobacter pylori is the major cause of type B antral gastritis and, in addition, plays an important role in the pathogenesis of peptic ulcer disease, especially duodenal ulcer disease. Its role in non-ulcer dyspepsia remains controversial [2]. Non-ulcer dyspepsia is an ill-defined clinical syndrome. The aetiology is unknown, but *H. pylori* as well as motility disorders are held responsible for the syndrome. The major problem facing the clinician is the fact that *H. pylori* gastritis does not cause a typical clinical syndrome. In other words, identification of the disease is not possible on the basis of the medical history. Another confounding factor is the fact that *H. pylori* gastritis also occurs in apparently healthy individuals.

The prevalence of *H. pylori* infection in healthy individuals increases to about 50% with advancing age. The overall prevalence is 25%. The prevalence of *H. pylori* gastritis in patients suffering from non-ulcer dyspepsia appears to be 50%–60%. If age cohorts of healthy individuals are compared with non-ulcer dyspepsia patients, it can be noticed that the prevalence of *H. pylori* infection is higher in every age cohort of non-ulcer dyspepsia patients (Fig. 1). Since studies done in healthy individuals are serological studies, this difference can be more impressive since an unknown number of individuals show the serological scar of past infection, implying that the actual figure of persons showing active *H. pylori* infection might be considerably lower.

This observation indicates that a significant number of non-ulcer dyspepsia patients actually suffer from *H. pylori* infection. Unfortunately these patients are not identifiable in the melee of upper abdominal complaints.

Since non-ulcer dyspepsia is a major problem in public health, identification of patients suffering from *H. pylori* infection should be the first goal in non-ulcer dyspepsia research.

One way to establish this is the performance of randomised therapeutic trials. The results of these studies have so far been disappointing. Some

Department of Internal Medicine, University Hospital Maastricht, PO Box 5800, 6202 AZ Maastricht, The Netherlands

J.M. Pajares et al. (Eds.)
H. pylori and Gastroduodenal Pathology

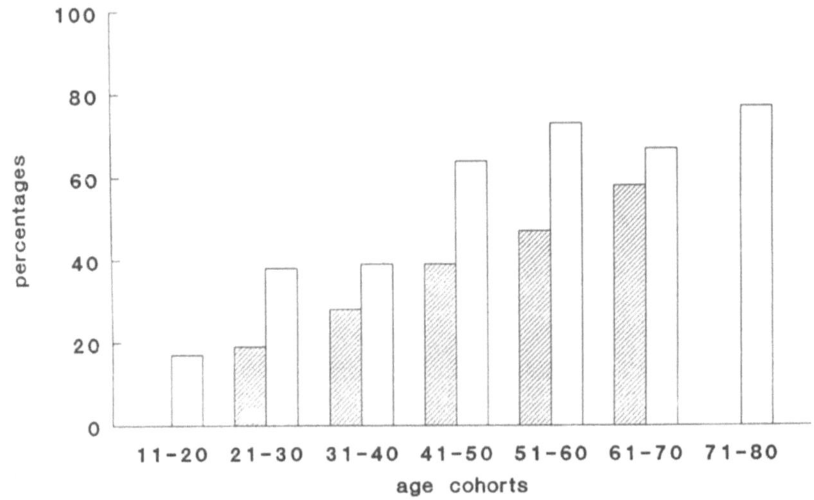

Fig. 1. Presence of *H. pylori* in different age cohorts of non-ulcer dyspepsia patients (*open columns*) and healthy blood donors (*hatched columns*). The non-ulcer dyspepsia patients were tested histologically and microbiologically. The donors were tested serologically

complaints show decline in severity or frequency, others do not; while in other studies no differences are noted between treated patients and placebo patients. The only valid conclusion from these trials can be that some patients benefit from therapy [3–7].

The other possibility to answer the question is the development of a very detailed medical history, linking complaints and/or combination of complaints with the results of histological and microbiological studies from antral biopsy specimens. The routine medical history remains a time-consuming and often unsatisfactory task. Therefore, it seems logical to use a computer-assisted system, for medical history taking.

Material and Methods

A computerised questionnaire is being developed as part of an on-line decision system in gastroenterology. The hardware consists of a Novell network with Olivetti personal computers. The programming language is Turbo Pascal 5.5, the software is Ashton Tate's DBASE III plus.

The questionnaire comprises 54 questions in a so-called branching structure. Twenty-two questions always have to be answered, while the remainder are refinement questions only to be answered if a main question is answered in the affirmitive. The answers are given as multiple choice options. The patient has the possibility to go back and change previously given answers. It is not possible to skip a question.

Table 1. Patients' complaints in the presence or absence of *H. pylori*

	HP present (%)	*HP absent* (%)
Epigastric pain	55	50
Retrosternal pain	52	43
Heartburn	48	29
Burping	84	67
Nausea	46	40
Vomiting	39	43
Anorexia	33	47
Bloating	54	54
Influence of stress	71	90

The questionnaire includes questions on upper abdominal complaints, as well as complaints pointing towards irritable bowel syndrome and symptomatic gall bladder disease. The severity of some complaints is scored, and additionally pain localisation is asked for. Data on medication, smoking habits and alcohol consumption are noted.

All patients referred for upper gastrointestinal endoscopy are encouraged to participate in this computerised history taking. A pilot study was done in patients suffering from non-ulcer dyspepsia. For the purpose of the study all patients referred for upper gastrointestinal endoscopy showing macroscopic abnormalities such as peptic ulceration or reflux oesophagitis were excluded. Patients with upper abdominal complaints with normal endoscopic investigation were included.

One antral biopsy specimen was taken from all patients for histological examination; in several patients one specimen was also used for the CLO test (Delta West Ltd., Western Australia). The sections were examined with the haematoxylin and eosin stain for scoring of gastritis according to Whitehead. *H. pylori* was detected with the modified Giemsa stain [8].

Results

Seventy-seven consecutive patients were included. Eighteen patients had to be excluded because they did not participate, hence no computer data were available. The remainder could be evaluated (31 men, 28 women; mean age 51 years, range 24–79). Gastritis was present in 35 patients (59%), all were positive for *H. pylori.* Twenty-four biopsy specimens (41%) revealed normal antral mucosa, nine of them were *H. pylori*-positive (37.5%), 15 were *H. pylori* negative (62.5%). *H. pylori* was present in 44 out of 59 patients (74.5%). The results regarding several upper abdominal complaints for *H. pylori* positive and -negative non-ulcer dyspepsia patients are shown in Table 1.

Discussion

This study was especially done to study the feasibility and the acceptability of a computerised system for non-ulcer dyspepsia. The results of this pilot study are promising. However, the system has some important shortcomings. At the present time, the number of patients is too small to permit any conclusions or to do statistical analysis. A significant number of patients are not able to participate, mostly because of visual problems or illness. On the other hand, the system needs to be explained to the patient, hence extra personnel. In times of great pressure on the department, it was not possible to instruct all patients.

To the question: "Were you able to express all your complaints?" 58% of the patients answered in the affirmative, while 25% answered in the negative. The remainder had no opinion.

For this system to work it is very important that all referred patients participate and that every patient is able to express their complaints to the full extent. This indicates that some amplifications and alterations of the system are necessary, and that, if active participation is not possible, the questionnaire should be used in the traditional oral way. This requires more attendance and explanation by the endoscopy staff. The computerised questionnaire was easily accepted by the participants. The majority found the computer work easy and interesting.

In this study only one antral biopsy specimen was taken for the evaluation of the histological appearance of the mucosa and presence of *H. pylori*. This is not the correct way of diagnosing *H. pylori* infection. At least two biopsy specimens should be taken from the antrum. However, owing to logistic problems this was not possible in this pilot phase.

After the necessary improvements and solution of the logistic problems, this questionnaire, together with the antral biopsy, seems to be a potentially valid instrument to identify patients suffering from *H. pylori* infection. It is our opinion that it will only be a matter of time before a large part of the puzzling problem of *H. pylori* and non-ulcer dyspepsia has been solved.

References

1. Marshall BJ, Warren JR (1984) Unidentified curved bacilli in the stomach of patients with gastritis and peptic ulceration. Lancet i: 1311–1315
2. Tytgat GNJ, Axon ATR, Dixon MF, Graham DY, Lee A, Marshall BJ (1990) Helicobacter pylori, causal agent in peptic ulcer disease. Working party reports 1990. World Congresses of Gastroenterology Sydney. Blackwell, Melbourne, pp. 36–45
3. McNulty CAM, Gearty JC, Crump B et al. (1986) Campylobacter pyloridis and associated gastritis: investigator blind placebo controlled trial of bismuth salicylate and erythromycin ethylsuccinate. Br Med J 293: 645–649
4. Glupczynski Y, Burette A, Labbe M, Deprez C, Reuck M, Deltenre M (1988) Campylobacter pylori associated gastritis: a double blind placebo controlled trial with amoxycillin. Am J Gastroenterol 83: 365–372
5. Loffeld RJLF, Potters HVPJ, Stobberingh E, van Spreeuwel JP, Flendrig JA, Arends JW (1989) Campylobacter associated gastrits in patients with non-ulcer dyspepsia: a placebo controlled double blind trial with colloidal bismuth subcitrate. Gut 30: 1206–1212

6. Oderda G, Holton J, Altare F, Vaira D, Ainley C, Ansaldi N (1989) Amoxycillin plus tinidazole for Campylobacter pylori gastritis in children: assessment by serum IgG antibody, pepsinogen I and gastrin. Lancet i: 690–692
7. Lambert JR, Borromeo M, Komann MG et al. (1987) Effect of colloidal bismuth (De-Nol) on healing and relapse of duodenal ulcers—role of Campylobacter pyloridis. Gastroenterology 92: 1489
8. Loffeld RJLF, Potters HVPJ, Arends JW, Stobberingh E, Flendrig JA, van Spreeuwel JP (1988) Campylobacter associated gastritis in patients with non-ulcer dyspepsia. J Clin Pathol 41: 84–88

VIII. *Helicobacter pylori* and Malignant Growth

Prevalence of *Helicobacter pylori* in Patients with Malignant Growth of Gastric Mucosa

J.M. Pajares, M. Crespo, J. Fraga, P. Sanchez Risco, A. Lancho, and M. Blanco

Introduction

Although the relation between *Helicobacter pylori* (HP), gastritis and peptic ulcer disease has already been well demonstrated [1–6], very little is still known about HP and its possible role in the pathogenesis of gastric cancer. Research carried out up to now on the aetiology of gastric cancer has contributed many data on different risk factors or circumstances that play a role in the pre-cancerous process.

Owing to the only recent discovery of HP as a cause of gastroduodenal pathology, researchers have aimed their work at trying to clear up the pathogenetic implications of HP in pathologies such as gastritis [7–9] and peptic ulcer [10–14]. Only when these implications had been sufficiently demonstrated, did they shift their interests towards another pathology, i.e., gastric cancer. The point of view that HP is an infectious agent and a possible "cancer promoter" [15], together with others already recognized as risk factors in gastric cancer, raises a new line of study and research. However, until now, there have only been a few studies published about HP and gastric neoplasia which have tried specifically to add data to this subject.

Aims

In our work we tried to carry out a study exclusively with patients with gastric neoplastic or preneoplastic pathology. The aims of our study were: (a) to establish the general prevalence of HP in patients with malignant growth of gastric mucosa; (b) to study HP colonization in different types of malignant growth of gastric mucosa, considering factors such as the histologic type of the neoplasia or its location.

Patients and Methods

A total of 69 patients were studied (43 men and 26 women). They were sent to the digestive endoscopy service in our hospital with clinical symptoms

Hospital de la Princesa, Universidad Autonoma de Madrid, Madrid, Spain

J.M. Pajares et al. (Eds.)
H. pylori and Gastroduodenal Pathology

suggesting gastric pathology, and they were all endoscoped and diagnosed, with later histopathologic confirmation, as having gastric cancer. The diagnoses were: 57 gastric adenocarcinomas, five gastric lymphomas, two Kaposi's sarcomas, two adenocarcinomas of the lower third of the esophagus, five gastric dysplasias. The patients were divided into two groups according to the method used to detect HP presence.

Histologic Detection of HP

A total of 49 patients were included. Several biopsies were obtained from each patient and these were processed in the pathology service of our hospital using conventional histologic methods. Hematoxylin and eosin stains were performed in all biopsies. We also used Giemsa stains on the biopsies of six patients. Slides were reviewed with a conventional optic microscope, and HP presence was confirmed in all cases with high-power fields ($\times$ 100 objective).

Microbiologic Detection of HP

Twenty patients were included. The urease test and CLO-test (Delta West Ltd., W.Australia) were performed, with the biopsies obtained from these patients, and samples were also sent to the microbiology service of the hospital for culture and Gram staining, using the appropriate methods and culture media. With reference to the microbiologic diagnostic criteria of HP infection, patients were considered as HP positive when at least two of the four diagnostic methods named were positive. The location of the neoplasia (body/antrum) was set, followed by the endoscopy criterion. Statistical analysis of the data was performed by the chi-square method, the results being considered as significant with p values less than 0.05.

Results

The general prevalence of HP in patients with malignant growth of gastric mucosa was 32% (22 HP positive from 69 patients). There is a significantly higher presence of HP in gastric adenocarcinomas growing from the antrum than in those growing from the body ($p < 0.05$).

It is interesting to observe, from the histologic point of view, that there was not a high level of HP colonization in any of the infected patients, and the highest concentration of HP was found in two of the three gastric lymphomas studied. With reference to the patients with histologic diagnosis of gastric epithelial dysplasia, HP was found only in those with an antral location (two HP positive from patients with dysplasia).

No HP presence could be demonstrated in patients with adenocarcinoma of the lower third of the esophagus or Kaposi's sarcoma. Patients with adenocarcinoma of the lower third of the esophagus were included in this study

Table 1. Helicobacter pylori (HP) colonization to type and location of malignant growth

Type	HP positive (*n*)	HP positive (%)	HP negative (*n*)	HP negative (%)	Total (*n*)
Adenocarcinoma of the esophagus	—		2		2
Gastric adenocarcinoma					
Body	5	18	23	82	28
Antrum	13	45	16	55	29
Gastric dysplasia					
Body	—		2		2
Antrum	2	66	1	33	3
Lymphoma	2	66	1	33	3
Kaposi's sarcoma	—		2		2
Total	22	32	47	68	69

due to the fact that many of these neoplasias grow from the so-called Barrett's esophagus, a gastric epithelial metaplasia that may be colonized by HP [16].

Considering the diagnostic method used to detect HP presence, results differ according to the method (histologic or microbiologic) used. In patients studied microbiologically, we obtained a higher percentage of HP positivity: 55% (11 HP positive from 20 patients) in relation to those studied histologically: 23% (11 HP positive from 69 patients).

However, the tendency towards a higher level of HP colonization in antral neoplasias in relation to those from the body subsits as much in the group of patients studied with histologic methods, as in those studied with microbiologic methods. The results of all patients are summarized in Table 1.

Discussion

The results of the study showed the relatively low level of HP colonization in patients with gastric neoplasia in relation to the levels of infection reported in other gastroduodenal pathologies (chronic gastritis [1, 3, 8, 10, 17, 18] and peptic ulcer [2, 18, 19]).

In a comparative analysis of the results from the different studies published which include patients with gastric cancer, the great difference in the prevalences obtained and the very broad range could be surprising [2, 18, 20–26], but it should be pointed out that not all of these studies were specifically designed to establish the relation between HP and gastric cancer, and consequently the number of patients with gastric neoplasia included is usually very low. Nevertheless, in general, the levels of HP colonization reported are never as high as in the rest of gastroduodenal pathologies, as we have confirmed in our study.

With reference to the greater sensitivity of detecting HP presence in patients studied using microbiologic methods in relation to those studied histologically, we observed that the reason could be that HP is never found colonizing neoplastic tissue but adjacent "normal" gastric mucosa with no tumorous infiltration, confirming observations that have been previously reported in other studies [18, 21]. In this sense, many of the biopsies reviewed histologically were nearly all tumorous tissue, in which no HP presence could be demonstrated.

We therefore recommend that biopsy samples should be taken at a sufficient distance from the neoplastic tissue if a study of the prevalence of HP in patients with gastric cancer is to be carried out.

It remains to be confirmed that HP colonization, which is a chronic infection that probably persists over the gastric mucosa for many years, may contribute together with other risk factors to the development of carcinomatous changes [15]. At the same time, the currently accepted pathogenic sequence, chronic gastritis – intestinal metaplasia – dysplasia – cancer, could be extended by the incorporation of HP presence in the first step (chronic gastritis).

Considering the significant difference in colonization between antral and body carcinomas, the interpretation and significance are still in doubt, although HP might be one of the most important risk factors in the antral location, while in body carcinomas other more important risk factors such as nitrosamines or other exogenous factors could have a more determining effect.

An interesting perspective arises in relation to gastric primitive lymphomas and the possible role HP could play in their pathogenesis as HP infection could be considered as a continuous antigenic stimulation over the gastric mucosa, with a more or less important role according to the level of colonization.

In summary, the prevalence of HP in gastric cancer is less than in other gastroduodenal pathologies (chronic gastritis and peptic ulcer). Nevertheless, HP could play an important role as a risk factor or carcinogen in the first stages of the neoplastic process.

This pathogenic role could be more or less important according to the circumstances such as tumor location (a greater implication of HP in antral carcinomas) or the histologic type of this carcinoma.

References

1. Marshall BJ, Warren JR (1984) Unidentified curved bacilli in the stomach of patients with gastritis and peptic ulceration. Lancet i: 1311–1315
2. Marchall BJ, McGechie DB et al. (1985) Pyloric campylobacter infection and gastroduodenal disease. Med J Aust 142: 439–444
3. Buck GE, Gourley WK et al. (1986) Relation of Campylobacter pyloridis to gastritis and peptic ulcer. J Infect Dis 153: 664–669
4. Goodwin CS, Armstrong JA et al. (1986) Campylobacter pyloridis, gastritis and peptic ulceration. J Clin Pathol 39: 353–365
5. Rathbone BJ, Wyatt JI et al. (1987) Possible pathogenetic pathways of Campylobacter pylori in gastroduodenal disease. Scand J Gastroenterol 23 [Suppl 142]: 40–43
6. Blaser MJ (1990) Helicobacter pylori and the pathogenesis of gastroduodenal inflammation. J Infect Dis 161 (4): 626–633
7. Marshall BJ (1986) Campylobacter pyloridis and gastritis. J Infect Dis 153: 650–657
8. Nedenskov-Sorensen P, Bjornelett A et al. (1988) Campylobacter pylori infection and its relation to chronic gastritis. An endoscopic, bacteriologic, and histomorphologic study. Scand J Gastroenterol 23: 867–874
9. McNulty CA (1989) Pathogenecity of Campylobacter pylori – a causative factor in gastritis. Scand J Gastroenterol [Suppl.] 160: 3–6

10. Andersen LP, Holk S et al. (1987) Campilobacter pyloridis and peptic ulcer disease. I. gastric and duodenal infection caused by C. pyloridis: histopathologic and microbiologic findings. Scnad J Gastroenterol 22: 219–224
11. Wyatt JI, Rathbone BJ et al. (1988) Campylobacter pylori and development of duodenal ulcer. Lancet i: 118–119
12. Graham DY (1989) Campylobacter pylori and peptic ulcer disease. Gastroenterology 96 [2 suppl]: 615–625
13. Graham DY (1989) Campylobacter pylori as a pathogenetic factor in duodenal ulcer: the case for. Scand J Gastroenterol [Suppl] 160: 40–52
14. Smith AC (1989) Duodenal ulcer disease: what role does Campylobacter pylori play? Scand J Gastroenterol [Suppl] 160: 41–18
15. Correa P, Ruiz B (1989) Campylobacter pylori and gastric cancer. In: Rathbone BJ, Heatley RV (eds) Campylobacter pylori and gastroduodenal disease. Blackwell, London, pp 39–146
16. Paull G, Yardley JH (1988) Gastric and esophageal Campylobacter pylori in patients with Barrett's esophagus. Gastroenterology 95: 216–218
17. Johnston BJ, Reed Pl et al. (1986) Campylobacter-like organisms in duodenal and antral endoscopic biopsies: relationship to inflammation. Gut 27: 1132–1137
18. Jiang SJ, Liu WZ et al. (1987) Campylobacter-like organisms in chronic gastritis, peptic ulcer and gastric carcinoma. Scand J Gastroenterol 22: 553–558
19. Coghlan JG, Gilligan D et al. (1987) Campylobacter pylori and recurrence of duodenal ulcers – a 12 month follow-up study. Lancet ii: 1109–1111.
20. Lambert RH, Leon-Barua R et al. (1985) Campylobacter-like organisms in human stomach. Gastroenterology 88: 1463 (abstract)
21. Gilman RH, Leon-Baura R et al. (1987) Campylobacter pyloridis fails to colonize sites of adenocarcinoma but not adjacent non-cancerous tissue in patients with gastric cancer. Gastroenterology 92: 1406 (abstract)
22. Cheng SCW, Sanderson CR et al. (1987) Campylobacter pyloridis in patients with gastric carcinoma. Med J Aust 147: 202–203 (letter)
23. Rathbone B, Wyatt J (1988) Campylobacter pylori and precancerous lesions. In: Reed Pl, Hill MJ (eds) Gastric carcinogenesis. Excerpta Medica, Amsterdam, pp 132–144
24. Cortés A, Carmona F et al. (1987) Campylobacter pyloris en biopsia de mucosa gástrica. Colombia Med 18: 101–109
25. Von Wulffen H, Grote HJ et al. (1988) Immunoblot analysis of immune response to Campylobacter pylori and its clinal associations. J Clin Pathol 41: 653–659
26. Feng Y, Wang Y (1988) Campylobacter pylori in patients with gastritis, peptic ulcer and carcinoma of the stomach in Lanzhou, China. Lancet i: 1055 (letter)

A Role for *Helicobacter pylori* in Gastric Carcinoma?

R.J.L.F. Loffeld[1], I. Willems[2], and J.W. Arends[2]

Introduction

The exact aetiology of gastric carcinoma is unknown. Many impetuses are known or suspected to play a role in the pathogenesis of gastric cancer. It is generally accepted in the literature that induction of gastritis is the initial event in the pathogenesis. If the inflammation persists for many years, this will eventually lead to the development of intestinal metaplasia and atrophia. Together with other impetuses, such as, for instance, toxic substances, this can finally lead to the development of gastric cancer. Of course this sequence is much more complicated and not unraveled to its full extent.

Helicobacter pylori is the major cause of type B antral gastritis, characterised by an acute and chronic inflammatory infiltrate in the superficial portion of the lamina propria and around groups of gastric glands. Krienitz [1] postulated that the spiral organisms observed in resection specimens from gastric cancers had some aetiological importance. Glands in which *H. pylori* are present display increased proliferative activity. It is thought that the bacterium acts as an irritant and induces excessive cell replication, thus acting as a cancer promotor [2]. Although only a minor portion of patients with type B gastritis will develop gastric cancer, it is interesting to know what the prevalence of the infection is in this condition.

Material and Methods

Gastric carcinomas diagnosed from 1980 to 1984 were retrieved from our files. The material consisted of 105 carcinomas (64 men and 41 women, mean age 76.2 years, range 41–97).

The presence of *H. pylori* was studied with a modified Giemsa stain, applied on all material available from the patients. Biopsy specimens as well as resection specimens were studied. Presence of *H. pylori* was scored as previously described [3].

---

*This paper has been published in a different form in *Histopathology* (1990) 17: 537–541.

[1]Department of Internal Medicine, University Hospital Maastricht, PO Box 5800, 6202 AZ Maastricht, The Netherlands

[2]Department of Pathology, University Hospital Maastricht, PO Box 5800, 6202 AZ Maastricht, The Netherlands

J.M. Pajares et al. (Eds.)
H. pylori and Gastroduodenal Pathology

Standard haematoxylin and eosin-stained sections were re-evaluated. Tumour differentiation was scored as poor or high, tumour type (intestinal or diffuse) was also noted. Presence of intestinal metaplasia was scored semiquantitatively: grade 0, none; grade 1, focal; grade 2, less than half of the epithelium affected; and grade 3, more than half of the epithelium affected.

The cancer patients were compared with seropositive blood donors and *H. pylori*-positive non-ulcer dyspepsia patients. A total of 415 healthy blood donors were tested serologically with a highly sensitive and specific enzyme-linked immunosorbent assay (ELISA), using a crude sonicate as antigen [4]. Two hundred patients suffering from non-ulcer dyspepsia were studied histologically and microbiologically for *H. pylori*. Both control groups have already been published [3, 5].

Results

Fourteen carcinomas had to be excluded because of the absence of normal epithelium in the sections. The remaining 91 carcinomas were available for evaluation.

As shown in Table 1 no significant difference was present in *H. pylori* presence between resection specimens or biopsy specimens from the same patient. However, a trend towards a higher bacterial load was observed in the biopsy specimens. If all available biopsy specimens and resection specimens were compared, a highly significant difference in bacterial load was present.

Table 2 shows the presence of *H. pylori* in poorly or highly differentiated tumours. No significant difference was seen, although *H. pylori* was more often present in tumours revealing a high differentiation grade.

No significant difference was seen in the presence of *H. pylori* between intestinal and diffuse types of carcinoma (Table 3). However, the number of *H.*

Table 1. Presence of *H. pylori* in resection and biopsy specimens

H. pylori score	Grade 0		Grade 1		Grades 2/3	
	(*n*)	(%)	(*n*)	(%)	(*n*)	(%)
Resection	33	54	17	28	11	18
Biopsy	26	43	17	28.5	17	28.5

Table 2. Presence of *H. pylori* in poorly and highly differentiated samples

H. pylori score	Grade 0		Grade 1		Grades 2/3	
	(*n*)	(%)	(*n*)	(%)	(*n*)	(%)
Poor differentiation	37	45	21	24	24	31
High differentiation	1	11	5	56	3	33

Table 3. Presence of *H. pylori* according to the type of carcinoma

H. pylori score	Grade 0		Grade 1		Grades 2/3	
	(*n*)	(%)	(*n*)	(%)	(*n*)	(%)
Diffuse	6	54	3	27	2	19
Intestinal	32	40	23	29	25	31

pylori positive patients was higher in the intestinal type. This figure did not reach statistical significance, probably because the majority of excluded carcinomas were of the diffuse type.

As could be expected, there was an inverse relation between the presence *H. pylori* and the presence of intestinal metaplasia (Table 4).

Figure 1 shows the comparison of *H. pylori* positivity in different age cohorts of gastric carcinoma patients, patients with non-ulcer dyspepsia and healthy blood donors. Presence of *H. pylori* in the age cohorts 41–50 years and 51–60 years was significantly higher than to be expected in the healthy population ($p < 0.001$ and $p < 0.05$, respectively). No difference was seen in the comparison with non-ulcer dyspepsia patients.

Table 4. Presence of *H. pylori* and intestinal metaplasia

Intestinal metaplasia score / *H. pylori* score	Grade 0		Grade 1		Grades 2/3	
	(*n*)	(%)	(*n*)	(%)	(*n*)	(%)
Grades 0/1	19	35	15	27	21	38
Grades 2/3	20	56	11	30	5	14

$p < 0.05$

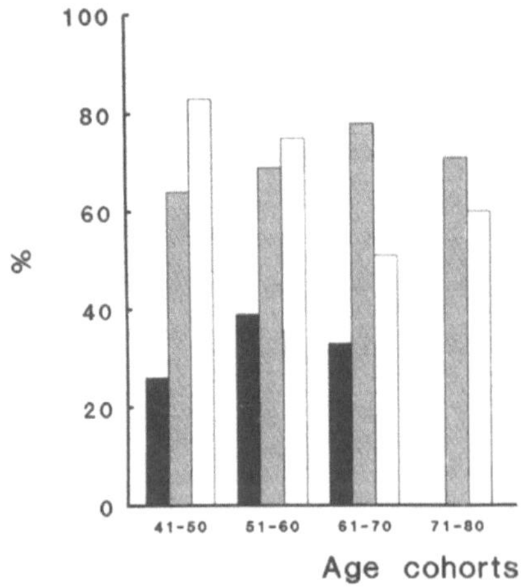

Fig. 1. Presence of *H. pylori* in different age cohorts from gastric cancer patients (*Open columns*) patients suffering from non-ulcer dyspepsia (*hatched columns*) and healthy blood donors (*solid columns*)

Discussion

The prevalence of *H. pylori* in our material was 59%.

This figure is in keeping with data in the literature.

The bacterial load was higher in biopsy specimens. The reason for this phenomenon is not obvious. A possible explanation could be autolysis and drying of the resection specimens after surgical removal. Although biopsy specimens are immediately fixed, resection specimens are not. Since *H. pylori* is susceptible to drying, this could explain a decrease in bacterial load and possibly the disappearance of the microorganism. Hence studies on *H. pylori* in gastric carcinoma should be done on freshly fixed biopsy material.

Helicobacter pylori was only seen overlying normal-looking epithelium adjacent to the tumour site. The bacterium was also absent in areas of intestinal metaplasia. *H. pylori* has a special affinity for gastric epithelium due to the presence of growth factors and a specific glycerolipid receptor in the epithelium [6]. It is tempting to assume that these points of recognition are lost in the epithelium adjacent to tumours of poor differentiation, explaining the lower *H. pylori* positivity of these tumours.

Although not statistically significant, *H. pylori* seems to be linked with the intestinal type of carcinoma. The reason for the absence of a significant difference is possibly the fact that the majority of the tumours of the diffuse type had to be excluded because no normal-looking epithelium was present in the sections.

On the other hand, the microorganism might not necessarily be present throughout the evolution of the tumour. This might explain the *H. pylori*-negative cases of intestinal-type carcinoma.

About half of the patients suffering from non-ulcer dyspepsia and half of the healthy blood donors, at a certain age, reveal *H. pylori* histologically and/or serologically. At first glance, the presence of *H. pylori* in gastric carcinoma patients seems to reflect the average *H. pylori* status in the normal population. However, an impressive observation was done when different age cohorts of cancer patients were compared with healthy individuals. *H. pylori* was present in 83% and 75% of cancer patients aged 41–50 years and 51–60 years, compared with 26% and 39% of the healthy population, respectively. In other words, the presence of *H. pylori* in these age groups is considerably higher than to be expected in the normal population, implying a role for the bacterium in this condition.

The presence of *H. pylori* in different age cohorts is not influenced by the presence of intestinal metaplasia.

It is concluded that *H. pylori* plays a role in the pathogenesis of gastric cancer. The infection occurs in more than half of the patients and seems to be linked to the intestinal tumour type with high differentiation. The presence in the younger age cohorts also suggests a pathogenetic role. Further studies to elucidate the exact pathophysiological mechanisms are, however, mandatory.

References

1. Krienitz W (1906) Ueber das Autreten von Mageninhalt bei Carcinoma Ventriculi. Dtsch Med Wochenschr 22:872
2. Correa P, Ruiz B (1989) Campylobacter pylori and gastric cancer. In: Rathbone BJ, Heatly RV (eds) Campylobacter pylori and gastroduodenal disease, Blackwell, London
3. Loffeld RJLF, Potters HVPJ, Arends JW, Stobberingh E, Flendrig JA, van Spreeuwel JP (1988) Campylobacter associated gastritis in patients with non-ulcer dyspepsia. J Clin Pathol 41:85–88
4. Loffeld RJLF, Stobberingh E, Flendring JA, van Spreeuwel JP, Arends JW (1989) Diagnostic value of an immunoassay to detect anti Campylobacter pylori antibodies in non-ulcer dyspepsia. Lancet i:1182–1185
5. Loffeld RJLF, Stobberingh E, Flendrig JA, Arends JW (1990) The prevalence of anti Campylobacter pylori antibodies in patients and healthy blood donors. J Med Microbiol 32:105–109
6. Lingwood CA, Pellizzari A, Law H, Sherman P, Drumm B (1989) Gastric glycerolipid as a receptor for Campylobacter pylori. Lancet ii:238–241

Increased Incidence of *Helicobacter pylori* in Gastric Cancer, as Shown by the Rapid Urease Test

D. Boixeda[1], A.L. San Roman[1], C. Martin de Argila[1], R. Cantón[2], C. Redondo[3], L.A. Gil Grande[1], and L. de Rafael[2]

Introduction

Nowadays it is widely accepted that *Helicobacter pylori* infection is the main etiologic factor in chronic gastritis type B (CGB). However, its contribution to the developement of gastric ulcer remains controversial. In vivo studies on gnotobiotic piglets [1, 2] and healthy humans [3, 4] have shown that a typical CGB picture can be reproduced by the ingestion of this microorganism. A further point supporting a relationship between *H. pylori* and CGB is the possibility of inducing a partial or complete histologic recovery of the disease after the clearance of bacteria by bismuth salts or antibiotics [5–8].

A correlation between gastritis and the possibility of developing gastric cancer (GC) has been consistently demonstrated, especially in areas where the prevalence of this neoplasms is high. However, this indirect relationship between *H. pylori* and gastric carcinogenesis has scarcely been investigated.

Our proposal has been to assess the prevalence of *H. pylori* colonization in patients with endoscopically diagnosed and histologically confirmed GC in an area with a high incidence of GC [9].

Patients, Methods, and Materials

Two cohorts of patients were studied. Group A included 30 cases of endoscopically diagnosed and histologically confirmed GC (25 adenocarcinomas, five undifferentiated neoplasms). Group B comprised 27 patients in whom no pathologic findings could be shown on gastroscopy nor on gastric biopsies. Twenty-four patients were male. Ages varied between 18 and 87 years. Both groups were similar as to age and sex distribution. In each case, gastric biopsies were processed following a classical routine. A further sample, obtained 2 cm proximal to the pylorus (or at 2 cm of the macroscopical tumor border, in cases

[1]Department of Gastroenterology, Hospital "Ramon y Cajal," Alcala de Henares University, Madrid, Spain
[2]Department of Microbiology, Hospital "Ramon y Cajal," Alcala de Henares University, Madrid, Spain
[3]Department of Pathology, Hospital "Ramon y Cajal," Alcala de Henares University, Madrid, Spain

J.M. Pajares et al. (Eds.)
H. pylori and Gastroduodenal Pathology

of pyloric invasion by neoplasia), was immediately placed in 1 cc of 10% urea solution in deionized water at pH 6.8 with two drops of phenol red, as described by Arvind [10]. A change in color from yellowish to red-pink at 3, 15, 30, and 180 min or at 24 h was considered a positive result. In our experience this technique had a 81.53% sensitivity and a 100% specificity [11].

Statistical Methods. The chi-square test was used to compare the presence of cancer with positivity or negativity of *H. pylori* in both groups.

Results

The rapid urease test was positive in 12 of 30 patients in group A (40%) but only in three of 27 normal controls (11.1%) ($p < 0.05$). All five patients with undifferentiated GC showed a negative Arvind's urease test.

Discussion

A possible relationship between GC and *H. pylori* has scarcely been studied. However, current results permit us to conclude that *H. pylori* colonization is rare in GC patients, as compared to patients with gastritis. This microorganism cannot usually be detected on neoplastic tissue [12, 13]. Our results show a prevalence of 40% in *H. pylori* detection in GC patients and are in good agreement with those previously published. Other authors, however, have found a figure reaching 100% [14, 15], although the number of patients in both studies was small [14, 15]. This low prevalence of *H. pylori* colonization in GC patients could be explained by microbial competence against intestinal flora which colonizes the hypochloridic neoplastic stomach. An accompanying diffuse intestinal metaplasia, a condition rarely associated with *H. pylori* [16] is found with increased frequency in GC patients. Scott et al. [17] recently reported on a family with a high prevalence of GC: five of eight siblings were shown to have at an early age (< 30 years) an *H. pylori*-associated chronic atrophic gastritis, and three of them also showed intestinal metaplasia. According to the authors, *H. pylori* would play a role in the metaplastic transformation of the gastric epithelium. This high prevalence of *H. pylori* detection has also been shown in patients with chronic atrophic gastritis living in high-risk areas for GC [15, 18].

In summary, a possible role for *H. pylori* in gastric carcinogenesis remains to be established. Epidemiologic studies focusing on young patients in both low- and high-risk areas, are urgently needed to elucidate any possible relationship to GC and/or to preneoplastic conditions.

References

1. Krakowka S, Morgan DR, Kraft WG, Leunk RD (1987) Establishment of gastric campylobacter pylori infection in the neonatal gnotobiotic piglet. Infect Immun 55: 2789–2796
2. Lambert JR, Borromeo M, Pinkard KJ, Turner H, Chapman CB, Smith ML (1987) Colonization of gnotobiotic piglets with Campylobacter pyloridis, an animal model? J Infect Dis 155: 1344
3. Marshall BJ, Armstrong JA, McGechie DB, Glancy RJ (1985) Attempts to fulfill Koch's postulates for pyloric Campylobacter. Med J Aust 142: 436–439
4. Morris A, Nicholson G (1987) Ingestion of Campylobacter pyloridis causes gastritis and raised fasting gastric pH. Am J Gastroenterol 82: 192–199
5. McNulty CAM, Gearty JC, Crump B, Davis M, Donovan IA, Melikian V, Lister DM, Wise R (1986) Campylobacter pyloridis and associated gastritis: investigator blind, placebo controlled trial of bismuth salicylate and erythromycin ethylsuccinate. Br Med J 293: 645–649
6. Rauws EAJ, Langenberg W, Houthoff HJ, Zanen HC, Tytgat GNJ (1988) Campylobacter pyloridis associated chronic active antral gastritis. A prospective study. Gastroenterology 94: 33–40
7. Morgan D, Kraft W, Bender M, Pearson A (1988) Nitrofurans in the treatment of gastritis associated with Campylobacter pyloridis, Gastroenterology 95: 1178–1184
8. Glupczynski Y, Burette A, Labbe M, Deprez C, De Reuck M, Deltenre M (1988) Campylobacter pylori associated gastritis. A double-blind placebo-controlled trial with Amoxycillin. Am J Gastroenterol 83: 365–372
9. Lopez-Abente G, Escolar A, Errezola M (1984) Atlas del cancer en España. Departamento de Sanidad del Gobierno Vasco, Vitoria, pp 37–44
10. Arvind AS, Cook RS, Tabaqchali S, Farthing MJG (1988) One minute endoscopy room test for Campylobacter pylori. Lancet 1: 704
11. Boixeda D, San Roman AL, Canton R, Erdozain JC, Redondo C, Moreira VF, De Rafael L (1990) Is rapid urease test sufficient for the diagnosis of Helicobacter pylori in an endoscopy unit? Rev Esp Enferm Dig 78 [Suppl 1]: 15
12. Jiang SJ, Liu WZ, Zhang DZ (1987) Campylobacter-like organisms in chronic gastritis, peptic ulcer and gastric carcinoma. Scand J Gastroenterol 22: 553–558
13. Feng YY, Wang Y (1988) Campylobacter pylori in patients with gastritis, peptic ulcer and carcinoma of the stomach in Lanzhou, China. Lancet 1: 1055–1056
14. Inouye H, Tonokatsu Y, Mikami J, Yamamoto I, Fukuda Y, Tamura K, Shimoyama T, Yamamoto Y, Tanaka T, Tamura T, Shoji K, Ishii E, Kishi T (1988) Campylobacter pylori in patients with gastric disease in Japan. I Workshop "Gastroduodenal pathology and campylobacter pylori", Bordeaux, p 129
15. Jaskiewcz K, Louwrens HD, Woodroof CW, Van Wyk MJ, Price SK (1989) The association of Campylobacter pylori with mucosal pathological changes in aq population at risk for gastric cancer. S Afr Med J 75: 417–419
16. Paull G, Yardley JH (1989) Pathology of Campylobacter pylori-associated gastric and esophageal lesions. In: Blaser MJ (ed.) Campylobacter pylori in gastritis and peptic ulcer disease. Igaku-Shoin, New York, pp 73–97
17. Scott N, Lansdown M, Rathbone B, Murday L, Wyatt JI, McMahon M, Dixon MF, Quirke P (1990) Helicobacter gastritis and intestinal metaplasia in a gastric cancer family. Lancet 335: 728
18. Fox JG, Correa P, Taylor NS, Zavala D, Fontham E, Janney F, Rodriguez E, Hunter F, Diavolitsis S (1989) Campylobacter pylori associated gastritis and immune response in a population at increased risk of gastric carcinoma. Gastroenterology 84: 775–781

Evaluation of the Presence and Amount of *Helicobacter pylori* in Early Gastric Cancer

C. Avellini[1], V. Cocchi[2], G.M. Paganelli[3], and G. Biasco[3]

Introduction

Careful consideration of several factors is necessary if we are to arrive at a reasonable hypothesis in the evaluation of the pathogenic role of bacterial agents in the gastric mucosa, regarding both chronic inflammation and the later stages of atrophic gastritis and dysplasia. In many populations, diffusion of inflammation in the stomach is accompanied by the presence of *Helicobacter pylori* (HP) [2]. This in itself does not demonstrate that the bacillus has a causal role, but suggests a close association between it and a pathological picture which, albeit in a limited number of cases, may constitute the first step in the pathogenetic sequence leading to neoplasia [1].

Adequate pharmacological treatment often allows drastic reduction in the bacterial population along with a corresponding improvement of the trophic conditions of the mucosa. Complete eradication of this bacillus, however, is much more difficult [3, 8]. Once again, although this does not represent an argument in favour of its playing a causal role, there is suggestion that within the context of the chemical and physical restructuring of the mucosa under the effects of inflammation, the bacillus is capable of invading and adapting itself to the foveolar surface [7].

The difficulty in observing aggressive morphological pictures of the bacillus with respect to the cell, together with its frequent perpendicular arrangement in relation to the epithelial layer (almost lining itself up along a chemical gradient in correspondence with the intercellular connections), without in any way wishing to relegate the role of HP to a saprophytic one, could suggest a mutual adaptation of both the cell and the bacillus in the inflammatory process [4].

Finally, it is thought that the degree of inflammation in the presence of HP does not so much depend on the varying aggressiveness of different strains, but rather on the reactivity of the host and his or her exposure to extrinsic irritating agents, that is to say to those environmental factors so often cited in reference to the pathogenesis of cancer [1]. On the one hand, the wide variety of cell types present in the gastric mucosa implies different types of adaptation and reaction

[1] Department of Pathology, University of Bologna, 40138 Bologna, Italy
[2] Department of Oncology "F. Addarii", S. Orsola Hospital, 40138 Bologna, Italy
[3] "G. Prodi" Center for Cancer Research, University of Bologna, 40138 Bologna, Italy

J.M. Pajares et al. (Eds.)
H. pylori and Gastroduodenal Pathology

to the inflammatory process; on the other hand, the hypothesis that progression of dysplasia towards neoplasia begins with multipotent elements of the foveolar neck partly redimensions this phenomenon.

The relationship of the bacillus with cell glycocalix and the pericellular microenvironment may provide the key for correctly interpreting the role of HP. Hence, there is a need to start long-term studies on the relationship between pre- and early neoplastic lesions of the gastric mucosa and HP. Since early gastric cancer (EGC) relies on the comprehension of the pathology and histogenesis of gastric cancer [5], we evaluated the presence and amount of HP in specimens of EGC and its surrounding mucosa.

Material and Methods

We studied 23 surgical specimens of EGC. Formalin-fixed, paraffin-embedded samples were taken both from the neoplastic tissue and from the surrounding mucosa. Specimens were stained with haematoxylin and eosin, alcian blue-PAS and Gram staining for histological diagnosis and HP evaluation. The presence of spiral bacteria, observed on four superficial and three glandular high-power fields ($\times 40$), was confirmed at $\times 100$ magnification. Their amount was evaluated from + to + + +: +, few bacteria after careful search; + +, groups of bacteria in most fields; + + +, many bacteria in most high-power fields. The histological type of neoplasia was described according to Lauren's criteria [6].

Results

Helicobacter pylori was not detected in any of the neoplastic tissue specimens. On the contrary, 16/23 (69.6%) specimens taken from the surrounding mucosa were HP-positive (eight cases score +, six cases score + +, two cases score + + +). As far as the histological type of neoplasia is concerned, HP was detected in specimens taken from the surrounding mucosa of both intestinal and diffuse-type gastric cancer (Table 1).

Table 1. Helicobacter pylori positivity in neoplastic tissue and in the surrounding mucosa according to Lauren's histological criteria

	Neoplasia HP positive (*n*)	Surrounding mucosa HP positive (*n*)
Histological type		
Intestinal (18 cases)	0	13
Mixed (1 case)	0	0
Diffuse (4 cases)	0	3

Discussion

The long-established ability of HP, when ingested in sufficient quantity, to provoke acute gastritis, together with the complex and indirect relationship that it assumes in the pathogenesis of chronic gastritis and peptic ulcer determines the direction of study to find the relationship between HP and gastric cancer. The most likely hypothesis seems to be that the bacillus represents one of the many elements which react with each other and with gastric mucosa and which are capable of producing chemical and physical modifications in the micro-environment by the alteration of epithelial cell proliferation and differentiation. Hence, the importance of the morphological relations between HP and the earliest lesions connected with the development of neoplasia.

The presence of a large number of bacilli on the surrounding mucosa in the majority of cases of EGC studied by us is a preliminary indication. Further investigation is required to determine how much this phenomenon could depend on the microenvironmental modifications brought about by the neoplasia and leading to greater sensitivity to infection in the peripheral gastric mucosa, and how much it could depend on the intensification and extension of a pre-existing inflammatory process.

To this end, it is also necessary to point out the absence of the bacillus in the neoplastic areas and, in the majority of cases, in the dysplastic mucosa. This could suggest that the bacillus effectively plays a role in the initiation of the neoplastic process. On the other hand, a high concentration of bacilli in healed peptic ulcers could represent not only a marker of risk of relapse, but could also be the effect of a reduction of mucosal defence mechanisms due to trophic alterations induced by a preneoplastic lesion.

Here again it is important to consider the hypothesis of a relationship between morphological alterations of the various cellular lines (glandular, foveolar, endocrine) and the presence of bacilli in early neoplastic lesions.

References

1. Correa P (1988) A human model of gastric carcinogenesis. Cancer Res 48:3554–3560
2. Forman D, Sitas F, Newell DG, Satcey AR, Berham J, Peto R, Campbell TC, Li J, Chen J (1990) Geographic association of Helicobacter pylori antibody prevalence and gastric cancer mortality in rural China. Int J Cancer 46:608–611
3. Graham DY (1989) Campylobacter pylori and peptic ulcer disease. Gastroenterology 96 [Suppl 2]:615–625
4. Hazell SL, Lee A, Brady L, Hennessy W (1986) Campylobacter pyloridis and gastritis: association with intercellular spaces and adaptation to an environment of mucus as important factors in colonization of the gastric epithelium. J Infect Dis 153:658–663
5. Johansen A (1981) Early gastric cancer. A contribution to pathology and to gastric cancer histogenesis. Pone Petri, Copenhagen
6. Lauren P (1965) The two histological main types of gastric carcinoma: diffuse and so-called intestinal-type carcinoma. Acta Pathol Microbiol Scand 64:31–49
7. Sidebotham RL, Baron JH (1990) Hypothesis: Helicobacter pylori, urease, mucus, and gastric ulcer. Lancet 335:193–195
8. Tytgat GN, Rauws EA, de Koster EH (1989) Campylobacter pylori. Diagnosis and treatment. J Clin Gastroenterol 11 [Suppl 1]:49–53

Helicobacter pylori Prevalence in Patients with Severe Dysplasia of Gastric Epithelium and Associated Pathology

H.K. Koch and K. Grillmeier

Introduction

Gastric cancer and dysplasia are often related to chronic gastritis [5]. *Helicobacter pylori* infection has been implied as the main cause of type B gastritis in recent years [8, 13]. This raises the question of a causal role of *H. pylori* infection in the metaplasia/dysplasia/carcinoma sequence suggested by Correa in 1988 [5]. Prevalence rates of *H. pylori* infection in patients with gastric cancer or gastric dysplasia have already been reported as being 25%–85% [1–4, 6, 7, 12]. In addition, the coincidence of *H. pylori* infection, early manifestation of gastritis, and metaplasia was recently reported in a "gastric cancer family" [11]. To evaluate the significance of *H. pylori* infection in patients with gastric dysplasia, we studied the prevalence of *H. pylori* in patients with severe gastric dysplasia with reference to coexisting mucosal pathology.

Material and Methods

Cases of severe gastric dysplasia from the years 1988 and 1989 were retrospectively analyzed. Only cases without already manifest carcinoma were studied. Gastric dysplasia was diagnosed and classified according to the consensus criteria laid down in 1984 [10]. A total of 120 patients with severe gastric dysplasia were evaluated for the presence of *H. pylori* by screening the slides with a high-power dry objective (×63) followed by oil immersion (×100) whenever a doubtful result was obtained. *H. pylori*-positive and -negative patients were further analyzed for the type of coexisting gastritis, intestinal metaplasia, lymphoid hyperplasia, gastric ulcer, or polypoid lesions. The *H. pylori* prevalence associated with these lesions in patients with severe gastric dysplasia was compared to prevalence rates associated with these lesions found in patients without dysplasia.

Results

Severe gastric dysplasia was observed in antral biopsies of 69 patients (57.5%, male = 39, female = 30), in fundic biopsies of 38 patients (31.7%, male = 25,

Institut für Patholgie, Rosastr. 9 / Postfach 1260, 7800 Freiburg im Breisgau, FRG

J.M. Pajares et al. (Eds.)
H. pylori and Gastroduodenal Pathology

female = 19) and in the operated stomach of 13 patients (10%, male = 9, female = 4) either at the anastomosis (ten patients) or in the gastric stump. *H. pylori* was microscopically demonstrated in 57 patients (47.5%). In another ten patients (8.3%) the presence of bacteria could not be ruled out since the biopsies displayed only dysplastic or metaplastic epithelia where *H. pylori* might not be able to grow. No difference was observed in the age or sex distribution of patients with severe gastric dysplasia with or without *H. pylori* infection (Table 1).

A higher rate of *H. pylori* infection in patients with severe gastric dysplasia was found associated with active gastritis, lymphoid hyperplasia, gastric ulcer, and intestinal metaplasia (Table 2). Intestinal metaplasia was mostly of the

Table 1. Mean age and sex distribution in patients with severe gastric dysplasia with or without *H. pylori* infection or where *H. pylori* infection could neither be ascertained nor ruled out

	Total patients		Male		Female	
	n	Mean age (years)	*n*	Mean age (years)	*n*	Mean age (years)
H. pylori positive	57	70	36	69	21	73
H. pylori negative	53	70	31	67	22	73
H. pylori ???	10	69	6	70	4	67
Total	120		73		47	

Table 2. *H. pylori* prevalence with regard to different lesions coexistent with severe dysplasia.

Associated lesion	Patients (*n*)	*H. pylori* positive *n*	(%)	*H. pylori* negative *n*	(%)	*H. pylori* ??? *n*
Active chronic gastritis	86	44	51.2	36	41.8	6
Inactive chronic gastritis	16	5		9		2
Metaplasia	63	33	50.8	30	46.2	2
Lymphoid hyperplasia	22	13	59.1	8	36.4	1
Gastric ulcer	34	18	52.9	14	41.2	2
Polypoid lesion	14	4		8		2

Table 3. Comparison of *H. pylori* prevalence associated with different lesions in patients with and without dysplasia (From [8])

Associated lesion	Patients with dysplasia			Patients without dysplasia		
	Total (*n*)	*H. pylori* positive (*n*)	(%)	Total (*n*)	*H. pylori* positive (*n*)	(%)
Active chronic gastritis	86	44	51.2	379	304	80.2
Inactive chronic gastritis	16	5	31.2	157	86	54.8
Metaplasia	65	33	50.8	186	127	68.3
Lymphoid hyperplasia	22	13	59.1	175	145	82.8
Gastric ulcer	34	18	52.9	134	88	65.7

incomplete type both in patients with ($n = 31$) and without ($n = 29$) *H. pylori* infection.

Comparing the *H. pylori* infection rate associated with different mucosal lesions in patients with or without gastric dysplasia, it was found that patients with severe gastric dysplasia had a substantially lower percentage of infection regardless of the type of lesion (Table 3).

Discussion

The *H. pylori* infection rate in gastric cancer or dysplasia has been reported to be 25% and 85% [4, 7]. More recent studies still differ considerably, stating prevalence rates from 32% [6], 59% [12], and 70% [1] in patients with gastric cancer. Our results gave an *H. pylori* infection rate of 47.5% in patients with severe gastric dysplasia which may be considered a precursor of gastric cancer [9, 10]. If those patients in whom neither exclusion nor verification of *H. pylori* infection could be confirmed were simply included, an *H. pylori* prevalence of 56% can be assumed.

Judging from epidemiological parameters such as age and sex distribution, patients with severe gastric dysplasia and *H. pylori* infection were not different from patients without *H. pylori* infection.

The patients with severe dysplasia and *H. pylori* infection more often had active gastritis, lymphoid hyperplasia, and gastric ulcer. However, when *H. pylori* prevalence for these lesions was compared in patients with or without gastric dysplasia, it was seen that the prevalence was always significantly lower in patients with severe gastric dysplasia. This was still true even when the patients with questionable *H. pylori* infection status were simply added to the *H. pylori* positive patients. Patients with severe gastric dysplasia (mean age 70 years) were older than the patients without gastric dysplasia cited for comparison [8] of prevalence rates (mean age 50 - 60 years).

Since the prevalence of *H. pylori* infection generally increases with age [13], the lower infection rates in patients with severe dysplasia are even more confounding. These results are not suggestive of a causative role of *H. pylori* infection in gastric carcinogesis. If *H. pylori* infection plays any role, it is probable that this effect lies with an early event and is followed by changes in the gastric milieu which may hamper the growth of the organisms.

In this context it is very interesting indeed that a declining prevalence of *H. pylori* infection may also be seen when cases of early and manifest advanced gastric cancer are compared [3].

References

1. Avellini C, Cocchi V, Paganelli GM, Biasco G (1990) Evaluation of presence and amount of Helicobacter pylori in early gastric cancer. Rev Esp Enferm Dig 78 [Suppl 1]: 92
2. Boixeda D, San Roman AL, Erdozain JC, Canton R, Redondo C, Gil Grande L, De Rafael L (1990) Increased incidence of Helicobacter pylori in gastric cancer as shown by the rapid urease test. Rev Esp Enferm Dig 78 [Suppl 1]: 92
3. Caruso ML, Fucci L (1990) Histological Identification of Helicobacter pylori in early and advanced gastric cancer. J Clin Gastroenterol 12: 601–602
4. Cheng SCW, Sanderson CR, Waters TE, Goodwin CS (1987) Campylobacter pyloridis in patients with gastric carcinoma. Med J Aust 147: 202–203
5. Correa P (1988) A human model of gastric carcinogenesis. Cancer Res 48: 3554–3560
6. Crespo M, Pajares JM, Lancho A, Lopez Brea M, Blanco M (1990) Prevalence of Helicobacter pylori in patients with malignant growth of gastric mucosa. Rev Esp Enferm Dig 78 [Suppl 1]: 92–93
7. Jiang SJ, Liu WZ, Zhang DZ, Shi Y, Xiao SD, Zhang ZH, Lu DY (1987) Campylobacter-like organisms in chronic gastritis, peptic ulcer, and gastric carcinoma. Scand J Gastroenterol 22: 553–558
8. Koch HK, Baumert B, Koch U, Oehlert M, Oehlert W (1990) Prevalence of Campylobacter pylori as demonstrated by histology or CLO-test in different types of gastritis. A study in 5 clinically predefined groups of patients. Pathol Res Pract 186: 154–158
9. Koch HK, Oehlert M, Oehlert W (1990) An evaluation of gastric dysplasia in the years 1986 and 1987. Pathol Res Pract 186: 80–84
10. Ming SC, Bajtai A, Correa P, Elster K, Jarvi OH, Munoz N, Nagayo T, Stemmerman GN (1984) Gastric dysplasia. Significance and pathologic criteria. Cancer 54: 1794–1801
11. Scott N, Lansdown M, Diament R, Rathbone B, Murday V, Wyatt JI, McMahon M, Dixon MF, Quirke P (1990) Helicobacter gastritis and intestinal metaplasia in a gastric cancer family. Lancer 335: 728
12. Talley NJ, DiMagno E, Zinsmeister AR, Perez-Perez G, Blaser M (1990) Helicobacter pylori and gastric cancer: a case control study. Rev Esp Enferm Dig 78 [Suppl 1]: 7–8
13. Wyatt JI, Rathbone BJ (1988) Immune response of the gastric mucosa to Campylobacter pylori. Scand J Gastroenterol 23 [Suppl 142]: 44–49

Gastric Cancer and *Helicobacter pylori*

P. Correa[1] and J. Fox[2]

Chronic atrophic gastritis has been identified as a precursor of gastric cancer [1]. Since *Helicobacter pylori* is a cause of chronic gastritis [2], the possible role of *H. pylori* infection in gastric carcinogenesis deserves attention. This has been investigated in populations with different risks of gastric cancer in Colombia and New Orleans (USA).

Material and Methods

In Colombia, sera were obtained from adult patients on whom gastric biopsies were done at the gastroenterology clinics of the Hospital Departamental of Pasto, Nariño, and the University Hospital of Cali. In New Orleans, serum samples were obtained from normal adults seen at Louisiana State University's (LSU) Dental School for routine dental care. These individuals represent the lower middle classes of the city. In New Orleans serum samples were also obtained from children receiving medical care for diseases unrelated to the gastrointestinal tract at Charity Hospital and Children's Hospital. Charity Hospital is a state institution providing medical care predominantly for black patients of low socioeconomic strata. Children's Hospital is a private institution which provides medical care for predominantly middle-class white and black patients.

In Colombia, patients from Nariño (Pasto) represent a population at very high risk of gastric cancer (over 100 per 100000) while patients from Cali represent a population at intermediate risk (around 50 per 100000 in males). In New Orleans, blacks have a cancer incidence rate which is approximately double that of whites (16 vs. 7.3 per 100000 males).

Immunoglobulin G anti-*Helicobacter* antibodies were assayed by an enzyme-linked Immunosorbent assay (ELISA) method previously published [3]. Gastric biopsies were evaluated according to histopathologic parameters previously published [4].

[1] Department of Pathology, Louisiana State University Medical Center, New Orleans, LA 70112, USA
[2] Division of Comparative Medicine, Massachusetts Institute of Technology, Cambridge, MA, USA

J.M. Pajares et al. (Eds.)
H. pylori and Gastroduodenal Pathology

Results

Table 1 shows the prevalence of *H. pylori*-positive patients 35–84 years of age, by locality and histologic diagnosis. There were no subjects with normal histology in this series. The cases with superficial gastritis had a similar prevalence of anti-*H. pylori* antibodies in Nariño and Cali (75%–70%). There was, however, a statistically significant difference between patients with multifocal atrophic gastritis (MAG) in Pasto (93%) compared to Cali (67%). MAG includes patients with gastric atrophy (gland loss), with or without intestinal metaplasia.

Table 2 shows the prevalence of *H. pylori* antibodies in "normal" subjects attending the University's dental school clinic. A significantly higher prevalence is seen in blacks. The rate in whites is somewhat higher than other series, probably reflecting a somewhat lower socioeconomic status in the New Orleans sample.

Table 3 shows the prevalence of *H. pylori* infection in New Orleans children. At Charity Hospital a relatively high proportion of children are born with *H. pylori* antibodies, indicating transplacental transfer of maternal antibodies. Older children attending Charity Hospital also shows a relatively high prevalence, more pronounced in blacks than in whites. These prevalence rates are

Table 1. Prevalence of IgG *H. pylori* antibody positivity in Colombian patients, 35-84 years of age

	SG			MAG		
	T	*P*		T	*P*	
	(*n*)	(*n*)	(%)	(*n*)	(*n*)	(%)
Cali	10	7	70	12	8	67
Nariño	4	3	75	15	14	93

SG, superficial gastritis; MAG, multifocal atrophic gastritis; T, total examined; P, positive for *H. pylori*

Table 2. Prevalence of IgG *H. pylori* antibody positivity in subjects aged 35-84 years attending LSU dental school, by race

	T	P	
	(*n*)	(*n*)	(%)
White	85	50	59
Black	85	67	79

T, total examined; P, positive for *H. pylori*.

Table 3. Prevalence of IgG *H. pylori* antibody positivity in New Orleans children, by age, race, and hospital

Age	Charity Hospital						Children's Hospital					
	White			Black			White			Black		
	T	P		T	P		T	P		T	P	
	(*n*)	(*n*)	(%)	(*n*)	(*n*)	(%)	(*n*)	(*n*)	(%)	(*n*)	(*n*)	(%)
0–5 months	9	4	44	38	16	48		–			–	
6 months–9 years	3	1	33	34	19	56	42	12	28	22	6	27
10–18 years	15	8	53	37	25	67	8	0	0	11	2	18

higher than those seen at Children's Hospital, especially when black children of the two institutions are compared.

Discussion

The intent of this investigation was to examine any possible correlation between *H. pylori* infection, as reflected in IgG serum antibodies, and the risk of gastric cancer. Serum samples were therefore obtained from populations with contrasting gastric cancer risks. The results clearly show that the prevalence of *H. pylori* infection was positively correlated with gastric cancer risk in Colombia and New Orleans. Histopathologic examination of gastric biopsy specimens from Pasto frequently revealed a massive infection, not seen in the New Orleans material.

A high prevalence of *H. pylori* infection has been reported from other populations with high gastric cancer risk such as Peru [5, 6] and China [7, 8]. In New Orleans a positive correlation was found between *H. pylori* infection prevalence and cancer risk when blacks were compared with whites. This correlation, however, appeared to be confounded by socioeconomic factors. Our findings indicate that socioeconomic conditions rather than race are the overriding forces influencing *H. pylori* infection, as clearly shown in the data for New Orleans black children. Similar conclusions can be reached with data from other countries such as Madagascar, where the blacks have lower socioeconomic conditions and higher *H. pylori* prevalence but low gastric cancer rates [9]. The prevalence rates reported here for white adults are much higher than other series from the United States, Australia, and Europe [10] in which the prevalence was around 30%. This is probably another indicator of the influence of socioeconomic factors, since our New Orleans white population represented a lower socioeconomic stratum. Low socioeconomic status is usually associated with higher gastric cancer risks.

Epidemiologic studies based on the detection of antibodies against *H. pylori* in serum samples have confirmed a high degree of correlation with gastric cancer rates. The prevalence of IgG antibodies against *H. pylori* in 1882 Chinese men was correlated [8] with cancer mortality rates of the 46 rural counties they represented. No significant correlation was found between *H. pylori* antibodies and cancer rates, with the only exception of gastric cancer, for which a significant ($2p = 0.02$) correlation coefficient of 0.34 was found. A cohort study in England [11] has shown that the prevalence of IgG anti-*Helicobacter* antibodies was 69% in subjects who years later developed gastric cancer, compared to 47% in matched controls, yielding a statistically significant relative risk of 2.77. Studies of seroepidemiology have consistently found that, in populations at high gastric cancer risk, the prevalence of the infection in children is higher than in populations of low risk [12]. Early infection with *H. pylori* may result from overcrowding. Housing patterns in childhood have been found positively correlated with gastric cancer death rates, leading the authors to state that overcrowding at home during childhood "might act by promoting the transmission of causative organisms" [13]. The influence of socioeconomic factors on *H. pylori* infection in childhood is shown by our study in New Orleans, where both cancer rates and *H. pylori* antibody prevalence are higher in blacks than in whites.

It is well known that *H. pylori* does not colonize cancer cells, and it follows that the prevalence in cancer cases represents colonization in the remaining foveolar epithelium of the stomach. A study of gastrectomy specimens from California [14] found that the prevalence of *H. pylori* infection in the mucosa surrounding gastric carcinomas of the intestinal (or "epidemic") type was 89%, compared to 32% for carcinomas of the "diffuse" type [14]. The difference was statistically highly significant. This strong association of *H. pylori* infection with the intestinal type of gastric cancer, which greatly exceeds the prevalence in normal populations in the United States provides additional support for a role of *H. pylori* in the gastric carcinogenic process.

The prevalence of *H. pylori* infection in children is infrequent, around 3% in England [12] and other countries with low gastric cancer rates [10]. It is, however, high in populations with a high gastric cancer risk [15]. This led to the suggestion that such infection may increase the risk of gastric cancer [16]. That *H. pylori* infection is not a sufficient cause of cancer is indicated by the high prevalence of infection in populations with a low gastric cancer risk such as Madagascar [9] and the Ivory Coast [15]. Additionally, patients with duodenal ulcer and diffuse antral gastritis, who have a very high prevalence of *H. pylori* infection, are not known to be at increased risk of gastric cancer. It thus appears that other factors are needed to induce gastric cancer. Genotoxic agents, whose role could be potentiated by factors which increase the rate of proliferation of the gastric mucosa, have been suggested. *H. pylori* is considered a causal agent in chronic gastritis in which epithelial hyperplasia is one of the characteristic histologic changes. Atypical hyperplasia observed in Peruvian patients with massive *H. pylori* infection regressed after bismuth therapy [5, 6]. Therefore *H. pylori* may be considered as adjuvant or co-factor in gastric carcinogenesis, and it may play an important role in populations where other etiologic factors are present. Given the interdependency of etiologic factors in the gastric carcinogenic process [1], it is possible that controlling *H. pylori* infection could decrease the effectiveness of other carcinogenic agents. Its role may be mediated through stimulation of epithelial hyperplasia (hyperproliferative state). A hyperproliferative state has been previously documented in precursor lesions of gastric cancer. It has also been shown that hyperproliferating cells express abnormal fetal antigens, indicators of partial loss of differentiation [17]. Therefore *H. pylori* may increase the gastric cancer risk in populations with a high prevalence of such precursor lesions as chronic atrophic gastritis, intestinal metaplasia, and dysplasia.

Acknowledgement. This work was supported by grant POI-CA-28842 from the National Cancer Institute.

References

1. Correa P (1988) A human model of gastric carcinogenesis. Cancer Res 48:3554–3560
2. Blaser MJ (1987) Gastric Campylobacter-like organisms gastritis and peptic ulcer disease. Gastroenterology 93:371–383
3. Fox JG, Correa P, Taylor NS et al. (1989) Campylobacter pylori-associated gastritis and immune response in populations at increased risk of gastric carcinoma. Am J Gastoenterol 84:775–781
4. Correa P (1988) Chronic gastritis. A clinicopathological classification. Am J Gastroenterol 83:504–9
5. Leon-Barua R, Recavarren S, Ramirez A et al. (1988) Reversal of gastric mucosal dysplasia associated with Campylobacter pylori using oral bismuth therapy (abstract). Gastroenterology 94:256
6. Recavarren S (1989) Histopatologia de la infeccion por Campylobacter pylori. In: Ramirez-Ramos A (ed) Campylobacter pylori y Patologia Gastroduodental Universidad Peruana Cayetano Heredia, Lima, Peru, pp 117–133 (Impresiones Santa Ana)
7. Feng Y, Wang Y (1988) Campylobacter pylori in patients with gastritis, peptic ulcer and carcinoma of the stomach in Lanzhou, China. Lancet 1:1055 (letter)
8. Forman D, Sitas F, Newell DG, Boreham J, Peto R, Li J, Chen J (1990) Geographic association of Helicabacter pylori anti-body prevalence and gastric cancer mortality in China. Int J Cancer 46:608–611
9. Olive C, Michault A, Brassens-Rabbe MP, Megruad F (1989) Seroepidemiology of Campylobacter pylori in Reunion Island (abstract). Klin Wochenschr 67:52
10. Parsonnet J (1989) The epidemiology of C. pylori. In: Blaser MJ (ed) Campylobacter pylori in gastritis and peptic ulcer disease. Igaku-Shou New York pp 51–60
11. Forman D, Newell DG, Fullerton F, Yarnell JWG, Stacey AR, Wald N, Sitas F (1991) An association between Helicobacter pylori infection and the risk of gastric cancer: evidence from a prospective investigation. Br Med J 302:1302–1305
12. Eldridge J, Jones DM (1989) Longitudinal study of Campylobacter pylori in school children. In: Megraude F, Lamouliatte H (eds) Gastroduodenal Pathology and Campylobacter pylori. Excepta Medica, Amsterdam, pp 419–421
13. Barker DJP, Coggon D, Osmond C, Wickham C (1990) Poor housing in childhood and high rates of stomach cancer in England and Wales. Br Cancer 61:575–578
14. Parsonnet J, Vandersteen D, Goates J, Sibley RK, Pritikin J, Chang Y (1991) Helicobacter pylori infection in intestinal and diffuse-type gastric adenocarcinoma. JNCI 83:640–643
15. Brassens-Rabbe MP, Mégraud F, Denis F et al. (1989) Seroepidemiology of Campylobacter pylori infection in four populations which differ in their ethnic and geographic characteristics. In: Megraud F, Lamouliatte H (eds) Gastrointestinal pathology and campylobacter pylori. Excerpta Medica, Amsterdam, pp 42–430
16. Rathbone B, Wyatt J (1988) Campylobacter pylori and precancerous lesions. In: Reed PI, Hill MJ (eds) Gastric carcinogenesis. Excerpta Medica, Amsterdam, pp 132–144
17. Lipkin M, Correa P, Mikol IB et al. (1985) Proliferate and antigenic modifications in human epithelial cells in chronic atrophic gastritis. JNCI 75:613–619

Role of *Helicobacter pylori* in Gastric Carcinogenesis: Future Prospects

P. Correa

The role of *Helicobacter pylori* in human disease seems to become more complex as the scientific scrutiny intensifies. The original skepticism about the pathogenicity of the bacteria has largely been overcome. Few investigators or clinicians today doubt that *H. pylori* is an important cause of chronic gastritis. Since gastritis has been linked to gastric carcinoma in several epidemiologic studies, the role of *H. pylori* in gastric carcinogenesis is under scrutiny. So far, the epidemiologic findings have been remarkably consistent in favoring a causal relationship between this rather recently recognized bacteria and one of the oldest and more fatal neoplastic diseases of mankind.

The epidemiologic enquiry has progressed from descriptive studies which reported positive correlations between prevalence of infection and cancer rates to analytical case–control studies which reported an elevated relative risk of gastric cancer in subjects infected with *H. pylori*, and finally to prospective studies in which the prevalence of infection was found excessive in subjects who, many years later, developed gastric carcinoma.

What lies ahead almost certainly is an enthusiastic scientific pursuit to explore the phenomenon further. The best proof of causality could be provided by intervention studies aimed at curing or preventing the infection, expecting a decrease in cancer rates. These studies might be almost impossible, given the very prolonged latency of the neoplastic process. As a substitute for a cancer end-point, intermediate end-points in the neoplastic process could be used, such as intestinal metaplasia and dysplasia of the gastric mucosa.

The possible mechanisms by which *H. pylori* may increase gastric cancer frequency will most probably be the subject of intensive scientific investigation. At the moment such mechanisms are unknown. Several apparent negative findings have been pointed out, such as the fact that the bacteria do not reduce nitrate to nitrite, a suspected carcinogen precursor. Another peculiar negative observation is that the bacteria do not colonize the intestinalized epithelium, which seems the most frequent site of origin of neoplastic cells. Possible pathogenic mechanisms which deserve scientific explanation are: (a) the role of *H. pylori* in increasing cell replication; (b) the release of mutagenic oxygen radicals by polymorphonuclear leukocytes; (c) the in situ formation of *N*-nitroso compounds. Research along those lines is in progress and should appear in future publications.

Department of Pathology, Louisiana State University Medical Center, New Orleans, LA 70112, USA

J.M. Pajares et al. (Eds.)
H. pylori and Gastroduodenal Pathology

IX. *Helicobacter pylori* in Children

Epidemiology of *Helicobacter pylori* Infection in Pediatrics: A Serologic Study

C. De Giacomo

Introduction

Helicobacter pylori infection has been demonstrated in children from all continents. Data regarding the prevalence of the infection in children are derived from endoscopic series [1]. Since children undergoing endoscopy represent a highly selected population, this is a rough indication of the occurrence of the organism in childhood. Noninvasive methods such as serology are more suited for epidemiologic studies.

To evaluate the prevalence of the infection we measured antibody response to *H. pylori* using an enzyme-linked immunosorbent assay (ELISA) in children and adolescents. To study the spread of the infection within families of infected children, we evaluated sera from relatives of children with *H. pylori* infection.

Material and Methods

Serum Samples

We evaluated the presence of antibodies against *H. pylori* in 150 children and adolescents (age range 6 months–16 years) regardless of the presence of chronic dyspepsia (defined as upper abdominal or retrosternal pain, discomfort, heartburn, and vomiting lasting for more than 4 weeks [2]). We evaluated sera from 47 relatives (19 mothers, 13 fathers, ten sisters, and five brothers) of 19 *H. pylori*-infected children.

Enzyme-Linked Immunosorbent Assay

An ELISA using a suspension of six isolates of *H. pylori* from different antral biopsies was developed as antigen [3]. Briefly, plates were coated with whole bacterial cells by means of methylglyoxal. Serum samples were added after 1:30000 dilution for IgG and 1:300 dilution for immunoglobulin A (IgA) assay, respectively. IgG- or IgA-specific peroxidase-conjugated rabbit anti-human immune globulins (DAKO) were added after preadsorbtion with a *H. pylori*

Clinica Pediatrica dell' Universita' di Pavia, IRCCS Policlinico S. Matteo, 27100 Pavia, Italy

J.M. Pajares et al. (Eds.)
H. pylori and Gastroduodenal Pathology

suspension. Colour reaction was obtained with a *O*-phenylenediamine solution and read at 492 nm. Antibody levels were expressed in standard units referring to a standard curve based on progressive dilutions of a serum with high specific antibody activity. To evaluate the specificity of the ELISA, sera were also tested against a pool of Gram-negative bacteria.

Cut-off points were established in testing serum samples from 47 children (22 with *H. pylori* infection) undergoing upper digestive endoscopy with antral biopsy. An efficacy of 87% and 85% for IgG and IgA, respectively, was obtained. Sensitivity and specificity were 82% and 92% for IgG and 75% and 92% for IgA.

Combination of IgG and IgA seropositivity allows an increase in the positive and negative predictive value of the assay: both IgG and IgA seropositivity in the same child represents a 100% positive predictive value [3].

Results

Immunoglobulin seropositivity was found in 18 of 150 children, while IgA seropositivity was present in 19. Increased titres of both anti-*H. pylori* IgG and IgA in the same child were found in ten children. The infection may be considered unusual before 10 years of age. The prevalence of seropositivity in children increases with age (Table 1).

Table 1. Distribution of sera with IgG, IgA, and both IgG and IgA antibodies specific to *H. pylori* in various age groups. Numerators indicate the numbers positive by ELISA while denominators indicate the numbers tested in each group.

	Total		Age groups (years)								
	(*n*)	(%)	< 2	< 4	< 6	< 8	< 10	< 12	< 14	< 16	< 18
IgG positivity	18/150	12	1/8	1/11	2/22	1/5	0/19	3/30	4/19	2/10	4/16
IgA positivity	19/150	12.6	1/8	1/11	0/22	0/15	2/19	3/30	3/19	4/10	5/16
IgG and IgA positivity	10/150	6.6	0/8	0/11	0/22	0/15	0/19	2/30	2/19	2/10	4/16

Table 2. Distribution of sera with IgG, IgA, and both IgG and IgA antibodies specific to *H. pylori* in parents and siblings over 6 years of 18 children with *H. pylori* infection

	Fathers ($n = 13$)	Mothers ($n = 19$)	Brothers ($n = 5$)	Sisters ($n = 10$)
IgG positivity	10	17	4	5
IgA positivity	8	18	3	5
IgG and IgA positivity	7	16	3	4
Absence of IgG and IgA	2	0	2	3

Of 32 parents 30 (94%) and 10 of 15 siblings (66%) had specific antibodies in their serum (Table 2). All mothers in our series showed IgG or IgA seropositivity. Two fathers were negative for specific IgG and IgA. Two of five brothers and three of ten sisters were seronegative. In total, 14 of 18 males and 26 of 29 females showed IgG or IgA seropositivity (Table 2).

Discussion

Helicobacter pylori colonization of the gastric antrum has been found in 10%–31% of children examined by different pediatric endoscopic services [4, 5]. Differences between the many studies are mainly due to the methodology of the survey (retro- or prospective) and the selection of patients [1]. In spite of these discrepancies, infection has been always found in chronic antral gastritis, in association or not with peptic ulcers. A true prevalence of the infection in the pediatric population may be available only by means of noninvasive techniques such as serology.

Complement fixation tests, passive hemagglutination, ELISA, and immunoblot techniques have been employed to evaluate the specific immune response against *H. pylori* [6]. Although ELISA was the most diffuse method, a difference in the antigenic preparation results in different sensitivity and specificity [6]. Two different studies indicate that *H. pylori* antibodies are present in 3%–5% of healthy children and adolescents [7, 8]. No specific antibodies were detected before 9 years of age in another series [9]. Evaluation of 10–19-year-old adolescents showed an overall rate of seropositivity of 15.8% [10]. We report here the distribution of seropositivity for IgG and IgA in various pediatric age groups. A 100% positive predictivity of infection was found in ten of 150 (6%) children and adolescents unselected for digestive complaints. The infection may be considered as unusual before 10 years of age. As in adults, the prevalence of seropositivity increases with age [7].

Serologic evaluation of the families with infected members have given opposing conclusions in different studies. Antibody response to *H. pylori* was found with increased prevalence in families of index children [11]. Successive studies on household contacts of infected patients did not confirm intrafamilial diffusion [12]. Recently, specific antibodies were detected in 73.5% of 34 parents and 82% of 22 siblings of colonized children [13]. Our study confirms that the prevalence of antibody in relatives of index children is significantly higher than in adult and pediatric series. These data provide further evidence that *H. pylori* is spread by close personal contact.

References

1. Mahony MJ, Littlewood JM (1989) Campylobacter pylori in pediatric populations. In: Rathbone BJ, Heatley RV (eds) Campylobacter pylori and gastroduodenal disease. Blackwell, London, pp 167–75

2. Anonymous (1988) Management of dyspepsia: report of a working party. Lancet 1:576–9
3. De Giacomo C, Lisato L, Negrini R, Licardi G, Maggiore G (1991) Serum immune response to Helicobacter pylori in children: epidemiologic, and clinical applications. J Pediatr 119:205–210
4. Drumm B, Sherman P, Cutz E, Karmali M (1987) Association of Campylobacter pylori on the gastric mucosa with antral gastritis in children. N Engl J Med 316:155–161
5. Cadranel S, Glupczynsky Y, Labbe M, De Prez C (1988) Campylobacter pylori in children. In: Menge H, Gregor M, Tytgat GNJ, Marshall BJ (eds) Campylobacter pylori. Springer, Berlin Heidelberg New York, pp 110–115
6. Newell DG, Stacey AR (1989) The serology of Campylobacter pylori infections. In: Rathbone BJ, Heatley RV (eds) Campylobacter pylori and gastroduodenal disease. Blackwell, London, pp 75–82
7. Jones DM, Eldridge J, Fox AJ, Sethi P, Whorwell PJ (1986) Antibody to the gastric Campylobacter-like organisms ("Campylobacter pyloridis") – clinical correlations and distribution in the normal population. J Med Microbiol 22:57–62
8. Thomas JE, Eastham EJ, Elliott TSJ, Nelson R (1988) Campylobacter pylori gastritis in children – a common cause of symptoms. Gut 29:A707
9. Perez-Perez GI, Dworkin BM, Chodos JE, Blaser MJ (1988) Campylobacter pylori antibodies in humans. Ann Intern Med 109:11–17
10. Dwyer B, Kaldor J, Tee W, Raios K (1989) The prevalence of Campylobacter pylori in human populations. In: Rathbone BJ, Heatley RV (eds) Campylobacter pylori and gastroduodenal disease. Blackwell, London, pp 190–196
11. Mitchell HM, Bohane TD, Berkowicz J, Mazell SL, Lee A (1987) Antibody to Campylobacter pylori in families of index children with gastrointestinal illness due to C. pylori. Lancet 11:681–682
12. Jones DM, Eldridge J, Whorwell PJ (1987) Antibodies to Campylobacter pyloridis in household contacts of infected patients. Br Med J 294:615
13. Drumm B, Perez-Perez GI, Blaser MJ, Sherman PM (1990) Intrafamilial clustering of Helicobacter pylori infection. N Engl J Med 322:359–63

Prevalence of Antibody to *Helicobacter pylori* in Children in Northern Nigeria

C. Holcombe[1], S. Tsimiri[1], J. Eldridge[2], and D.M. Jones[1]

Introduction

Helicobacter pylori infection is uncommon in children in the West [1], but it is important to remember that *H. pylori* infection has a worldwide distribution. The social, cultural and environmental influences on populations in the developing world are very different and have a profound effect on the infection profile.

In a random serological survey we established that 178 of 212 (85%) adult subjects in northern Nigeria had IgG antibodies to *H. pylori* [2]. These levels of infection were confirmed endoscopically in 40 asymptomatic volunteers, 31 (77%) of whom were infected by *H. pylori* [3]. In contrast to the infection profile in the West, the prevalence of infection did not vary with age, suggesting that infection is acquired before the age of 20. As a result of these initial studies, we went on to measure antibodies to *H. pylori* in children to determine the age at which infection is acquired and thereby to define the infection profile of this population.

In considering the aetiological significance of *H. pylori*, it is vital to consider the prevalence of infection in the population and the age at which it is acquired. If most people are infected from early childhood and yet remain asymptomatic, it is difficult to postulate a significant pathological role for *H. pylori* in this environment.

Background

Maiduguri lies in the Sahel of north-eastern Nigeria, just south of the Sahara desert. The climate is hot and dry, most of the population are Muslim, rural, subsistence farmers. People live in extended families in what would be considered cramped conditions by Western standards, often without running water or adequate facilities for disposal of sewage.

[1]Academic Department of Surgery, University of Maiduguri, Northern Nigeria
[2]Public Health Laboratory, Manchester, UK

J.M. Pajares et al. (Eds.)
H. pylori and Gastroduodenal Pathology

Table 1. Comparison of patients with and without antibodies to *H. pylori*

Diagnosis	H. pylori positive (*n*)	(%)	H. pylori negative (*n*)	(%)
Pneumonia	10		5	
Unknown	7		0	
Parasites	4		0	
Malaria	4		2	
Diarrhoea and vomiting	4	8	6	31
UTI	2		1	
Sickle cell disease	2		0	
Glomerular nephritis	0		2	
Marasmus	1		1	
Miscellaneous	14		2	
Total	48		19	
Sex Male	29		13	
Female	19		6	
Antibiotic intake	16	33	7	36
Abdominal pain	3	6	1	5

Methods

In a random serological survey serum was taken from children aged 10–19 years. Further samples were taken from children aged 0–10 years who were inpatients at the University of Maiduguri Teaching Hospital and had been admitted with a variety of common paediatric conditions (Table 1), each had a standard proforma completed and 5 ml serum was taken.

Immunoglobulin G (IgG) antibodies to whole cell *H. pylori* antigen were measured using a standard enzyme-linked immunosorbant assay (ELISA) previously validated [1]. Sera were absorbed with *Campylobacter jejuni* prior to testing for antibodies to *H. pylori*. The optical density for validated negatives was determined on the basis of optical densities obtained in patients known to be *H. pylori* positive on the basis of culture and histological examination of gastric mucosal biopsies.

Results

Of the 43 randomly chosen children aged 10–19 years, 39 (91%) had IgG antibodies to *H. pylori*. Seven of these children (16%) had experienced dyspepsia in the preceding 6 months, all of these seven had IgG antibodies to *H. pylori*.

One hundred children under the age of 10 years were tested and 69 (69%) had IgG antibodies to *H. pylori*. There is a trend to increasing levels of antibody with increasing age, but this does not reach statistical significance (Table 2), even 22 of 38 (58%) subjects under 1 year of age had antibodies to *H. pylori*.

Table 2. *H. pylori* infection according to age

Age	Positive		Negative		Total
	(*n*)	(%)	(*n*)	(%)	(*n*)
0– 5 months	6	50	6	50	12
6–11 months	16	61	10	39	26
1 year	10	71	4	29	14
2 years	6	86	1	14	7
3 years	5	71	2	29	7
4 years	1	100	0		1
5 years	3	75	1	25	4
6 years	6	75	2	25	8
7 years	4	67	2	33	6
8 years	4	77	2	33	6
9 years	8	89	1	11	9
Total	69	69	31	31	100
10–19 years	39	91	4	9	43

Complete clinical details are available in 67 of those aged 0–9 years (Table 1). There were no significant differences in the medical diagnoses, antibiotic intake, amount of diarrhoea or abdominal pain between subjects who did and did not have antibodies to *H. pylori* (Table 1).

Discussion

These results are strikingly different from the West, with a prevalence of antibodies to *H. pylori* of 69% under 10 years compared to a prevalence of only 3% in the West [4]. In northern Nigeria 91% of teenagers have antibodies to *H. pylori* compared to only 18% in the West [1] (Fig. 1). In no population has a higher prevalence of antibodies in children been recorded, and this is all the more important in view of the apparent lack of pathological effects.

There is a distinctive infection profile for *H. pylori* in Africa, with high levels of antibodies acquired at a young age in all populations studied [4–6]. The source of *H. pylori* and mode of transmission is unknown, although circumstantial evidence suggests that person-to-person spread is most likely [7]. Most children in northern Nigeria are members of an extended family, often in crowded conditions, without running water or adequate facilities for sewage disposal, conditions in which infection could easily spread from person to person.

This high prevalence of antibodies has been shown by endoscopic studies to be due to active *H. pylori* infection and, as in the West, is associated almost invariably with antral gastritis in both dyspeptic patients [5, 6] and asymptomatic controls [3].

There were no significant differences between those with and without antibodies to *H. pylori.* In those with antibodies only three (4%) had diarrhoea, whereas five (26%) of those without antibodies had diarrhoea. There was only

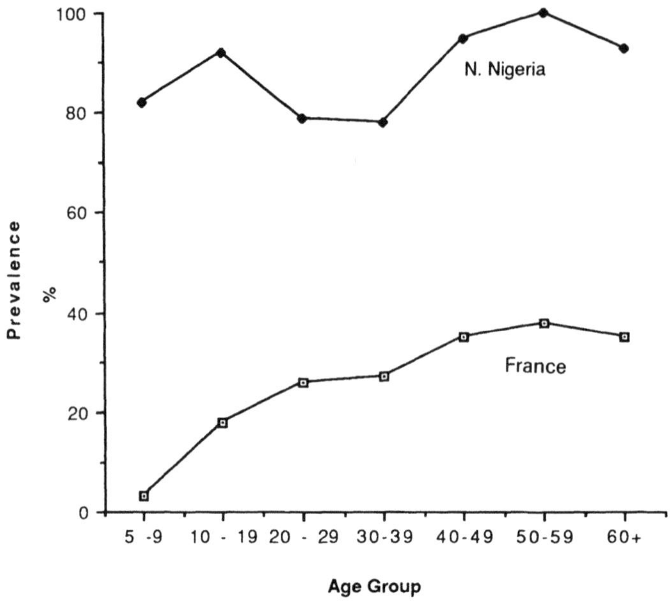

Fig 1. Prevalence of antibodies to *H. pylori* according to age. *Black squares*, N. Nigeria [2] ; *white squares*, France [4]

one patient with malnutrition in each group. Antibiotics had been taken by 18 (20%) of those with antibodies and seven (39%) of those without. Three (6%) of those with antibodies had abdominal pain as did one (5%) without antibodies. Our overwhelming impression has been of very high levels of *H. pylori* infection in every group tested and at every age, whether patients with peptic ulcer and non-ulcer dyspepsia [8, 9], asymptomatic controls [3] or children.

These children do not seem to present with adverse effects caused by their *H. pylori* infection, and treatment in this environment would be rarely, if ever, justified. Not only is the organism difficult to eradicate [10] and the treatment expensive, but the re-infection rate is unknown. However, it is likely to be high when other family members are infected, as demonstrated by Collins et al. [11].

Conclusion

The clinical significance of *H. pylori* infection is unclear; however, it is clear that over three quarters of the population in northern Nigeria are infected by *H. pylori* for most of their lives with apparently few consequences. This is all the more striking as this study was carried out in an area where both peptic ulcer [12] and gastric cancer [13] are uncommon.

References

1. Jones DM, Eldridge J, Fox AJ, Sethi P, Whorwell PJ (1986) Antibody to the gastric Campylobacter-like organism ('Campylobacter pyloridis') – clinical correlations and distribution in the normal population. J Med Microbiol 22: 57–62
2. Holcombe C, Omotara BA, Eldridge J, Jones DM (1992) The commonest bacterial infection in Africa. Helicobacter pylori: A random serological survey. Am J Gastroenterology 87 (1): 28–30
3. Holcombe C, Kaluba J, Lucas SB (1991) Non-ulcer dyspepsia in Nigeria: a case control study. Trans R Soc Trop Med Hyg 4: 553–555
4. Megraud F, Brassens-Rabbe MP, Denis F, Belbouri A, Hoa DQ (1989) Seroepidemiology of Campylobacter pylori in various populations. J Clin Microbiol 27 (8) 1870–1873
5. Rouvroy D, Bogaerts J, Nsengiumwa O, Omar M (1987) Campylobacter pylori, gastritis and peptic ulcer disease in central Africa. Br Med J 295: 1174
6. Weir WRC, Goodgame R, Kiire CF, Lucas SB (1988) Campylobacter like organisms and gastritis in Africa. Trans R Soc Trop Med Hyg 82: 172
7. Mitchell HM, Lee A, Bohane TD (1989) Evidence for person to person spread of Campylobacter pylori. In: Rathbone BJ, Heatley RV (eds) Campylobacter pylori and gastroduodenal disease. Blackwell, London, pp 197–202
8. Holcombe C, Lucas SB, Umar H, Abba A (1990) Helicobacter pylori (= Campylobacter) in Africa. Trans R Soc Trop Med Hyg 84 (2): 294–296
9. Holcombe C, Lucas SB, Thom C (1990) Helicobacter pylori and Non-ulcer dyspepsia in Northern Nigeria. Rev Rsp Enferm Dig 78 (1): 71
10. Holcombe C, Thom C, Kaluba J, Lucas SB (1991) Helicobacter pylori clearance in the treatment of non-ulcer dyspepsia. Ailmentary Pharmacology & Therapeutics 6 (1): 119
11. Collins R, Patchett S, Keane C, O'Norain C (1991) Reinfection with Helicobacter pylori due to intrafamilial clustering of the organism. Rev Esp Enferm Dig 78 (1): 9
12. Tovey FI, Tunstall M (1975) Duodenal ulcer in black populations in Africa south of the Sahara. Gut 16: 564–576
13. Holcombe C, Babayo U (1991) The pattern of malignant disease in North East Nigeria. Trop Geogr Med 43 (2): 189–192

Prevalence of Immunoglobulin G Antibodies to *Helicobacter pylori* in Mothers of Newborns with Upper Gastrointestinal Bleeding

O. Sangaletti[1], R. Tornaghi[2], M. Petrillo[1], M. Fontana[2], and G. Bianchi Porro[1]

Introduction

In recent years, a reduction in the diameter of the fiberoscope has made possible endoscopic exploration of the digestive tract in very small infants [1]. Neonatal gastrointestinal endoscopy has shown the occurrence of ulcerated lesions of the stomach and duodenum as a cause of upper gastrointestinal haemorrhage in newborns [2–4]. With few exceptions – severe stress caused by sepsis, asphyxia, etc – the aetiology of ulcerated lesions in newborns is unknown. *Helicobacter pylori* (HP) infection has been recognized as the cause of type B gastritis in adults [5] and in children [6], and also as a possible pathogenetic factor of peptic ulceration, being found in the gastric antrum of 90%–100% of patients with duodenal ulcer [7] and in 60%–80% of those with gastric ulcer [8].

The determination of immunoglobulin G (IgG) antibody levels against HP has been shown to provide a good and non-invasive method to evaluate the presence of HP infection [9, 10]. To explore the possibility that maternal HP infection might be a risk factor for the development of upper gastrointestinal, lesions in newborns, we looked for IgG antibodies to HP in a group of mothers of newborns with gastrointestinal bleeding and in a control group of mothers of healthy newborns.

Patients and Methods

The sera of 30 newborns and their respective mothers were tested for IgG against HP. Seventeen newborns had an upper gastrointestinal haemorrhage within 24 h of birth: the endoscopical examination showed the presence of multiple superficial erosions and ulcers of the gastric body and fundus. The control group included 13 healthy newborns and their mothers. In both groups, samples of sera were taken simultaneously from mothers and newborns within 24 h of birth.

Immunoglobulin G levels were measured by enzyme-linked immunosorbent assay (ELISA) (G.A.P. Test, BIO-RAD Laboratories, Segrate, Milan, Italy). All of the samples were diluted 1:200 and assayed simultaneously: the antibody

[1]Gastrointestinal Unit, L. Sacco Hospital, Via G.B. Grassi 74, 20157 Milan, Italy
[2] Pediatric Unit, University of Milan, Milan, Italy

J.M. Pajares et al. (Eds.)
H. pylori and Gastroduodenal Pathology

titres were determined at an optical density (OD) of 405 nm. The sensitivity and the specificity of the test were 80% and 100%, respectively. The antibody activity was expressed as OD and samples with OD above 0.150 were considered as positive.

Statistical analysis was performed by means of the chi-square test, Wilcoxon's sum rank test and Spearman's rank correlation as and where appropriate.

Results

Table 1 shows the prevalence of HP antibodies in the two groups examined: Nine of 17 (52.9%) mothers of newborns with upper gastrointestinal bleeding and seven of 13 (53.8%) mothers of healthy newborns had antibodies against HP. The difference was not statistically significant. No differences were found when the levels of HP antibodies of mothers of newborns with or without upper gastrointestinal haemorrhage were compared (Table 1): the median values were 0.569 and 0.431, respectively ($p > 0.05$). Only the newborns of mothers with the presence of serum IgG to HP showed antibodies (Fig. 1), and there was a good

Table 1. Prevalence and amount of Helicobacter pylori antibodies in mothers of newborns with upper gastrointestinal bleeding and in mothers of healthy newborns

	Prevalence (*n*)	Prevalence (%)	Optical density (median)
Mothers of newborns with gastrointestinal bleeding	9/17	52.9	0.569
Mothers of healthy newborns	7/13	53.8	0.431
	p = NS		p = NS

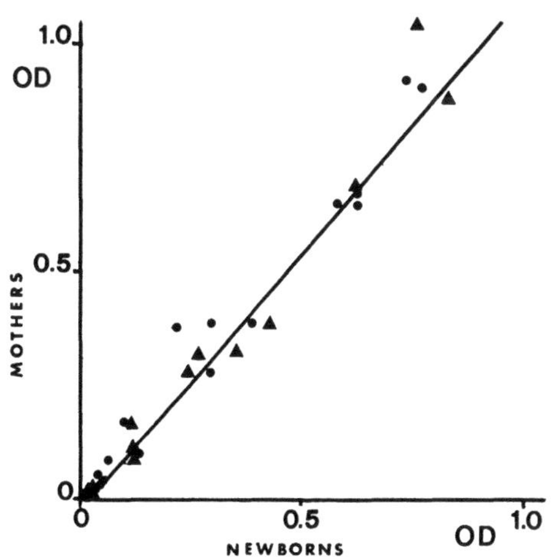

Fig. 1. Correlation between antibody titres of mothers and newborns. *Circles*, upper gastrointestinal bleeding group; *triangles*, control group; *OD*, optical density

correlation between the antibody titres of mothers and newborns ($r = 0.97$; $p < 0.001$).

Discussion

Our results suggest that maternal HP infection is not a risk factor for the development of upper gastrointestinal bleeding. The prevalence of IgG antibodies against HP and antibody titres were similar in both study groups and do not differ from the prevalence found in a normal population [11]. As in newborns, the antibody response was poor, the antibodies found probably being acquired from the mother.

Nevertheless, a definitive exclusion of maternal HP infection as a risk factor for gastrointestinal bleeding in newborns needs further investigation to evaluate the presence of HP on gastric biopsies and a follow up of antibody titres in newborns.

References

1. Olives JP (1989) Neonatal gastrointestinal endoscopy. Front Gastrointest Res 15:65–73
2. Sherman MJ, Clatworth JW (1967) Gastro-intestinal bleeding in neonates. A study of 94 cases. Surgery 62:614–617
3. Liebman WM, Thaler MM, Bujanover Y (1978) Endoscopic evaluation of upper gastrointestinal bleeding in the newborn. Am J Gastroenterol 69:607–608
4. Walker-Smith JA, Hamilton JR, Walker WA (1983) Gastrointestinal haemorrhage. In: Walker JA, Hamilton JR (eds) Practical pediatric gastroenterology. Butterworths, London pp 49–56
5. Goodwin CS, Armstrong JA, Marshall BJ (1986) Campylobacter pyloridis, gastritis and peptic ulceration. J Clin Pathol 39:353–365
6. Drumm B, Sherman P, Cutz E, Karmil M (1987) Association of campylobacter pylori on the gastric mucosa and antral gastritis in children. N Engl J Med 316:1557–1617
7. Wyatt JI (1989) Campylobacter pylori, duodenitis and duodenal ulceration. In: Rathbone BJ, Heatley RV (eds) Campylobacter pylori and gastroduodenal disease. Blackwell, Oxford, pp 117–124
8. O'Connor HJ, Axon ATR (1989) Campylobacter pylori, gastric ulceration and the postoperative stomach. In: Rathbone BJ, Heatley RV (eds) Campylobacter pylori and gastroduodenal disease. Blackwell, Oxford, pp 125–138
9. Rathbone BJ, Wyatt JI, Worsley BW, Shires SE, Trejdosiewicz LK, Hearley RV, Losowsky MS (1986) Systemic and local antibody responses to gastric Campylobacter pyloridis in non-ulcer dyspepsia. Gut 27:643–647
10. Goodwin CS, Blincow E, Peterson G, Sanderson C, Cheng W, Marshall BJ, Warren JR, McCulloch R (1987) Enzyme-linked immunosorbent assay for Campylobacter pyloridis: correlation with presence of Campylobacter pyloridis in the gastric mucosa. J Infect Dis 155:488–494
11. Dwyer B, Kaldor J, Tee W, Raios K (1989) The prevalence of Campylobacter pylori in human populations. In: Rathbone BJ, Heatley RV (eds) Campylobacter pylori and gastroduodenal disease. Blackwell, Oxford, pp 193–196

Helicobacter pylori Infection in Children

U. Blecker, A. Franckx-Goossens, S. Lauwers, and Y. Vandenplas

Introduction

Helicobacter pylori is a recently discovered bacterium that has quickly awakened an enormous interest when a possible link was discovered between this microorganism and a number of diseases of the upper gastrointestinal tract [10, 11, 14]. For this reason *H. pylori* has been the subject of numerous clinical trials since its discovery 8 years ago. *H. pylori*-associated gastritis and peptic ulcer were initially reported in adult patients [17]. However, there has recently been increasing interest in the incidence of this microorganism in the paediatric age group [3, 4, 6, 9, 10, 12, 14].

Materials and Methods

Over a period of 7 months, between June and December 1989, we investigated 48 paediatric patients aged 1–15 years (mean age 8 years 11 months; 20 boys/28 girls). All patients were admitted because of aspecific abdominal complaints. The complaints consisted mostly of chronic abdominal or epigastric pain, sometimes combined with nausea or vomiting. Our aim was to examine the prevalence of an *H. pylori* infection in this group of patients. Furthermore, we were interested in knowing whether there were any "typical" *H. pylori* complaints.

All 48 patients underwent an upper gastrointestinal tract endoscopy during which biopsies were taken at different locations of the antral mucosa.

In addition to culture, the presence of *H. pylori* was also investigated histologically (Gram and Giemsa stains) and with a rapid urease test (CLO test, Delta West Ltd., Western Australia). Furthermore, in order to exclude other possible causes for the complaints, all patients (with the exception of two who presented with acute haematemesis) also underwent a lactose H_2–breath test, a histology, stool culture, abdominal X-ray, and dietary history. None of the investigated patients was taking any medication at the time of examination.

Academic Children's Hospital, Free University of Brussels, Laarbeeklaan 101, 1090 Brussels, Belgium

J.M. Pajares et al. (Eds.)
H. pylori and Gastroduodenal Pathology

Results

In ten patients of the 48 (20.8%) an infection with *H. pylori* was demonstrated. The results of the different diagnostic tests are outlined in Table 1. This shows that all these tests are complementary and it stresses the importance of using more than one test for the diagnosis of an infection with this microorganism.

In all ten *H. pylori*-positive patients a type B active chronic gastritis was histologically present. During endoscopy a clear macroscopical abnormality of the gastric antral mucosa (nodular gastritis) was only observed in three out of the ten *H. pylori*-positive patients. The various diagnoses of the other 38 patients are shown in Table 2.

As mentioned earlier, the 48 patients presented with a variety of symptoms. Nausea, vomiting, abdominal and/or epigastric pain, anorexia, and dyspepsia are just some of the symptoms the patients complained of. Typical *H. pylori* complaints could not be detected, although almost all *H. pylori*-infected patients complained of chronic abdominal pain (Table 3). This reflects the findings of other authors [1, 5, 10].

All patients with an *H. pylori* infection were treated with amoxycillin (50 mg/kg per day three times daily) and colloidal bismuth subcitrate (60 mg three times daily) over a period of 6 weeks [7]. All patients had fewer complaints shortly after starting the treatment. Four to eight weeks after the end of the treatment we performed a control endoscopy in order to evaluate the eradication of the bacteria from the stomach. This eradication was achieved in seven (70%).

To obtain an idea of the familial incidence of *H. pylori* infection, immunoglobulin G (IgG) anti-*H. pylori* titres (last generation serological test) were

Table 1. The results of the different diagnostic techniques

Diagnostic technique	Positive results (%)
Rapid urease test	70
Histology	80
Culture	80

Table 2. Diagnoses of the *H. pylori*-negative patients

Diagnosis	(*n*)	(%)
Campylobacter jejuni enteritis	2	5.3
Constipation	6	15.8
Psychosomatic symptoms	13	34.2
Parasitic infection	3	7.9
Lactase deficiency	7	18.4
Viral gastritis (cytomegalovirus)	1	2.6
Peptic ulcer, not *H.pylori* related	1	2.6
Unknown	5	13.2

Table 3. Symptoms of the *H. pylori* positive patients in comparison to all examined children (*n*:48)

Symptom	Total (*n*)	*H. pylori* Positive (*n*)	(%)
Nausea	14	3	21.4
Chronic epigastric pain	8	3	37.5
Chronic abdominal pain	30	5	16.7
Diarrhoea	5	1	20.0
Fatigue	3	1	33.3
Vomiting	2	2	100.0

determined in 35 family members (parents, brothers and sisters older than 10 years) of ten newly diagnosed *H. pylori*-positive patients from January 1990. In 17 family members the IgG titres were positive. The IgG level was normal in one of our patients with a positive culture for *H. pylori.*

All family members with a positive titre underwent a gastric endoscopy. An *H. pylori* infection was diagnosed in seven (by means of histology, culture, and/or rapid urease test). Retrospectively, two of these family members had minimal digestive complaints.

Discussion

The study reflects the results of a prospective evaluation with regard to clinical and pathological findings in *H. pylori*-positive and -negative children that underwent an upper gastrointestinal tract endoscopy/gastroscopy because of abdominal complaints. These results show that one should not depend on the clinical symptoms to diagnose an *H. pylori* infection because there are no typical complaints for an infection with this bacterium. This coincides with the findings of other authors [1, 2, 10].

Furthermore, we confirmed the correlation between the presence of an *H. pylori* infection and a histological type B active chronic gastritis. In fact, all *H. pylori*-positive patients presented this kind of gastritis. In addition, the presence of an *H. pylori* infection was demonstrated in approximately 20% of the examined children. This percentage reflects an important incidence of *H. pylori* infection in the sphere of the vague abdominal complaints in children. Corresponding percentages of *H. pylori*-positive gastritis in children with abdominal complaints have been reported by other authors [4, 10, 12, 14]. Considering the fact that the prevalence of *H. pylori* infection increases with age [13], one can, seeing the relatively young age of our patients, conclude that the incidence of *H. pylori* infection we found reflects the importance of *H. pylori* in the pathogenesis of certain diseases of the upper gastrointestinal tract.

Furthermore, we would like to stress the importance of family screening in all *H. pylori*-positive patients. As demonstrated in other clinical trials [8, 15, 16], we were also able to show a clear familial incidence in families with an *H. pylori*-positive member.

References

1. Blecker U, Renders F, Lanciers S, Vandenplas Y (1991) Syncopes leading to the diagnosis of a Helicobacter pylori positive chronic active haemorrhagic gastritis. Eur J Pediatr 150:560–561
2. Börsch G, Schmidt G, Wegener M, Sandman M, Adamek R, Leverkus F, Reitemeyer E (1988) Campylobacter pylori: prospective analysis of clinical and histologic factors associated with colonization of the upper gastrointestinal tract. Eur J Clin Invest 18:133–138
3. Bujanover Y, Konikoff F, Baratz M (1990) Nodular gastritis and Helicobacter pylori. J Pediatr Gastroenterol Nutr 11:41–44
4. Czinn SJ, Speck WT (1989) Campylobacter pylori: a new pathogen? J Pediatr 114:670–672
5. Czinn SJ, Dahms BB, Kaplan B, Rothstein FC (1986) Campylobacter-like organisms in association with symptomatic gastritis in children. J Pediatr 109:80–83
6. Drumm B, Sherman P, Cutz E, Karmali M (1987) Association of Campylobacter pylori on the gastric mucosa with antral gastritis in children. N Engl J Med 316:1557–1561
7. Drumm B, Sherman P, Chiasson D, Karmali M, Cutz E (1988) Treatment of Campylobacter pylori-associated antral gastritis in children with bismuth subsalicylate and ampicillin. J Pediatr 113:908–912
8. Drumm B, Perez-Perez GI, Blaser MJ, Sherman PM (1990) Intrafamilial clustering of Helicobacter pylori infection. N Engl J Med 322:359–363
9. Eastham EJ, Elliot SM (1987) Campylobacter pyloridis in children. Arch Dis Child 62:652
10. Glassman MS, Schwarz SM, Medow MS, Beneck D, Halata M, Berezin S, Newman LJ (1989) Campylobacter pylori-related gastrointestinal disease in children. Incidence and clinical findings. Dig Dis Sci 34:1501–1504
11. Johnston BJ, Reed PI, Ali MH (1988) Prevalence of Campylobacter pylori in duodenal and gastric mucosa-relationship to inflammation. Scand J Gastroenterol 23:69–75
12. Kilbridge PM, Dahms BB, Czinn SJ (1988) Campylobacter pylori-associated gastritis and peptic ulcer disease in children. Am J Dis Child 142:1149–1152
13. Kosunen TU, Höök J, Rautelin HI, Myllylä G (1989) Age-dependent increase of Campylobacter pylori antibodies in blood donors. Scand J Gastroenterol 24:110–114
14. Mahoney MJ, Wyatt JI, Littlewood JM (1988) Campylobacter pylori gastritis. Arch Dis Child 63:654–655
15. Mitchell HM, Bohane T, Berkowicz J, Hazell SL, Lee A (1987) Antibody to Campylobacter pylori in families of index children with gastrointestinal illness due to Campylobacter pylori. Lancet ii:681–682
16. Oderda G, Ansaldi N, Boero M, Ponzetto A, Bellis D (1988) Campylobacter pylori in families of children with relapsing gastroduodenal disease due to *C. pylori* infection. Am J Gastroenterol 83:1437–1438
17. Warren JR, Marshall BJ (1983) Unidentified curved bacillion gastric epithelium in active chronic gastritis. Lancet i:1273–1275

Histological and Serological Examination of *Helicobacter pylori* in Estonian Children with Abdominal Complaints

H.-I. Maaroos[1], I. Nilsson[2], T. Vorobjova[1], T. Rägo[1], and T. Wadström[2]

Chronic gastritis has been found to be common in Estonian children with abdominal complaints (61%), and in 85% of cases it was caused by *Helicobacter pylori* (HP) [1]. Infected adult patients elicit a signicicant circulating antibody response against antigens of HP [2]. The systemic antibody response to HP in children has not been studied to the same extent. Some authors have found that HP antibody levels in children mirror investigations in adult patients [3, 4], others claim that HP antibodies are not a common finding in children [5, 6]. The aim of our study was to determine HP-IgG antibody frequency in children with histologically proven but slight gastritis and to examine the correlation between HP-IgG antibodies and HP colonisation intensity in the antral and body mucosa. Glycin extract antigen and the enzyme-linked immunosorbent assay (ELISA) method were used in parallel in two laboratories (in Tartu and in Lund) on the same children's sera.

Material and Methods

Thirty-nine children aged 4–15 years (mean age 10 ± 3 years, 15 boys and 24 girls) were examined. All of them were endoscoped mainly for upper abdominal pains. Patients with peptic ulcer were not included in the group. None of the subjects had received any antibiotics within 4 weeks before the endoscopy.

During the endoscopy two specimens were taken from the antral and two from the body mucosa. The state of the gastric mucosa was histologically examined in haematoxylin and eosin-stained specimens by P. Sipponen [1]. Normal mucosa, slight, moderate or severe superficial gastritis both in the antral and body mucosa were diagnosed by chronic round cell infiltration. Acute neutrophil infiltration was also graded. No atrophic changes were found in children's gastric mucosa [1]. HP intensity was graded as follows: 0, no bacteria; 1, fewer than 20 bacteria; 2, 20–60 bacteria; and 3, more than 60 bacteria per high-power microscopic field (objective, 40, ocular 10). HP-IgG antibody response was examined using the HP strain NCTC 11637 glycine extract as an antigen [7] using an ELISA method both in the Lund and Tartu laboratories. The results were expressed as the relative antibody activity (RAA) by absorbance values. Commercial IgG (Kabi AB, Stockholm, Sweden) was used on each

[1] Tartu University, EE2400 Tartu, Estonia
[2] Lund University, Lund, Sweden

J.M. Pajares et al. (Eds.)
H. pylori and Gastroduodenal Pathology

Table 1. Distribution of patients by HP histological state, HP-IgG antibodies and gastric mucosal morphology

			Age (years)		Sex		Mucosal morphology		
								Chronic gastritis	
		No. of patients	Range	Mean ± SD	M	F	Normal	antral or corpus	pan-gastritis
Helicobacter pylori IgG antibodies	– –	17	6–14	9 ± 3	7	10	11	6	0
Helicobacter pylori IgG antibodies	+ –	6	4–14	10 ± 4	3	3	2	3	1
Helicobacter pylori IgG antibodies	+ +	16	7–15	11 ± 2	5	11	2	3	11
Total	2	39	4–15	10 ± 3	15	24	15	12	12

Table 2. Distribution of patients by HP-IgG RAA levels, age and HP score

RAA HP-IgG level			Age		Antral HP score				Corpus HP score			
	No. of patients	%	Range	Mean ± SD	0	1	2	3	0	1	2	3
< 25 units	20	51%	4–14	10 ± 3	17	1	1	1	16	1	0	3
25–40 units	3	8%	13–14	14 ± 1	2	0	0	1	2	1	0	0
> 40 units	16	41%	7–14	11 ± 2	1	3	4	8	3	5	5	3

ELISA plate as a positive control for 100 units of RAA values. The results were regarded as negative below 25 units, RAA values between 25 and 40 units were intermediate, and at over 40 units they were positive for HP-IgG antibodies.

The sera with intermediate values were examined by immunoblot analysis.

The results were statistically analysed using the "Statgraphics" package, the Spearman rank correlation test and linear regression analysis. The Scheffe method was used for the differences reliability of unpaired data.

Results

Chronic gastritis was found in 24 (62%), HP-positive cases by histological examination were 22 (56%), HP-IgG antibodies were detected in 16 patients out of 39 (41%) (Table 1). The distribution of patients by HP-IgG antibody RAA levels, age and HP score in the antral and body mucosa is given in Table 2. There were no differences in the mean ages of groups where the number of RAA units was lower than 25 and higher than 40 ($p > 0.05$). In patients with negative HP-IgG antibodies HP in the gastric mucosa was predominantly negative and in patients with HP-IgG antibody it was positive.

The mean HP-IgG antibody RAA level in the HP-negative group was 10.6 $\pm$ 9.5 and it differed significantly from the groups that had HP only in the antral region or HP both in the antral and body regions ($p < 0.001$), but no differences were found from the group with HP only in the body mucosa (Table 3).

The RAA mean levels of HP-IgG antibody in patients with different HP scores are shown in Table 4. The group with the highest HP score in the antral mucosa had the highest RAA mean level of HP-IgG antibody, and the difference from the groups with negative HP or from the groups with scores 1 and 2 was statistically significant ($F = 19.007$ $p < 0.001$, $F = 21.331$ $p < 0.001$ and $F = 29.217$ $p < 0.001$, respectively) (Table 4). A highly significant correlation was established between HP-IgG antibody RAA values and the antral mucosa by the Spearman rank correlation test ($r = 0.69$, $p < 0.001$), the body mucosa HP scores ($r = 0.46$, $p < 0.001$) and the sum of the scores of the antral and the body mucosa HP ($r = 0.65$, $p < 0.001$). The distribution of HP-IgG antibody RAA values and HP scores in the antral and the body mucosa is given in Figs. 1 and 2.

Table 3. Mean HP-IgG RAA levels in groups with different HP colonization pattern

Group	HP in histological specimens		RAA			
	Antrum	Corpus	No.	Mean $\pm$ SD	Range	p
I	–	–	17	10.6 $\pm$ 9.5	1-100	I-II $p < 0.001$
II	+	–	4	64.8 $\pm$ 29.0	19-100	
III	–	+	3	39.0 $\pm$ 43.1	8-100	
IV	+	+	15	61.4 $\pm$ 30.5	8-130	I-IV $p < 0.001$

Table 4. HP-IgG antibody mean RAA levels and HP scores in antral and corpus mucosa

Helicobacter pylori	RAA				
Scores	*n*	Mean ± SD	Range	*F*	*p*
Antrum					
0	20	14.8 ± 21.9[a]	1.0-100.0	19.007[a,d]	< 0.001
1	4	52.3 ± 24.1[b]	19.0- 52.3	21.331[b,d]	< 0.001
2	5	55.8 ± 26.5[c]	9.0- 70.0	29.271[c,d]	< 0.001
3	10	69.1 ± 36.2[d]	11.0-130.0		
Corpus					
0	21	20.9 ± 8[a]	1.0-100.0	13.983[a,d]	< 0.001
1	7	54.0 ± 29.3[b]	8.0-100.0	17.175[b,d]	< 0.001
2	5	78.0 ± 30.3[c]	50.0-130.0	11.8[c,d]	< 0.002
3	6	44.8 ± 42.0[d]	9.0-110.0		

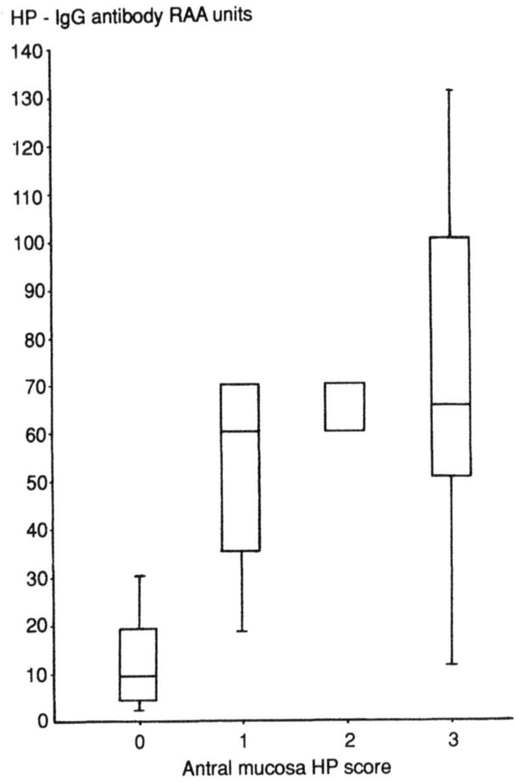

Fig. 1. Distribution of HP-IgG antibody RAA values with different HP scores in antral mucosa. *Boxes*, the middle 50% of the data values; *whiskers* extend to the minimum and maximum; *central line*, median

The screening of HP-negative and -positive patients by histological HP determination and by detection of HP-IgG antibody gave the same results in 33 cases (85%) (Table 1). HP was found in six children only by histological examination; five of these had HP-IgG RAA values below 25 units and one had

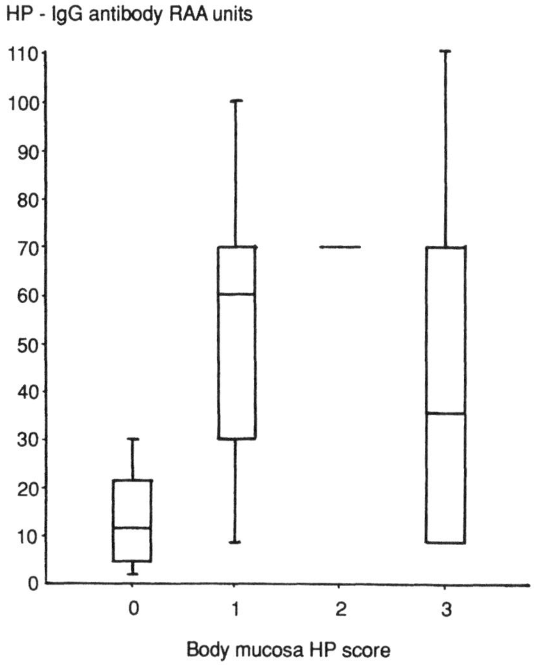

Fig. 2. Distribution of HP-IgG antibody RAA values with different HP scores in body mucosa. *Boxes*, the middle 50% of the data values; *whiskers* extend to the maximum and minimum; *central line*, median

an intermediate value (30 units); four out of the six had acute epigastric pain and histologically there was acute infiltration of the antral and the body mucosa. Two patients had no antral mucosa involvement and HP only in the body mucosa.

Sera with intermediate RAA values (Table 2) were analysed by immunoblot, and two of the three were positive for HP-IgG antibodies. They both had HP in the antral or body mucosa.

Comparing the HP-IgG antibody determination with histological HP detection results, we found specificity to be 100% and sensivity 73%, with a positive predictive value of 100% and a negative predictive value of 74% in the serological examination. ELISA determination of HP-IgG antibodies when using both the same antigen and method in Tartu and in Lund had a significantly high correlation ($r = 0.87$, $p < 0.001$).

Discussion

Immunoglobulin G antibodies to HP antigens are produced in cases of HP infection in adults, and circulating IgG antibodies are clearly of major importance in the immune defence against invasion of microorganisms [2]. In children's studies some authors have rarely found HP-IgG antibody in children below 14 years [5, 6]. The low frequency of HP-IgG antibody in these studies is obviously connected with a low incidence of HP infection in these regions. In our children's group, 56% of the children were infected with HP, and HP-IgG antibodies were found in 73% of the infected subjects. There were no age differences between HP-IgG antibody-positive and -negative groups ($p > 0.05$). Our results confirm that HP infection elicits a systemic HP-IgG antibody response in children as previously reported [3, 4]. Therefore HP colonisation is a true infection in children [3]. We made the semiquantitative evaluation of HP colonisation intensity in the gastric mucosa and found a correlation between the HP intensity and HP-IgG antibody RAA level. Our results confirm the opinion that the antral mucosa is more reactive to HP infection with a systemic antibody production [2] than the body mucosa since HP-IgG antibody RAA is more influenced by the antral mucosa HP scores than the body mucosa HP scores. Taking into consideration the different phenotypes of the antral and body HP topography, it was shown that the higher RAA level was connected with antrum colonisation. In our opinion, such a correlation is typical for children. Their systemic immune reaction to infection is not influenced by other factors as is the case, in adults, for example, by some medicines, patchiness of mucosal changes and age. Therefore the determination of HP-IgG antibodies allows the evaluation of the HP colonisation intensity, and antibody level may be helpful in following the response to the treatment of HP-associated gastritis and peptic ulcer in children [4].

In our series we had some patients whose HP infection was detected by histological examination, but the HP-IgG antibodies were under the lower limit and evaluated as negative. Some of these children may in fact have an active HP infection as they are clinically acutely ill, and their gastric mucosal morphology showed an acute infiltration with or without a chronic one. Two of these patients had HP only in the body mucosa without an antral involvement. There is a possibility that acute HP infection could be missed in some cases when using HP-IgG antibody determination and an additional analysis of other immunoglobulins should be made [2, 8].

For children, the limit of HP-IgG antibody is 25 units to evaluate the positive and negative results. All the intermediate values between 25 and 40 were additionally analysed by immunoblot, and it appeared that they were positive for HP. We had no false-positive tests using the serological examination. Therefore the antigen and the method we used had a high specificity for the detection of HP infection in children. The sensitivity of HP-IgG antibody determination was lower than the histological detection of HP. In their study Hirsche et al. [9] compared four antigen preparations; the glycine extract had the same sensitivity (78%) and specificity (98%) as we found. Therefore one reason for the low sensitivity of our test might be the acute HP infection in children; another reason could be the characteristic sensitivity of the HP glycine extract [9]. The reliability of our examinations is also shown by the very high correlation between results from the two laboratories.

References

1. H-I Maaroos, Rägo T, Sipponen P, Siurala M (1991) Helicobacter pylori and gastritis in children with abdominal complaints. Scand J Gastroenterol 26 (Suppl 186): 95–99
2. Rathbone BJ, Wyatt JI, Worsley BW, Shires SE, Trejdosiewicz LK, Heatley RV, Losowsky MS (1986) Systemic and local antibody responses to gastric Campylobacter pyloridis in non-ulcer dyspepsia. Gut 17:642–647
3. Czinn S, Carr H, Sheffler L, Aronoff S (1989) Serum IgG antibody to the outer membrane proteins of Campylobacter pylori in children with gastroduodenal disease. J Infect Dis 159:586–589
4. Oderda G, Vaira D, Holton J, Dowsett Jf, Ansaldi N (1989) Serum pepsinogen I and IgG antibody to Campylobacter pylori in non-specific abdominal pain in childhood. Gut 30:912–916
5. Bolton FJ, Hutchinson DN, Hinchliffe PM, Holt AV (1989) Distribution in various clinical groups of antibody to C.pylori detected by enzyme-linked immunosorbent assay, complement fixation and microagglutination tests. Serodiagn Immunother Infect Dis 3:41–50
6. Jones DM, Eldrige J, Fox AJ, Sethi P, Whorwell PJ (1986) Antibody to the gastric campylobacter-like organism ("Campylobacter pyloridis")-clinical correlations and distribution in the normal population. J Med Microbiol 22:57–62
7. Schalen C, Guruge JL, Ljungh A, Tyszkiewicz T, Wadström T (1989) Evaluation of Campylobacter pylori serology. Klin Wochenschr 67 [Suppl XVIII]: 61
8. Reiff A, Hellerich U, Grener P, Müller H, Jacobs E, Kist M (1989) Campylobacter pylori associated gastritis in childhood. Serological, morphological, and clinical aspects. Klin Wochenschr 67 [Suppl XVIII]: 58–59
9. Hirsch AM, Rathbone BJ, Wyatt JJ, Berger J, Rotter ML (1989) Four different antigen preparations for serodiagnosis of Campylobacter pylori infection. Klin Wochenschr 67 [Suppl XVIII]: 30–31

Helicobacter pylori in Children with Recurrent Abdominal Pain and Other Conditions

C. Camarero[1], I. Alvarez[2], L. de Rafael[3], M. Rodríguez[1], H. Escobar[1], and L. Suárez[1]

Introduction

In the last few years an association has been found between gastric colonization by *Helicobacter pylori* (HP) and some gastric conditions such as chronic antral gastritis [9, 10, 14] and peptic ulcer [4]. There have been numerous studies conducted with adult patients [5], but there is still some paucity of reports in the pediatric literature [1, 2, 6, 8, 11, 12, 15, 16].

We have investigated the presence of this microorganism in children suffering from different digestive conditions and specially in those with recurrent abdominal pain (RAP).

Patients and Methods

A total of 93 patients (47 boys, 46 girls) were studied. Their ages ranged from 4 to 17 years (mean age 10.2 years). The indications for upper digestive endoscopy were: RAP (56), epigastric pain (EP; 20), upper digestive hemorrhage (UDH; four), gastroesophageal reflux (GER; five), and other conditions (portal cavernomatosis one, caustic stenosis two, common variable immunodeficiency one, and gastrectomy control one).

Mucosa samples of the gastric antrum were obtained in the endoscopy unit and submitted for light-microscopic examination and culture. Samples for culture were processed within 1 h after being obtained. The biopsy sample was streaked onto a plate containing sheep blood (5%), Columbia agar supplemented with polymyxin B (5000 units per liter), amphotericin B (2 mg/l), trimethoprim (15 mg/l), and vancomycin (15 mg/l). Plates were incubated for 7 days at 37°C under an atmosphere containing 5% O_2, 10% CO_2, and 85% N_2. Suspected colonies were identified on the basis of positive reactions for oxidase, catalase, and rapid urease.

[1] Department of Pediatrics, Hospital Ramón y Cajal, 28034 Madrid, Spain
[2] Department of Pathology, Hospital Ramón y Cajal, 28034 Madrid, Spain
[3] Department of Microbiology, Hospital Ramón y Cajal, 28034 Madrid, Spain

J.M. Pajares et al. (Eds.)
H. pylori and Gastroduodenal Pathology

Results

Culture of HP

A positive culture was obtained in 30 patients (19 boys and 11 girls); the mean age was 10.4 years. In the 63 HP culture-negative patients the corresponding distribution was 27 boys and 36 girls with a mean age of 10.1 years. The indication for endoscopy in the HP-positive patients was: RAP four, EP eight, UDH one, GER one, and miscellaneous three. In the HP-negative patients the corresponding indication was: RAP 42, EP 13, UDH three, GER one, miscellaneous four (Tables 1, 2).

Endoscopy

Culture-Positive Patients: The mucosa had a normal appearance in 17. Findings consistent with chronic gastritis, duodenitis, and erosive gastritis were observed in nine, three and one patient, respectively.

Culture-Negative Patients. The mucosa had a normal appearance in 44 patients. Findings consistent with chronic gastritis, duodenal ulcer, and esophagitis were observed in 11, one, and one patient, respectively (Table 3).

Pathological Investigation of the Gastric Mucosa

Culture-Positive Patients. Mild to moderate chronic active gastritis was found in 12, mild follicular chronic active gastritis in six, and severe follicular chronic

Table 1. Culture of *H. Pylori*: sex distribution in our pediatric patients

	Boys		Girls	
	(*n*)	(%)	(*n*)	(%)
Culture-positive patients	19	63	11	36
Culture-negative patients	27	42.8	36	57.1

Table 2. Culture results of *H. Pylori* in our pediatric patients

Condition	Positive		Negative	
	(*n*)	(%)	(*n*)	(%)
Recurrent abdominal pain	14		42	
Epigastric pain	8		13	
Upper gastrointestinal bleeding	1		3	
Gastroesophageal reflux	4		1	
Miscellaneous	3		4	
Total	30	32.3	63	67.8

Table 3. Endoscopic findings in our pediatric patients

Endoscopy	*H. Pylori* culture Positive (*n*)	Negative (*n*)
Normal	17	44
Chronic gastritis	9	11
Duodenitis	3	–
Erosive gastritis	1	–
Duodenal ulcer	–	1
Esophagitis	–	1

active gastritis in five patients. One patient had erosive gastritis, and the gastritic mucosa was normal in four patients.

Culture-Negative Patients. Forty-nine patients were normal; follicular chronic gastritis was found in one and chronic active gastritis in seven patients. Six samples were inadequate for diagnosis.

Patients with RAP

Culture-Positive Patients. There were fourteen patients (ten boys, four girls) in this group. The endoscopic diagnosis in this group was duodenitis in five, antral hyperemia in one, and normal in ten patients. The pathological examination of the biopsy samples revealed: chronic antral gastritis in seven, follicular gastritis in three, and normal in three patients. One biopsy sample was inadequate for diagnosis.

Culture-Negative Patients. There were 42 patients (17 boys, 25 girls) in this group. The endoscopic findings were antral hyperemia in five, normal in 31, and miscellaneous in six patients. The pathological examination of the biopsy samples revealed 37 with a normal appearance, one with follicular chronic gastritis, and three with active chronic gastritis. One sample was inadequate (Table 4).

Table 4. Patients with recurrent abdominal pain: culture results and sex distribution

Patients	*H. pylori* culture Positive		Negative	
	(*n*)	(%)	(*n*)	(%)
Boys	10	71.4	17	40.4
Girls	4	28.5	25	59.5
Total	14	25	42	75

Discussion

Our findings are in good agreement with those of previous reports in which the presence of HP was documented in the gastric mucosa of children. In a high percentage of children this presence is associated with chronic antral gastritis. The percentage observed in our patients (76%) is similar to that observed in adults (> 70%) [5, 7, 17].

In this study a higher percentage of boys with HP was observed. This finding is relevant as the number of boys and girls at the beginning of the study was similar. This greater involvement of boys has not been reported in other pediatric series. The higher incidence of peptic ulcer in males is a well-known fact, and there is increasing evidence that HP can be involved in the origin of the ulcerous damage [13]. If it is true, gastric colonization by HP (more common in boys in our series) may be a plausible explanation for the higher incidence of the ulcerous damage in the male later in life. Most of the children studied were in the prepubertal age, and one might therefore, think that sex-linked hormonal factors do not play an important role in the gastric colonization by the microorganism.

The percentage of children with RAP and a positive HP culture (25%) associated with chronic gastritis is not to be neglected. One could suggest the existence of a subset of patients in whom this syndrome could be attributed to this association. However, these findings have been observed in asymptomatic patients with different percentages by several authors [3, 10]. Since this investigation was not carried out in our children we cannot accurately establish a cause–effect relationship. This issue will probably be elucidated in future studies with an efficient therapy in the eradication of HP from the gastric mucosa.

The mean age of HP-positive children was 10.4 years, similar to that of HP-negative patients. This observation reflects an early age for the possibility of colonization and opens many questions regarding the epidemiology of this microorganism.

References

1. Aleksandrova NZ, Minaev UI, Korsunskii AA, Kushel UR, Rytikov FM (1989) Detection of Campylobacter pylori in children with chronic gastritis. Lab Delo 6:54–56
2. Barbosa AJ, Queiroz DM, Mendes EN, Rocha GA, Carvalho AS, Roquete ML (1989) Campylobacter pylori associated acute gastritis in a child (letter). J Clin Pathol 42:779
3. Cohen H, Gramisu M, Fitzgibbons P, Appleman M, Skoglund M, Valenzuela JE (1989) Campylobacter pylori: associations with antral and fundic mucosal histology and diagnosis by serology in patients with upper gastrointestinal symptoms. Am J Gastroenterol 84:367–371
4. Dooley CP, Cohen H (1988) The clinical significance of Campylobacter pylori. Ann Intern Med 108:70–79
5. Dooley CP, Cohen H, Fitzgibbons PL, Bauer M. Appleman MD, Pérez-Pérez GI, Blaser MJ (1989) Prevalence of Helycobacter pylori infection and histologic gastritis in asymptomatic persons. N Engl J Med 321:1562–1566
6. Glassman MS, Schwarz SM, Medow MS, Beneck D, Halata M, Berezin S, Newman LJ (1989) Campylobacter pylori-related gastrointestinal disease in children. Incidence and clinical findings. Dig Dis Sci 34:1501–1504

7. Gregson DB, Low DE, Cohen MM, Cooter NB, Connon JJ, Wolman SL, Simor AE (1989) The prevalence of Campylobacter pylori gastritis among asymptomatic adults. Can Med Assoc J 140: 1449–1453
8. Kilbridge PH, Dahms BB, Czinn SJ (1988) Campylobacter pylori associated gastritis and peptic ulcer disease in children. Am J Dis Child 142: 1149–1152
9. Lanza FL, Skoglund ML, Rack MF, Yardley JH (1989) The effect of bismuth subsalicylate on the histologic gastritis seen with Campylobacter pylori: a placebo-controlled, randomized study. Am J Gastroenterol 84: 1060–1064
10. McNulty CA (1989) Pathogenicity of Campylobacter pylori. A causative factor in gastritis ? Scand J Gastroenerol [Suppl] 160: 3–6
11. Oderda G et al. (1989) Campylobacter pylori gastritis and peptic ulcer in children. Am J Dis Child 143: 877 (letter)
12. Oderda G, Dell 'Olio D, Morra I, Ansaldi N (1989) Campylobacter pylori gastritis: long term results of treatment with amoxycillin. Arch Dis Child. 64: 326–329
13. Rauws EAJ, Tytgat GNJ (1990) Cure of duedenal ulcer associated with eradication of Helycobacter pylori. Lancet 335: 1233–1235
14. Rauws EAJ, Langenberg W, Houthoff HJ, Zanen HC, Tytgat GNJ (1988) Campylobacter pyloridis-associated chronic active antral gastritis. A prospective study of its prevalence and the effects of antibacterial and antiulcer treatment. Gastroenterology 94: 33–10
15. Sullivan PB, Thomas JE, Wight DGD, Neale G, Eastham EJ, Corrah T Lloyd-Evans N, Greenwood BM. (1990) Helycobacter pylori in Gambian children with chronic diarrhoea and malnutrition. Arch Dis Child 65: 189–191
16. Tam PK, Saing H (1989) The use of H2-receptor antagonist in the treatment of peptic ulcer disease in children. J Pediatr Gastroenterol Nutr 8: 41–46
17. Wagner S, Freise J, Bar W, Fritsch S, Schmidt FW (1989) Epidemiology and therapy of Campylobacter pylori infection. Dtsch Med Wochenschr 114: 407–413

Helicobacter pylori-Associated Gastritis in Children

F. Gottrand[1], K. Francillette[1], D. Turck[1], P. Vincent[2], M. Lecomte[3], and J.P. Farriaux[1]

Introduction

Helicobacter pylori (HP)-associated gastritis and peptic ulcer were initially reported in adult patients [1]. More recently, a similar association has also been demonstrated in the pediatric age group [2, 3]. During the last year, HP-associated gastritis was found in 14 out of 55 children undergoing upper gastrointestinal endoscopy and gastric biopsies for suspected HP infection.

Patients and Methods

The 14 children presenting with HP infection were aged 2.5 months to 18.5 years (mean 11.2 years). The sex ratio was 1. All but one child were symptomatic: abdominal pain ($n = 9$), vomiting ($n = 7$), anorexia ($n = 3$), weight loss ($n = 2$), hematemesis ($n = 2$). The delay between symptoms and endoscopy varied from 3 days to 7 years (mean 1.3 years).

Endoscopy was performed using an Olympus XP 10 or XP 20 endoscope. During endoscopy, at least two antral biopsies were taken for histology (including Giemsa staining) and bacteriology. HP was identified in antral biopsies by at least one of the following microbiological techniques in four of six patients: Gram staining in three of six, urease production in two of six, or culture in four of six, or by histological techniques in 12 of 14 patients (Giemsa staining).

Results

Endoscopy always showed abnormalities; the macroscopic diagnosis was: gastritis ($n = 12$), esophagitis ($n = 6$), hiatal hernia ($n = 5$), gastric ulcer ($n = 1$). All but one child had chronic gastritis: 11 had interstitial inflammation of antral mucosa and one had chronic atrophic hypochlohydric gastritis. Only one child (age: 2.5 months) had acute superficial gastritis.

[1]Service de Pédiatrie, Gastroentérologie Pédiatrique et Génétique Médicale, [2]Laboratoire de Bactériologie, [3]Laboratoire d' Anatomie Pathologique, CHU Lille, 59037 Lille Cédex, France

J.M. Pajares et al. (Eds.)
H. pylori and Gastroduodenal Pathology

Discussion

The incidence of HP among children undergoing endoscopy for upper gastrointestinal symptoms ranges between 11% and 40% [4, 5]. In our study, HP was identified in 25% of the patients undergoing endoscopy for a variety of nonspecific symptoms.

Two points can be emphasized from this study. First, two children were less than 4 months old (Table 1). To the best of our knowledge, HP gastritis in young infants has not yet been reported in the literature; the youngest child previously described was 2 years old [6]. Secondly, five children were mentally retarded and institutionalized (Table 2). Three of them had gastroesophageal reflux, and endoscopy was done for suspected esophagitis. Two of them did not have any upper gastrointestinal symptoms and presented with weight loss

Table 1. HP gastritis in two children aged less than 4 months

Age (months)	Symptoms	Diagnosis delay	Endoscopic diagnosis	Histology	HP diagnosis
2.5	Abdominal pain Vomiting	3 days	Gastric ulcer	CG	Giemsa positive Bacteriology positive
3.5	Abdominal pain Hematemesis	5 days	Gastritis	ASG	Giemsa positive Bacteriology ND

CG, chronic gastritis; ASG, acute superficial gastritis; ND, not done.

Table 2. HP gastritis in five mentally retarded and institutionalized children

Age (years)	Symptoms	Diagnosis delay	Endoscopic diagnosis	Histology	HP diagnosis
15.5	Abdominal pain Vomiting	1.5 years	Gastritis Esophagitis Hiatal hernia	CG	Giemsa positive Bacteriology ND
18.5	Anorexia Weight loss	1 month	Esophagitis	CG	Giemsa positive Bacteriology ND
10.5	Weight loss	1 month	Gastritis Esophagitis	CG	Giemsa positive Bacteriology ND
10.5	–[a]	–	Gastritis	CG	Giemsa positive Bacteriology negative
16	Vomiting Anorexia	2 months	Gastritis Esophagitis	CG	Giemsa positive Bacteriology ND

[a] Endoscopy was done because of suspected caustic ingestion.
CG, chronic gastritis; ND, not done.

($n = 1$) and suspected caustic ingestion ($n = 1$). Esophagitis was found in four children; one of them had no gastritis at endoscopy. This high frequency of HP infection in mentally retarded and institutionalized children (five of 14) suggests a person-to-person transmission of HP [7]

Conclusions

First, HP gastritis can be found in infants aged less than 4 months; it is then associated with acute lesions. Secondly, HP gastritis seems to be frequent in mentally retarded institutionalized children. HP gastritis should be evaluated in these patients, even if symptoms are not specific and even if there is a story of gastroesophageal reflux or esophagitis.

References

1. Warren JR, Marshall BJ (1983) Unidentified curved bacilli on gastric epithelium in active chronic gastritis. Lancet 1:1273–1275
2. Czinn SJ, Dahns BB, Jacobs GH, Kaplan B, Rothstein FC (1986) Campylobacter-like organisms in association with symptomatic gastritis in children. J Pediatr 109:80–83
3. Cadranel S, Gossens H, De Boeck M, Malengreau A, Rodisch P, Butzler JP (1986) Campylobacter pyloridis in children. Lancet 1:735–736
4. Mahong MS, Wyatt J, Littlewood JM (1988) Campylobacter associated gastritis in children. Arch Dis Child 63:654–655
5. Bujanover Y, Konikoff F, Baratz M (1990) Nodular gastritis and Helicobacter pylori. J. Pediatr Gastroenterol Nutr 11:41–44
6. Czinn SJ, Carr H (1987) Rapid diagnosis of Campylobacter pyloridis-associated gastritis. J Pediatr 110:569–570
7. Berckowicz J, Lee A (1987) Person-to-person transmission of Campylobacter pylori. Lancet 1:680–681

Helicobacter pylori Gastritis in Children: Wide Spectrum of Symptoms

G. Oderda, E. Poli, P. Lerro, D. Dell'Olio, and N. Ansaldi

Introduction

Helicobacter pylori (HP) gastritis in adult patients can present with a wide variety of symptoms, but it is not yet clear whether gastritis is related to particular symptoms and which symptoms are most frequently associated.

A relationship between serum antibodies to HP and history of dyspepsia and peptic ulcer has been described in blood donors [1], but in most patients HP gastritis is a clinically silent condition [2]. In children the infection is described in association with symptomatic gastritis [3], chronic diarrhea and malnutrition [4], or protein-losing enteropathy [5]. HP infection is found in 16.8%–24% of children undergoing endoscopy [6–9]. Primary gastritis is associated with the infection in 70%–88% of children [10, 11] and in almost 100% of adults. The aim of this study was to report the clinical manifestations of 153 HP-positive children undergoing endoscopy for various complaints. The relationship between gastric infection and symptoms was reanalyzed 6 months after stopping antimicrobial therapy.

Patients and Methods

Between September 1982 and September 1990, 153 children (M/F 95/58; mean age 10.9 age, range 1–18 years) were diagnosed as having HP-associated gastritis: 31 of them were diagnosed retrospectively by reviewing slides of antral mucosa (1982–1987), 122 were diagnosed prospectively. HP had to be seen on microscopy (Giemsa staining) to diagnose the infection [12].

Gastritis was defined according to the Whitehead criteria [13]. A diffuse mononuclear inflammatory cell infiltrate of the lamina propria was regarded as "quiescent" chronic superficial gastritis (SCG); when polymorphonuclear cells were present in the lamina propria SCG was considered to be "active" [14].

Clinical charts of all children referred for upper gastrointestinal endoscopy were always completed by same investigators (either G.O. or N.A.) a few days before or on the morning of endoscopy. We analyzed: (a) indication for

Gastroenterology Section, Department of Pediatrics, University of Turin, P.za Polonia 94, 10126 Turin, Italy

J.M. Pajares et al. (Eds.)
H. pylori and Gastroduodenal Pathology

endoscopy – the clinical manifestation or symptom the family practitioner requested endoscopy for; (b) most prominent symptom – the symptom children or mothers were complaining of as the most severe or interfering with activity; (c) all symptoms reported; (d) initial symptom; (e) duration of symptoms prior to endoscopy; (f) presence of a sudden acute onset of dyspepsia, (g) family history of peptic ulcer or dyspepsia.

A symptom score was defined as previously reported [17]. Relationships between severity, kind and duration of symptoms, and severity of endoscopic and histologic changes were analyzed.

Antimicrobial therapy was suggested to all, but only 85 of those who complied with it consented to a follow-up endoscopic control 6 months after stopping treatment either with amoxicillin alone (17 children) or with amoxicillin plus tinidazole (68 children). Clinical charts and laboratory findings were again analyzed 6 months after stopping treatment in relation to eradication or persistence of the infection.

Paired and unpaired parametric data were analyzed by the appropriate *t* test, correlations by Spearman's rank and Pearson's linear correlation coefficient, and differences between groups by chi-square test. Results are given as mean $\pm$ 1 SD. A *p* value of less than 0.05 was accepted as significant.

Results

Indication for endoscopy was not always the most prominent symptom. Although recurrent abdominal pain was present in 85% of children, it was an indication for endoscopy in only 43%. Nevertheless, abdominal pain was the most common indication for endoscopy and was reported to be nocturnal and upon waking in 52 children. The second indication for endoscopy was to obtain a duodenal biopsy because of a failure to thrive and suspected celiac disease (14%). Endoscopy was performed in 18 siblings of infected children in a family study, even though ten of them were asymptomatic (Table 1).

The majority of children complained of more than one symptom, only 28 children complained of pain without additional symptoms. Most common associations of symptoms were abdominal pain with vomiting mimicking gastroesophageal reflux (66 children: 14 of them were also failing to thrive or suffered weight loss), or with recurrent diarrhea mimicking irritable bowel syndrome (23 children), or with anorexia (eight children). Anorexia, which was so severe as to be associated with failure to thrive or weight loss in 31 children, was associated with easy satiability and post-prandial epigastric fullness in 35. Recurrent diarrhea and failure to thrive mimicking celiac disease were seen in nine children. Hematemesis was associated with recurrent vomiting in four or with abdominal pain in four children. No symptoms were reported by 14 children even though one of them had duodenal ulcer, and seven severe active SCG.

A sudden acute onset of dyspepsia (abdominal pain, vomiting, and diarrhea of 2–4-day duration) was detected in 26 children 1–96 months prior to endoscopy (median 5.5).

Table 1. Clinical manifestations of 153 children with *Helicobacter pylori* gastritis

	Indication for endoscopy		Most prominent symptom		All symptoms reported		Initial symptoms	
	(*n*)	(%)	(*n*)	(%)	(*n*)	(%)	(*n*)	(%)
Recurrent abdominal pain	66	43	96	63	129	84	89	58
Failure to thrive	21	14	12	8	31	20	9	6
Recurrent vomiting	15	10	11	7	75	49	30	20
Hematemesis	7	4	3	2	8	5	2	1
Anorexia	3	2	6	4	66	43	4	3
Family study	18	11						
Others: follow up for	Gastroesophageal reflux	14	Abdominal distension	4	Nausea	45	Diarrhea	2
	Gluten challenge	5	Headache	2	Halitosis	37	Asthma	1
	Food allergy	2	Asthma	3	Headache	28	Headache	1
	Protein-losing enteropathy	1	Asthenia	1	Diarrhea	27	Eczema	1
	Abdominal distension	1	Nausea	1	Burping	18		

Mean symptom duration was 36.7 ± 27 months (range 1–156), and it was significantly related to the severity of symptoms ($r = 0.317$, $p < 0.01$) but not to age. Symptoms were significantly less severe in older children (Fig. 1); patients without symptoms were the oldest, and those with diarrhea and failure to thrive or severe anorexia were the youngest (13.4 years vs. 7.2; $p < 0.001$). Children with recurrent vomiting were significantly younger (9.8 years vs. 12.5 for nonvomiting children; $p < 0.05$). A positive family history of dyspepsia or peptic ulcer was found in 72% of children.

Table 2. Endoscopic and histological findings in 153 children with *Helicobacter pylori* gastritis

	Esophagus		Stomach		Duodenum	
	Finding	(*n*)	Finding	(*n*)	Finding	(*n*)
Endoscopy						
	Normal	68	Normal	45	Normal	91
	Hyperemia	41	Nodularity	61	Ulcers	26
	Erosions	17	Hyperemia	28	Hyperemia	17
	Nodularity	8	Ulcers	10	Erosions	12
	Sliding hernia	8	Erosions	9	Nodularity	7
	Barrett's esophagus	5				
	Ulcers	6				
Histology[a]						
	Normal	92	Normal	1	Normal	56
	Peptic esophagitis	30	Active SCG	52	Duodenitis	27
	Columnar metaplasia	5	Quiescent SCG	100	Gastric metaplasia	14

[a]Biospy specimens of esophageal or duodenal mucosa were taken only when some changes were seen at endoscopy

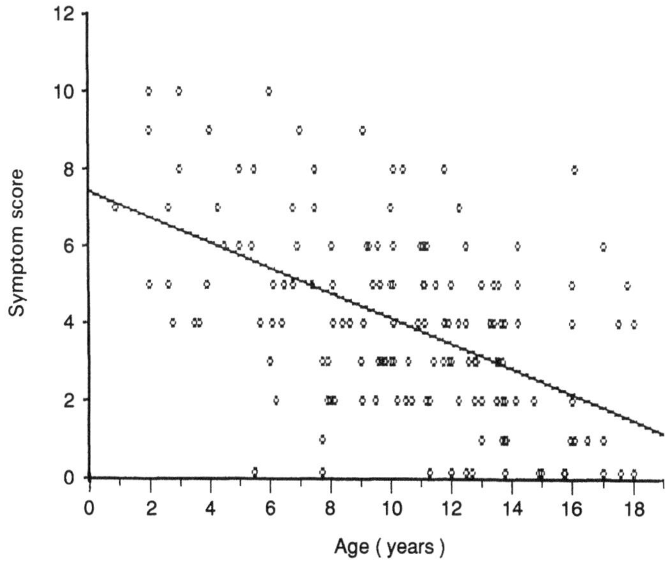

Fig. 1. Symptom score according to age. Symptoms were significantly less severe in older children ($r = -.497$, $p < .001$)

Endoscopy showed that the most common mucosal lesion of the stomach was nodularity of the antral area, but the stomach was frequently of normal appearance; ulcers were seen in the antral area in seven and at the greater curvature in three patients. The esophagus showed a wide range of lesions in half of the children. The duodenum was damaged less frequently, ulcers at the duodenal bulb were single in 17 and multiple in nine patients. Histology showed SCG in 152 out of 153 children; it was active in 52. Peptic esophagitis and duodenitis were infrequent (Table 2). The duration of symptoms was significantly shorter in active SCG (mean 20 ± 29 months vs. 50 ± 39 months in quiescent SCG; $p < 0.001$). No relationship was found between severity, duration, or kind of symptoms and endoscopic findings.

Six Months after Therapy

Helicobacter pylori was eradicated in 72 children (84.7%), all symptoms had disappeared in a mean of 12 ± 10 days in 61 out of 72; recurrence of mild pain without additional symptoms was reported by the others. Out of 13 children with persistent infection, symptoms were improved in five, unchanged in three, worsened in two; in three no symptoms were reported either before or after therapy.

Endoscopy showed a normal antrum in all HP-negative children. A nodular antrum or erosive gastritis were unchanged in children with persistent infection; relapse of duodenal ulcer was seen in one of them. Histology showed the disappearance of active SCG in all HP-negative children; mild quiescent SCG was still present in 61% of them. In HP-positive children, SCG was improved in 38% and was unchanged in the others.

Discussion

About one third of children undergoing endoscopy for abdominal pain are HP carriers [9, 10]. Although abdominal pain or symptoms of dyspepsia are reported in children with HP gastritis [3, 4, 7, 18], the prevalence and type of symptoms in large series of children are not yet known.

The majority of our patients were symptomatic: they complained of abdominal pain of very long duration, which was frequently associated with other symptoms. Associations of pain with recurrent vomiting or recurrent diarrhea were common, mimicking gastroesophageal reflux or irritable bowel syndrome. They were not considered an indication for endoscopy until they became severe or long enough to interfere with activity. The most prominent symptom and indication for endoscopy were not always the same due to different way of evaluating them by mothers and physicians.

Our data showed that symptoms were significantly less severe in older children: asymptomatic children were the oldest, whereas those of 4 years of age or younger had the most severe clinical picture. This suggests that with increasing age a sort of clinical tolerance develops to HP gastritis, which is a clinically silent condition in 62% of adult blood donors with serum antibodies against HP [1]. In children the positivity of serum antibodies against HP is related to symptoms in 50% [19]. Children reported to have severe symptoms such as protein-losing enteropathy [5] were 4 years old or younger. Chronic diarrhea and failure to thrive were more frequent in younger children. Malnutrition was usually mild, it was moderate only in four children with concomitant celiac disease and it was presumably due to the latter. In Gambian children diarrhea and malnutrition are frequently associated with HP infection [4], but it is not known whether malnutrition is caused by or predisposes to HP colonization.

With regard to the relationship between type of gastritis and duration of symptoms, we found a significantly shorter duration in children with active SCG. This suggests that SCG, starting as active, progresses in time to quiescent SCG.

A relationship between HP gastritis and symptoms was further suggested by their disappearance soon after treatment when HP was eradicated. We chose to analyze clinical data after 6 months, even though more children had an endoscopic control 1 month after treatment, because the symptoms they complained of were usually recurrent and 1 month could have been too soon to evaluate a true disappearance. Eradication of HP was associated with resolution of symptomatology in the majority of children.

References

1. Wyatt JI, Rathbone BJ (1989) The role of serology in the diagnosis of Campylobacter pylori infection. Scand J Gastroenterol 24 [Suppl 160:23–34
2. Andersen LP, Elsborg L, Justesen T (1988) Campylobacter pylori in peptic ulcer disease. III. Symptoms and paraclinical and epidemiologic findings. Scand J Gastroenterol 23:347–350
3. Czinn SJ, Dahms BB, Jacobs GH, Kaplan B, Rothstein FC (1986) Campylobacter-like organisms in association with symptomatic gastritis in children. J Pediatr 109:80–83
4. Sullivan PB, Thomas JE, Wight DGD et al. (1990) Helicobacter pylori in Gambian children with chronic diarrhoea and malnutrition. Arch Dis Child 65:189–191
5. Hill ID, Sinclair-Smith C, Lastovica AJ, Bowie MD, Emms M (1987) Transient protein losing enteropathy associated with acute gastritis and Campylobacter pylori. Arch Dis Child 62:1215–1219
6. Glassman MS, Schwarz SM, Medow MS et al. (1989) Campylobacter pylori-related gastrointestinal disease in children. Incidence and clinical findings. Dig Dis Sci 34:1501–1504
7. Kilbridge PM, Dahms BB, Czinn SJ (1988) Campylobacter pylori-associated gastritis and peptic ulcer disease in children. Am J Dis Child 142:1149–1152
8. Mahony MJ, Wyatt JI, Littlewood JM (1988) Campylobacter pylori associated gastritis. Arch Dis Child 63:654–655
9. Hill R, Pearman J, Worthy P et al. (1986) Campylobacter pyloridis and gastritis in children. Lancet i:387
10. Drumm B, Sherman P, Cutz E, Karmali M (1987) Association of Campylobacter pylori on the gastric mucosa with antral gastritis in children. N Engl J Med 316:1557–1561
11. Drumm B, Perez-Perez GI, Blaser MJ, Sherman PM (1990) Intrafamilial clustering of Helicobacter pylori infection. N Engl J Med 322:359–363
12. Barthel JS, Everett ED (1990) Diagnosis of Campylobacter pylori infections: the "gold standard" and the alternatives. Rev Infect Dis 12 [suppl 1]:S107–S114

13. Whitehead R, (1985) Mucosal biopsy of the gastrointestinal tract. In: Bennington JL (ed.) Major problems in pathology, vol 3. Saunders, Philadelphia, pp 47–58
14. Oderda G, Vaira D, Dell'Olio D, et al. (1985) Serum Pepsinogen I and gastrin concentrations in children positive for Helicobacter pylori. J Clin Pathol 1990 43:762–765
15. Oderda G, Altare F, Dell'Olio D, Ansaldi N (1988) Prognostic value of serum pepsinogen I in children with peptic ulcer. J Pediatr Gastroenterol Nutr 7:645–650
16. Oderda G, Vaira D, Holton J, Dowsett JF, Ansaldi N (1989) Serum Pepsinogen I and IgG antibody to Campylobacter pylori in non-specific abdominal pain in childhood. Gut 30:912–916
17. Oderda G, Forni M, Farina L, Dell'Olio D, Ansaldi N (1987) Duodenitis in children: clinical, endoscopic and pathological aspects. Gastrointest Endosc 33:366–369
18. Cadranel S, Goossens H, De Boek M et al. (1986) Campylobacter pyloridis in children. Lancet i:735–736
19. Thomas J, Eastham EJ, Elliott TSJ, Dobson CM, Jones DM (1988) Campylobacter pylori gastritis in children-a common cause of symptoms? Gut 29:A707

Childhood Peptic Ulcer Can Be Cured by Eradicating *Helicobacter pylori*

G. Oderda[1], M. Forni[2], I. Morra[2], F. Altare[2], D. Dell'Olio[1], and N. Ansaldi[1]

Introduction

Since fiberoptic endoscopes of small caliber have been available in the last 2 decades, peptic ulcers are diagnosed more frequently in children. Little is known on the long-term prognosis for pediatric peptic ulcer disease. A high incidence of morbidity persisting into adulthood has been reported, with a 47% endoscopically proven relapse rate [1] and a 70% recurrence of symptoms [2]. In adult patients the history of peptic ulcer is now changing after several reports showed the possibility of decreasing relapse rate by *Helicobacter pylori* eradication [3–6]

The aim of this study, begun in 1982, was to evaluate the healing rate of peptic ulcer in children after ranitidine and the endoscopically proven relapse rate after stopping the drug. After 1987 antibiotic treatment was added to ranitidine where ulcers were associated with *H. pylori* infection.

Patients and Methods

Since 1982 all children with gastric or duodenal ulcer diagnosed by endoscopy in our unit were treated with ranitidine (5–10 mg/kg in two divided doses) for 8 weeks, and endoscopy was repeated. In children with healed ulcers ranitidine was discontinued. The relapse rate was evaluated by repeating endoscopy every 6 months for 5 years or when symptoms recurred. In children with nonhealed peptic ulcer, ranitidine was continued at the same dosage for four more months and endoscopy repeated. When ulcers were healed, ranitidine was discontinued and endoscopy repeated every 6 months or at symptom recurrence. Relapses were managed by a second 8-week course of ranitidine, followed by continous maintenance with ranitidine at half dosage (2.5–5 mg/kg at bedtime).

During endoscopy biopsy specimens of esophageal, antral, and duodenal mucosa were taken for histologic examination after hematoxylin and Eosin staining. Gastritis was defined according to the Whitehead criteria. Blood was drawn, after an overnight fast, on the morning of endoscopy and serum

[1] Gastroenterology Section, Department of Pediatrics, University of Turin, P.za Polonia 94, 10126 Turin, Italy
[2] Histopathology Department, "Regina Margherita" Children's Hospital, 10126 Turin, Italy

J.M. Pajares et al. (Eds.)
H. pylori and Gastroduodenal Pathology

pepsinogen I and gastrin were measured by radio-immunoassay (RIA) as previously described [7].

In 1987 the rapid urease test (CLO test, Delta West Ltd., Western Australia) and culture media for *H. pylori* became available in our unit, and antral biopsies (Giemsa staining) were evaluated for *H. pylori* presence. Where antral gastritis and *H. pylori* infection was found, a course of amoxycillin (50 mg/kg in two divided doses) plus tinidazole (20 mg/kg in two divided doses) were added to ranitidine for the first 6 (21 patients or 4 weeks (15 patients) [8]. Antibiotics were also given to patients with peptic ulcer diagnosed before 1987 when *H. pylori* gastritis was seen retrospectively by reviewing slides of antral mucosa taken at first endoscopy.

Results

From 1982 to 1990 a total of 65 children with peptic ulcer were enrolled (M/F 43/22; median age 9.3 years, age range 1–18 years; Table 1).

Before 1987 the healing rate after 8 weeks of ranitidine was 94% (36 out of 38); after 1987, ranitidine with antibiotics in *H. pylori*-positive children or without antibiotics in *H. pylori*-negative children healed 96% of patients (26 out of 27) without any differences between *H. pylori*-positive or *H. pylori*-negative children (Table 2). In the remaining three patients, ulcers healed with ranitidine given for four more months. We did not find any nonresponders to ranitidine. In children with peptic ulcer healed with ranitidine, who were followed for 1 year or longer before 1987, the relapse rate was 47% (14 out of 30 patients), 11 relapses

Table 1. Type of peptic ulcers diagnosed in 65 children between 1982 and 1990

Sex	Duodenal		Gastric		Duodenal + gastric		Total	
	HP +	HP −	HP +	HP −	HP +	HP −	HP +	HP −
Males	20	10	5	6	2		27	16
Females	6	9	3	3		1	9	13
Total	45		17		3		36	29

HP +, *H. pylori* positive; HP −, *H. pylori* negative

Table 2. Type of ulcers, healing, and relapse rate of 58 children followed by endoscopy for 1 year or longer (mean follow-up time 3 years)

Sex	Healed at 8 weeks	*H. pylori* eradicated	Relapsed at 1 year: *H. pylori* Persistent	Eradicated	Negative
Males	41/43	23/27	2/4	0/22	0/13
Females	21/22	8/9	0/1	0/9	0/10
Total	62/65	31/36	2/5	0/31	0/23

were symptomless. No patients with gastric ulcer relapsed, but four patients with duodenal ulcer on admission had gastric ulcers after relapse. A second course of ranitidine followed by continuous maintenance prevented further relapses. Relapses were significantly more frequent in patients with *H. pylori* seen retrospectively in antral mucosa ($p < 0.001$). Mean time before relapse was 17 months (range 4–36 months). After 1987, antibiotics were given to relapsed patients and ranitidine maintenance discontinued. Antibiotics were given also to seven *H. pylori*-positive non-relapsed children and to 16 *H. pylori*-positive patients identified among 27 new peptic ulcer patients diagnosed since 1987. *H. pylori* was eradicated in 86%. Out of 58 children followed for 1 year or longer (mean follow-up time 3 years) duodenal ulcer relapsed in two out of five children with persistent *H. pylori* infection (Table 2). No relapses were seen in *H. pylori*-negative children.

Serum pepsinogen I levels on admission were significantly higher in *H. pylori*-positive patients (81.7 ± 28.9 vs. 47.4 ± 21.8 ng/ml in *H. pylori*-negative ones) and decreased significantly after *H. pylori* eradication ($p < 0.001$), whereas they were unchanged after ranitidine therapy in *H. pylori*-negative children. Gastrin levels were similar in the *H. pylori*-positive and -negative children (38.6 ± 18 μU/ml) before treatment, but significantly decreased in *H. pylori*-positive children after *H. pylori* eradication ($p < 0.01$). In all *H. pylori*-positive children histology showed superficial chronic gastritis of the antral mucosa which was active (with polymorphonuclear infiltration) in 42% of them. Mild superficial chronic gastritis without polymorphonuclear infiltration was seen in 30% of *H. pylori*-negative patients. Six months after *H. pylori* eradication, chronic gastritis was improved in 74% and healed in the others; it was unchanged in patients with persistent infection.

Discussion

The first part of our study (1982–1987) showed that ranitidine was very effective in peptic ulcer disease in children. The healing rate was 96% at 8 weeks and 100% at 6 months. Nonresponders to ranitidine were not found in our pediatric population. The relapse rate after ranitidine was discontinued was 47%, lower than it is in adults after H2 receptor antagonists [9], and no relapses were seen on maintenance therapy. It is interesting to note that four patients with duodenal ulcers at first endoscopy developed gastric ulcers when relapsed, suggesting that in the same patient *H. pylori* can cause both gastric or duodenal ulcers.

When we looked for factors predictive of relapse, we found that, where ulcers relapsed, children had higher pretreatment levels of serum pepsinogen I [7], and superficial chronic gastritis of the antral mucosa was more frequent and more severe than in nonrelapsed children. Since we had found that children with *H. pylori*-positive gastritis had higher serum levels of pepsinogen I than *H. pylori*-negative children [10], we reviewed the slides of the antral mucosa and found that all children who relapsed were *H. pylori* positive. Therefore we suspected that infection favored relapse. We tried to eradicate the infection in all *H. pylori*-positive children, both relapsed and nonrelapsed, discontinued ranitidine maintenance and repeated endoscopy every 6 months.

The second part of our study (1987–1990) showed that *H. pylori* eradication minimized the relapse rate. No relapses were seen in *H. pylori*-negative children, but out of five patients with persistent or recurrent infection two male patients with duodenal ulcer relapsed after 1 and 3 years, respectively. *H. pylori*-positive duodenal ulcers were more frequent in males, whereas in females duodenal ulcers were more frequently *H. pylori* negative and none relapsed. This finding can explain why in young females the clinical history of peptic ulcer is usually more favorable.

Our results are similar to those reported in adults. After *H. pylori* eradication with colloidal bismuth subcitrate (CBS) plus antibiotics, the relapse rate has been reported to be very low [3–5], but, since in adult studies CBS was always used, it is not yet clear whether this result is due to *H. pylori* eradication or to CBS-induced stimulation of prostaglandin production. In our study CBS was not used due to its possible toxicity in this age group and we can conclude that *H. pylori* eradication by itself minimized the relapse rate and cured peptic ulcer disease.

References

1. Murphy MS, Eastham EJ (1987) Peptic ulcer disease in childhood: long term prognosis. J Pediatr Gastroenterol Nutr 6:721–724
2. Murphy MS, Eastham EJ, Jimenez M, Nelson R, Jackson RH (1987) Duodenal ulceration: a review of 110 cases. Arch Dis Child 62:554–558
3. Coghlan JG, Gilligan D, Humphries H et al. (1987) Campylobacter pylori and recurrence of duodenal ulcers-a 12 month follow-up study. Lancet ii:1109–1111
4. Marshall BJ, Goodwin CS, Warren JR et al. (1988) Prospective double-blind trial of duodenal ulcer relapse after eradication of Campylobacter pylori. Lancet ii:1437–1441
5. Raws EAJ, Tytgat GNJ (1990) Cure of duodenal ulcer with eradication of Helicobacter pylori. Lancet 335:1233–1235
6. Tatsuda M, Ishikawa H, Iishi H, Okuda S, Yokota Y (1990) Reduction of gastric ulcer recurrence after suppression of Helicobacter pylori by cefixime. Gut 31:973–976
7. Oderda G, Altare F, Dell'Olio D, Ansaldi N (1988) Prognostic value of serum Pepsinogen I in children with peptic ulcer. J Pediatr Gastroenterol Nutr 7:645–650
8. Oderda G, Vaira D, Holton J et al. (1989) Amoxycillin plus tinidazole for Campylobacter pylori gastritis in children: assessment by serum IgG antibody, pepsinogen I and gastrin levels. Lancet i:690–692
9. Jones DB, Howden CW, Burget DW, Kerr GD, Hunt RH (1987) Acid suppression in duodenal ulcer: a meta-analysis to define optimal dosing with antisecretory drugs. Gut 28:1120–1127
10. Oderda G, Vaira D, Holton J, Dowsett JF, Ansaldi N (1989) Serum pepsinogen I and IgG antibody in non-specific abdominal pain in childhood. Gut 30:912–916

Helicobacter pylori Infection and Gastroduodenal Pathology in Children

G. Oderda

Immediately after Warren and Marshall's description in 1983 of an association between spiral bacteria and antral gastritis in adults, the disease was reported in children [1–3]. Subsequent studies showed that *Helicobacter pylori* was associated specifically with primary or unexplained gastritis [4], suggesting that the organism was a cause of gastric inflammation and was not just colonizing an already inflamed gastric mucosa. Several reports also suggested a strong association between *H. pylori* gastritis and duodenal ulcer in children [5–7].

Children represent a particularly important population group who may help to clarify our understanding of *H. pylori* infection. They seldom smoke, are not heavy alcohol drinkers, and their use of steroids or nonsteroidal anti-inflammatory drugs is usually modest. All these factors can worsen endoscopic and histologic pictures of the upper gastrointestinal tract and complicate our understanding of the mechanisms by which *H. pylori* causes gastric inflammation. Nodular antritis, for example, is a peculiar endoscopic picture reported in children. It does not seem to be present later in life as it has never been reported in adults. Nodules may have various diameters so that the mucosa may appear pseudopolypoid or micronodular. This endoscopic picture is quite specific for *H. pylori* infection: up to 90% of patients with nodular antritis are colonized by the organism [1, 3, 8]. Histologic examination of nodular mucosa reveals a high degree of inflammatory cell infiltrate consisting mainly of mononuclear cells, neutrophils, eosinophils, and an increased number of lymphoid follicles. This may be the early stage of gastric inflammation after colonization.

Children do not usually carry the infection for decades, therefore to study the distribution of the infection in this age group in different countries may help us to understand the actual distribution of the infection. Indeed, studies carried out in children have already shown that the hygienic conditions of the country are crucial factors in the spread of the infection. In Western countries the prevalence of children colonized with *H. pylori* is low and varies from 6% to 13% [6, 9], but a higher proportion of infected children are reported from developing countries. In Gambia 46% of children have high titers of serum antibodies to *H. pylori* by the age of 5 years [10]. In Peru the ^{13}C-urea breath test showed a prevalence of infection in 48% of children by the age of 10 years and was highly correlated with the socioeconomic status. The highest prevalence of infection was found among children living in households served by water from a community well, showing that water could be the source of the infection [11]. Once acquired, the

Gastroenterology Section, Department of Pediatrics, University of Turin, Turin, Italy

J.M. Pajares et al. (Eds.)
H. pylori and Gastroduodenal Pathology

infection appears to persist probably for life, but the mode of transmission to humans is still unknown. Useful informations came from studies in families of infected children. Despite the variability of the method used or the different countries where the studies were performed, infection rates among parents and siblings of infected children are significantly higher than those found in the normal population in the same age range [6, 12, 13]. Intrafamilial clustering of *H. pylori* infection suggests that person-to-person transmission within the family is likely.

Studies in children have also shown a correlation between *H. pylori* infection and symptoms, this is not yet well defined in adults. The organism is found in 10%–24% of children undergoing endoscopy for various complaints [14, 15], but the prevalence of the infection increases to 62% in children undergoing endoscopy for recurrent abdominal pain [16].

Studies on peptic ulcer disease in children showed some important differences between *H. pylori*-associated duodenal ulcer and ulcer without the infection [17, 18]. Treatment of peptic ulcer with an antibiotic combination but without a bismuth compound significantly lowered the relapse rate among children in whom *H. pylori* was eradicated [7]. These data helped to clarify the fact that the low relapse rate is due to *H. pylori* eradication rather than to bismuth-induced stimulation of prostaglandin production.

Furthermore, important information that could clarify the involvement of *H. pylori* in the pathogenesis of gastritis and/or duodenal ulcer came from studies performed in children. Gastric glycerolipid (as a receptor for the microorganism) has been shown to be present in smaller amount in the infant stomach [19]; bacterial adhesion to specialized attachment sites represents an important aspect of pathogenicity. A decreased histamine concentration in oxyntic mucosa was found in children with *H. pylori* gastritis compared to controls. The concentration was even lower in children with *H. pylori*-associated duodenal ulcer and was supposed to be due to increased mobilization of endogenous mucosal histamine with store depletion [20].

Histamine has long been recognized as a gastric secretagogue, and its increased release may lead to a low intraduodenal pH, thus preventing mucosal reparative processes.

Stimulation of gastric acid secretion due to increased gastrin release has been suggested to be another mechanism linked to *H. pylori* in the pathogenesis of duodenal ulcer disease, and it has been shown that the serum gastrin concentration in children with *H. pylori* gastritis fell after eradication of *H. pylori* by antibiotics [21].

Pepsin I is known to be increased in duodenal ulcer patients, and pepsinogen I levels are increased in their serum. This phenomenon was believed to have a genetic origin, but studies in children have shown that serum pepsinogen I levels are related to the severity of gastric inflammation associated with *H. pylori* [22] and decrease after eradication [21]. A family study in children with peptic ulcer has shown that increased serum pepsinogen I levels in relatives are related to *H. pylori* infection acquired by the relatives themselves [18]. Recently *H. pylori* has been shown to stimulate pepsin secretion from rabbit gastric glands [23], but this study did not clarify whether there is an actual increase in pepsinogen secretion or rather a diversion of normal secretion, and whether *H. pylori* eradication would be followed by normalization of pepsinogen secretion. Again, the answer to these questions came from studies on children. The concentration of pepsin in antral mucosa of children with *H. pylori* gastritis is significantly lower as compared to controls, but pepsinogen stores are similar. Serum pepsinogen I levels are inversely correlated with mucosal pepsin concentration, suggesting an abnormal leakage of pepsin from damaged chief cells, without an actual increase in pepsinogen production. Serum pepsinogen I and mucosal pepsin return to control levels after *H. pylori* eradication (personal observation).

In conclusion, children are a good model to study *H. pylori* gastritis. Even if the infection is less frequent in this age group and the series of children studied are usually small, their gastroduodenal mucosa is seldom damaged by factors other than *H. pylori* infection, and mechanisms by which *H. pylori* causes gastritis are easier to study in this population group.

References

1. Czinn SJ, Dahms BB, Jacobs GH, Kaplan B, Rothstein FC (1986) Campylobacter-like organisms in association with symptomatic gastritis in children. J Pediatr 109:80–83
2. Hill R, Pearman J, Worthy P et al. (1986) Campylobacter pyloridis and gastritis in children. Lancet i:387
3. Cadranel S, Goossens H, De Boeck M et al. (1986) Campylobacter pyloridis in children. Lancet i:735–6
4. Drumm B, O'Brian A, Cutz E, Sherman P (1987) Campylobacter pyloridis associated with primary gastritis in children. Pediatrics 80:192–195
5. Kilbridge PM, Dahms BB, Czinn SJ (1988) Campylobacter pylori associated gastritis and peptic ulcer disease in children. Am J Dis Child 142:1149–1152
6. Drumm B, Perez-Perez GI, Blaser MJ, Sherman PM (1990) Intrafamilial clustering of Helicobacter pylori infection. N Engl J Med 322:359–363
7. Oderda G, Forni M, Dell'Olio D, Ansaldi N (1990) Cure of peptic ulcer associated with eradication of Helicobacter pylori. Lancet 335:1599
8. Oderda G, Lerro P, Poli E, Tavassoli K, Ansaldi N (1988) Childhood nodular antritis and Campylobacter pylori. Endoscopy 20 [Suppl II]:86
9. Thomas JE, Eastham EJ, Elliott TSJ, Dobson CM, Jones DM (1988) Campylobacter pylori gastritis in children – a common cause of symptoms? Gut 29:A707
10. Sullivan PB, Thomas JE, Wight DGD et al. (1990) Helicobacter pylori in Gambian children with chronic diarrhoea and malnutrition. Arch Dis Child 65:189–191
11. Klein PD, GI Physiology Working Group, Graham DY et al. (1991) Water source as a risk factor for Helicobacter pylori infection in Peruvian children. Lancet 337:1503–1506
12. Oderda G, Boero M, Ponzetto A, Bellis D, Ansaldi N (1988) Campylobacter pylori in families of children with relapsing gastroduodenal disease due to C pylori. Am J Gastroenterol 1988;83:1437–1438
13. Malaty HM, Graham DY, Klein PD et al. (1991) Transmission of Helicobacter pylori infection. Studies in families of healthy individuals. Scand J Gastroenterol 26:927–932
14. Glassman MS, Schwartz SM, Medow MS et al. (1989) Campylobacter pylori related-gastrointestinal disease in children. Am J Dis Child 34:1501–1504
15. Mahony MJ, Wyatt JI, Littlewood JM (1988) Campylobacter pylori gastritis. Arch Dis Child 63:654–655
16. Oderda G, Vaira D, Holton J, Dowsett JF, Ansaldi N (1989) Serum Pepsinogen I and IgG antibody to Campylobacter pylori in non-specific abdominal pain in children. Gut 30:912–916
17. Hassal E, Dimmick JE (1991) Unique features of Helicobacter pylori disease in children. Dig Dis Sci 36:417–423

18. Oderda G, Vaira D, Holton J et al. (1991) Helicobacter pylori in children with peptic ulcer and their families. Dig Dis Sci 36 : 572–576
19. Lingwood CA, Law H, Pellizzari A, Sherman P, Drumm B (1989) Gastric glycerolipid as a receptor for Campyolobacter pylori. Lancet 11 : 238–241
20. Querioz DMM, Mendes EN, Rocha GA, Barbosa AJA, Carvalho AST, Cunha-Melo JR (1991) Histamine concentration in gastric mucosa in Helicobacter pylori positive and negative children. Gut 32 : 464–466
21. Oderda G, Vaira D, Holton J et al. (1989) Amoxycillin plus tinidazole for Campylobacter pylori gastritis in children: assessment by serum IgG antibody, pepsinogen I, and gastrin levels. Lancet i : 690–692
22. Oderda G, Vaira D, Dell'Olio D et al. (1990) Serum Pepsinogen I and gastrin concentrations in children positive for Helicobacter pylori. J Clin Pathol 43 : 762–765
23. Cave TR, Cave DR (1991) Helicobacter pylori stimulates pepsin secretion from isolated rabbit gastric glands. Scand J Gastroenterol 26 [Suppl 181]: 9–14

X. Treatment

Solubility Characteristics of Different Bismuth Salts in Gastric Juice

G. Bode[1], P. Malfertheiner[1], S. Härle[1], V. Krivan[2], and H. Ditschuneit[1]

Bismuth preparations are used for the treatment of various gastrointestinal disorders [1–3]. Various bismuth salts have recently become available for the treatment of *Helicobacter pylori* related gastroduodenal diseases [4–6]. Despite several beneficial effects of bismuth on the gastroduodenal mucosa, the antibacterial effect against *H. pylori* appears the most important. Relatively little is known about the mechanisms of bismuth in vivo [7, 8].

Most bismuth salts are insoluble in water and when ingested they are converted in the stomach into different insoluble bismuth complexes such as bismuth oxide, bismuth hydroxide, bismuth oxychloride, and to a variety of yet unknown complexes [9, 10]. The precise metabolism of these bismuth salts is not known. Part of the compounds generated seem to act topically on the inflamed tissue [11], whereas other bismuth complexes may interfere with *H. pylori* by inhibiting growth [12].

Little is known on the nature and fate of bismuth salts after exposure to the gastric juice. We therefore investigated the solubility of various bismuth salts in human gastric juice in vitro and tried to characterize the generated bismuth complexes by ultrastructural means.

Material and Methods

Gastric Juice Preparation

Gastric juice was sampled and pooled from patients during routine gastroscopy and freed from coarse particles by centrifugation (10 min; 3000 rpm, 4 °C). As the application of an antifoaming detergent normally used during endoscopic manipulation influenced the solubility of bismuth salts in an unexpected manner, this compound was omitted in the presented study.

Solubility Experiments and Bismuth Analysis

Gastric juice samples were incubated with an oversaturation of bismuth salts for different time intervals (15, 30, 45, 60, and 90 min) and at different pH values (pH

[1] Department of Internal Medicine II, University of Ulm, Robert-Koch-Str. 8, 7900 Ulm, FRG
[2] Department of Physical Chemistry, University of Ulm, Robert-Koch-Str. 8, 7900 Ulm, FRG

J.M. Pajares et al. (Eds.)
H. pylori and Gastroduodenal Pathology

1.5–7.5) at 37°C in a waterbath with continuous shaking. The pH level was adjusted to the desired value with aqueous sodium hydroxide. The following bismuth salts were studied: bismuth subcitrate (BSC_1), bismuth subsalicylate (BSS), bismuth subgallate (BSG), bismuth subcarbonate.(BSC_2), bismuth aluminate (BA), and bismuth subnitrate (BSN). After incubation, samples were centrifuged and passed through a microfilter (0.22 μm). Bismuth concentrations in gastric juice samples were analyzed using an atomic absorption spectrophotometer equipped with a specific bismuth light source after digestion of the sample with a HNO_3 solution in an appropriate concentration to avoid destabilization of samples. Calibration was performed by the addition of a stock solution of defined bismuth concentration to the gastric juice. The lower limit of detection of bismuth by this method is about 1.0 ng/ml. The coefficient of variation was within 5%.

Electron Microscopy

Bismuth salts were investigated by scanning electron microscopy before and during incubation experiments with a Philips EM 500 after air drying and sputter coating. Additionally the negative staining technique (2% phosphotungstic acid, pH 7.0) was used to visualize colloidal bismuth particles in gastric juice after incubation with bismuth salts under the described conditions.

Results

The various bismuth salts are characterized by their different crystal structure as revealed by electron microscopy (Fig. 1) : prismatic structures (BSS, Fig. 1a), as well as tetragonal (BSC_2, Fig. 1d) or concentric formations (BSC_1, Fig. 1e) are detected.

All salts investigated showed a specific pH optimum where the highest absolute solubility (ppm) was detected (Table 1). BSC_1 showed highest solubility of all salts covering a broad pH range (3.0–7.0). BSG and BSC_2 showed highest solubility at normal pH of gastric juice (pH 1.5–2.0), whereas others are more soluble at higher pH values (BSS 4.5–5.5 and BSN 3.0–5.0).

Electron microscopy revealed that BSC_1 generates rather homogenous colloidal particles (5–10 nm) in gastric juice with a tendency to form aggregates at higher pH values (Fig. 2a). The other bismuth salts showed similar particles in size and shape. However, the amounts of these particles differ, depending on the solubility of the salt (Fig. 2b, c).

Using scanning electron microscopy we found morphological alterations only with the BSC_1 preparation after contact with gastric juice. The mean diameter of the spherical crystal particles is smaller after incubation. In addition, small colloidal particles were detected.

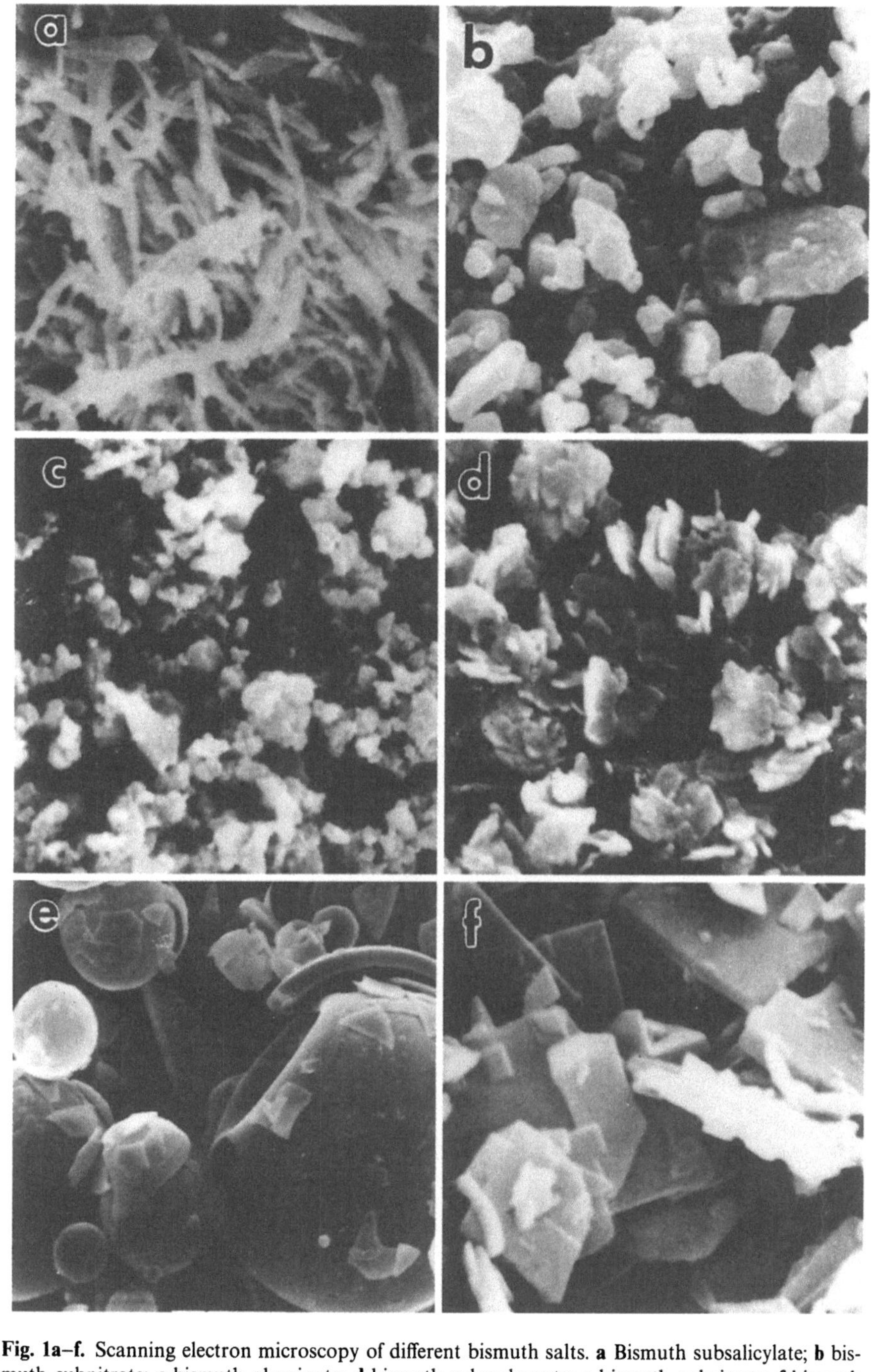

Fig. 1a–f. Scanning electron microscopy of different bismuth salts. **a** Bismuth subsalicylate; **b** bismuth subnitrate; **c** bismuth aluminate; **d** bismuth subcarbonate; **e** bismuth subcitrate; **f** bismuth subgallate. **a–d, f** × 15000; **e** × 1000

Table 1. Solubility characteristics of different bismuth salts in gastric juice

Salt	pH optimum	Solubility at the pH optimum (ppm)	Solubility at pH 1.5–2.0 (ppm)	Solubility at pH 6.5–7.0 (ppm)
BSC_1	3.0–7.0	10 000	10 000	10 000
BSS	4.5–5.5	55	10	25
BSG	1.5–2.0	50	50	2
BSC_2	1.5–2.0	45	45	4
BA	3.0–5.0	22	5	10
BSN	3.0–5.0	10	10	4

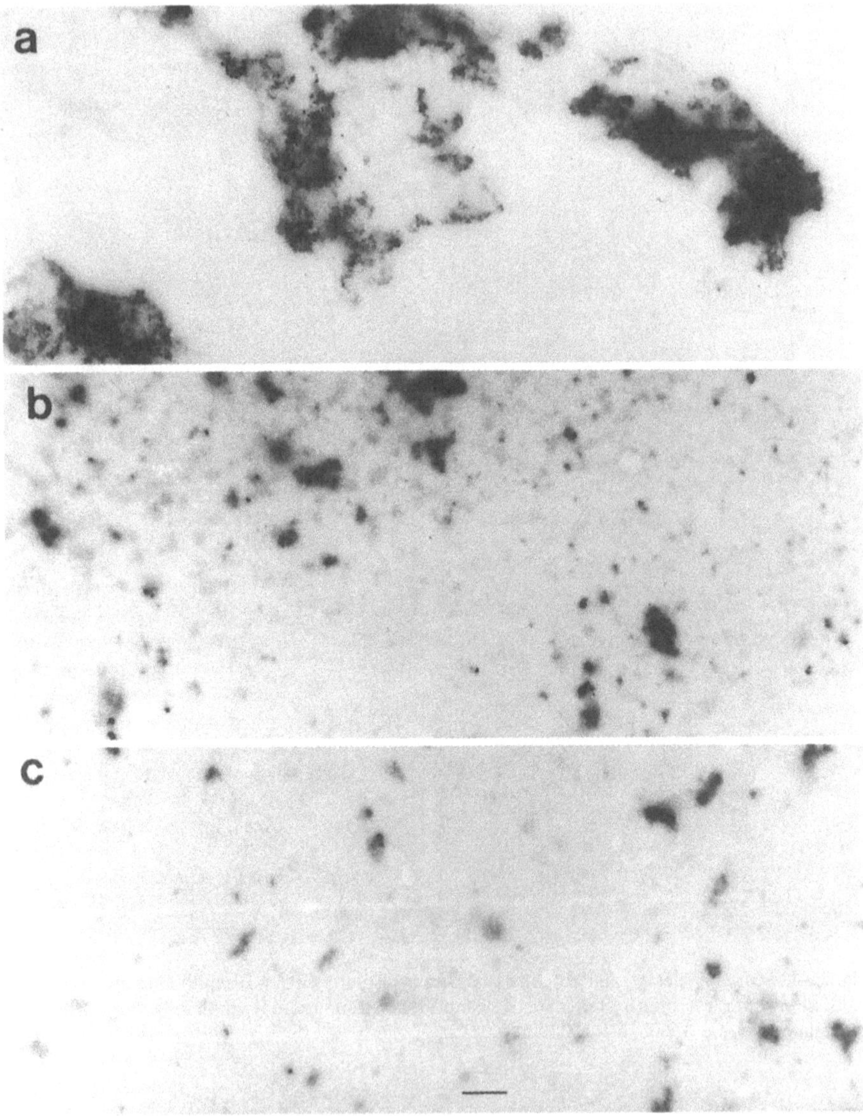

Discussion

These results show that all the investigated bismuth salts have their individual pH optimum in gastric juice regarding solubility. This pH for optimal solubility ranges from BSG, requiring the lowest pH, to BSS with the most alkaline pH.

The reason for the different behavior of bismuth salts is likely to be due to the complexity of chemical structures bismuth compounds are able to form: the formation of polymeric particles depends on the pH of gastric juice and the formation of complexes on the presence of other charged molecules in the gastric juice [13–15]. One part of the generated complexes may react with proteins, peptides, and free amino acids, leading to larger complexes and to precipitation on epithelial lesions. This kind of bismuth coating develops a local antipeptic effect at the ulcer site. This property seems to be strongly pH dependent because precipitated bismuth complexes are partly redissolved if the pH falls below 2. This is supported by investigations of Koo et al. [11], who showed that bismuth citrate has a selective coating affinity for the ulcer base and that this effect is hardly detectable with other commonly used bismuth compounds. The antiulcer effect of bismuth salts is far more complex than simple precipitation within the ulcer crater. The fact that small colloidal bismuth particles are found within *H. pylori* organisms and within epithelial surface cells demonstrates the possiblity that these particles can penetrate the bacterial cell wall and cell membranes. It is assumed that only Bi^{3+} ions are able to pass membranes and accumulate at certain membrane structures where they precipitate with sulfated groups of carbohydrates [16] or with the ATP/ATPase complex [15]. These colloidal bismuth particles may represent the biological active component exerting a bacteriostatic or bactericidal effect on *H. pylori*.

The direct comparison of different bismuth salts in vitro shows marked differences in the minimal inhibitory concentration (MIC) values and may be explained by their different capacity to dissolve. It is also possible that different crystalline structures generated in the gastric juice may also play a role in this respect [17].

The detection of a distinct complex with a well-defined size of 5–10 nm in all investigated salts seems to represent a characteristic property of all investigated bismuth salts. The same colloidal complex is found within the cell membrane of *H. pylori* after contact with these bismuth complexes in vitro (Fig. 3) and may thus represent a common mechanism inherent to bismuth salts.

Some properties of bismuth salts in vitro such as the solubility and the different minimal inhibitory concentrations which were detected with these salts against *H. pylori* in vitro [7] may be explained by the ability to form colloidal complexes.

The efficacy of a bismuth salt against *H. pylori* in gastric juice depends on the availability of such bactericidal bismuth complexes, and further studies will show if the efficacy of these salts can be improved by enhancing their solubility.

Fig. 2a–c. Negative contrast electron microscopy of bismuth particles in gastric juice. **a** Bismuth subcitrate; **b** bismuth subsalicylate; **c** bismuth subnitrate. ×114 000

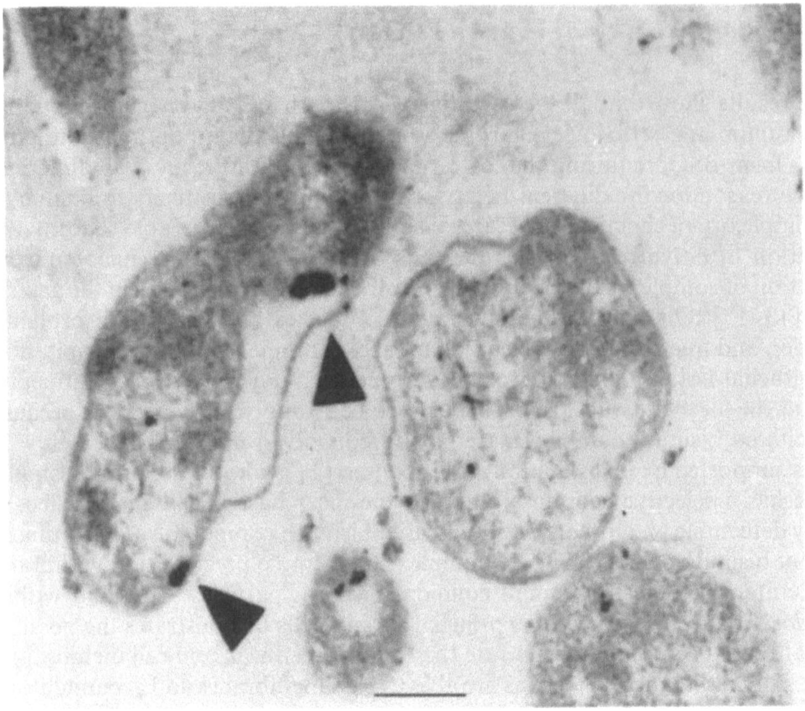

Fig. 3. Ultrathin section of *H. pylori* after exposition to bismuth subsalicylate (32 μg/ml) × 144 000

References

1. Dekker W, Reismak (1979) Double-blind controlled trial with colloidal bismuth subcitrate in the treatment of symptomatic duodenal ulcers, with special references to blood and urine bismuth levels. Ann Clin Res 11:94–97
2. Ericsson CD, Tannenbaum C, Charles TT (1990) Antisecretory and antiinflammatory properties of bismuth subsalicylate. Rev Infect Dis 12:16–20
3. Gorbach SL (1990) Bismuth therapy in gastrointestinal diseases. Gastroenterology 99:863–875
4. Elder JB (1986) Recent experimental and clinical studies on the pharmacology of colloidal bismuth subcitrate. Scand J Gastroenterol 21 [Suppl 122]: 14–16
5. McNulty CAM (1987) The treatment of Campylobacter-associated gastritis. Am J Gastroenterol 82:245–247
6. Kang JY, Tay HH, Wee A, Guan R, Math MV, Yap I (1990) Effect of colloidal bismuth subcitrate on symptoms and gastric histology in non-ulcer dyspepsia. A double blind placebo controlled study. Gut 31:476–480
7. Armstrong JA, Wee SH, Goodwin CS, Wilson DH (1987) Response of Campylobacter pyloridis to antibiotics, bismuth and an acid-reducing agent in vitro – an ultrastructural study. J Med Microbiol 24:343–350
8. Cornick NA, Silva M, Gorbach SL (1990) In vitro antibacterial activity of bismuth subsalicylate. Rev Infect Dis 12:9–10
9. Billon JP, Gernez G, Gourdin JC, Martin C, Palliere M (1976) Quelques caracteristiques physico-chimiques de sels de bismuth utilises en pharmacie. Ann Pharm Fr 34:161–171

10. Pellerin F, Goulle J-P, Dumitrescu D (1977) La formation de complexes et chelates solubles de bismuth avec divers medicaments et aliments: detection par absorption atomique. Ann Pharm Fr 35:281–286
11. Koo J, Ho J, Lam SK, Wang J, Ong GB (1982) Selective coating of gastric ulcer by tripotassium dicitrato bismuthate in the rat. Gastroenterology 82:864–870
12. Sox TE, Olson CA (1989) Binding and killing of bacteria by bismuth subsalicylate. Antimicrob Agents Chemother 33:2075–2082
13. Winship KA (1983) Toxicity of bismuth salts. Adverse Drug React Acute Poisoning Rev 2:103–121
14. Stemmer KL (1976) Pharmacology and toxicology of heavy metals: bismuth. Pharmacol Ther A 1:153–155
15. Paradies HH (1990) Bismuth – from the element to the tablet: the general chemistry of bismuth with relevance to pharmacy and medicine. In: Malfertheiner P, Ditschuneit H (eds) Helicobacter pylori, gastritis and peptic ulcer. Springer, Berlin Heidelberg New York, pp 409–426
16. Konger G, Thomas DJ, Weinhardt F, Hoyer S (1976) Disturbed oxidative metabolism in organic brain syndrome caused by bismuth in skin creams. Lancet II:485–487
17. Bode G, Malfertheiner P, Ditschuneit H (1990) Ultrastrukturelle Untersuchungen an verschiedenen Wismutsalzen. In: Malfertheiner P, Hotz J, Ditschuneit H (eds) Wismut-Therapiekonzept bei Helicobacter pylori, Gastritis und Ulkuskrankheit. Zuckschwerdt, München pp 59–66

Assessment of Susceptibility of *Helicobacter pylori* to Tinidazole, Amoxicillin and Tetracycline Among 555 Patients

D. Famerée[1], B. Ramdani[2], and V. Lamy[2]

Introduction

Most antibiotics are effective in vitro against *Helicobacter pylori* [1–3], but clinical failures have often been observed in our own practice as well as in the literature [4]. We therefore tested the susceptibility of the strains of *Helicobacter pylori* to three currently used antibiotics: tinidazole, amoxicillin, and tetracycline.

Material and Methods

Between February 1988 and October 1990, 721 antral biopsies for *H. pylori* culture were obtained from 555 patients referred to the endoscopic unit. The isolated *H. pylori* strains were subsequently checked for sensitivity to antibiotics. The average age of patients was 50.8 ± 15.1 years (53.2 ± 15.5 years for females, 48.9 ± 15.1 years for males) and the sex ratio male/female was 1.3.

Transport of the biopsies to the laboratory was carried out in a saline solution as soon as possible. Tissue biopsies were crushplated on chocolate agar and Martin-Lewis agar (Becton Dickinson) and cultured at 37 °C under microaerophilic conditions (Anaerocult C mini, Merck). As soon as the culture was positive, susceptibility to antibiotics was checked by a disk diffusion test. The inoculum had a density of no. 0.5 MacFarland Standard and the medium was Muller Hinton medium supplemented with 5% horse blood. Only one antimicrobial disk (Neosensitabs, Rosco) was tested on each plate. Plates were incubated for 3 days at 37 °C under microaerophilic conditions. Inhibition zones were measured and recorded. All the strains were tested for tinidazole susceptibility. Resistant strains were further tested for amoxicillin and tetracycline.

Results

From the initial 555 patients (245 females and 310 males), 75.5% were sensitive to tinidazole (76.7% of females, 74.2% of males).

[1] Microbiological Unit, Hôpital Civil de Jumet, 73 rue de Gosselies, 6040 Charleroi, Belgium
[2] Gastroenterological Unit, Hôpital Civil de Jumet, 73 rue de Gosselies, 6040 Charleroi, Belgium

J.M. Pajares et al. (Eds.)
H. pylori and Gastroduodenal Pathology

Biopsy cultures from 136 tinidazole-resistant patients were further tested for amoxicillin and tetracycline susceptibility; 132 (97%) were sensitive to amoxicillin and 135 (99%) were sensitive to tetracycline. A total of 125 patients (22.5%) have been biopsied several times (two to six). The mean follow-up time was 22 weeks (4–88 weeks). Among these patients 51 (40.8%) showed a change in susceptibility: 43 (34.4%) tinidazole-sensitive patients became resistant when controlled (16 females and 27 males). The mean follow-up time was 20 weeks (4–76 weeks). Eight (6.4%) tinidazole-resistant patients became sensitive when controlled (six women and one man).

Discussion

Our study seems to confirm that the susceptibility of *H. pylori* to the routinely used antibiotics is good. However, resistance seemed to be raised especially to tinidazole, probably because it is the most widely used in our country. Therefore we think it is important to test the susceptibility of *H. pylori* to antibiotics before each treatment. Onset of resistance to tinidazole in the treated patients makes it necessary to repeat control of susceptibility as often as possible. This resistance appearing in patients not treated by this drug is surprising, and may possibly due to reinfection by a different strain. However, most of our patients are referred to the endoscopic unit and we have no details of their prior treatment.

On the other hand, we have observed strains previously resistant becoming sensitive to tinidazole.

We can imagine that *H. pylori* with no contact to tinidazole recovers susceptibility to this drug. In our series differences between the sexes is not statistically significant, mainly with regard to tinidazole resistance.

Even if amoxicillin and tetracycline are the most efficient on *H. pylori* in vitro, we obtained the same discouraging results as in the literature [4–6] with regard to the eradication of *H. pylori.*

Conclusion

More research is needed to investigate which factors, such as a previous intake of antibiotics, could induce a resistance. It is worthwhile checking *H. pylori* susceptibility before starting any antibiotic treatment, mainly tinidazole.

References

1. McNulty CA, Dent JC (1988) Susceptibility of clinical isolates of Campylobacter pylori to 21 antimicrobial agents. Eur J Clin Microbiol Infect Dis 74: 566–569
2. Glupczynski Y, Delmée M, Bruck C, Labbe M, Avezani V, Burette A (1988) Susceptibility of clinical isolates of Campylobacter pylori to 24 antimicrobial and anti-ulcer agents. Eur J Epidemiol 42: 154-157

3. Andreasen JJ, Andersen LP (1987) In vitro susceptibility of Campylobacter pyloridis to cimetidine, sucralfate, bismuth and 16 antibiotics. Acta Pathol Microbiol Scand B 952: 147–149
4. Glupczynski Y, Burette A, Nyst JF, Deprez C, De Koster E, Deltenre M. (1989) Campylobacter pylori-associated gastritis: attempts to eradicate the bacteria by various antibiotics and anti-ulcer regimens. Acta Gastroenterol Belg 51(4/5): 329–337
5. Axon AT (1989) Campylobacter pylori: therapy review. Scand J Gastroenterol [Suppl 160]: 35–38
6. Tytgat GN, Rauws EA, De Koster EH (1989) Campylobacter pylori: diagnosis and treatment. J Clin Gastroenterol 11 [Suppl 1]: S49–S53

In Vivo Acquired Resistance to Fluorquinolones in *Helicobacter pylori* During Therapy with Ciprofloxacin Alone or Ciprofloxacin plus Amoxicillin Clavulanate

R. Cantón[1], L. de Rafael[1], D. Boixeda[2], M.T. Serrano[1], M. Garcia-Gonzalez[2], and F. Baquero[1]

Introduction

In vitro *Helicobacter pylori* is susceptible to several antimicrobial agents [1–3]. However, the long-term eradication of this organism from the gastric mucosa after a 3- or 4-week therapy seems rather difficult. In most cases a single-drug therapy is remarkably ineffective, and even the association of two or three substances achieves only the clearance or temporal eradication of *H. pylori* from the colonised gastric mucosa in humans. Relapse following what appears to be a successful treatment is common and probably due to incomplete eradication of this organism [4]. The conventional therapy includes not only the antimicrobial agents, but also gastric inhibiting substances [5].

In the present work, therapy with ciprofloxacin plus ranitidine and ciprofloxacin plus amoxicillin clavulanate plus ranitidine was evaluated, both in microbiological and clinical terms. These agents were selected after a routine susceptibility evaluation with different *H. pylori* strains isolated in our laboratory.

Materials and Methods

Patients and Biopsy Specimens. A total of 266 patients (166 men and 100 women) with ages ranging from 15 to 79 years (mean age 47.4 years) and clinical symptoms related to gastroduodenal pathology were studied. They underwent oral endoscopy. When possible, four samples from each patient were obtained by endoscopy (duodenal bulb, gastric antrum, body and fundus) to investigate the presence of *H. pylori.* An additional ulcer sample was obtained from patients with this disease.

Microbiological Study. Biopsy specimens were processed within 1 h of being obtained. Each biopsy specimen was smeared onto a culture plate and then rubbed onto a slide to be Gram stained. The culture plate consisted of

[1] Servicio de Microbiologia, Hospital Ramón Y Cajal, Carretera de Colmenar Km 9,100, 28034 Madrid, Spain
[2] Servicio de Gastroenterologia, Hospital Ramón Y Cajal, Carretera de Colmenar Km 9, 100, 28034 Madrid, Spain

J.M. Pajares et al. (Eds.)
H. pylori and Gastroduodenal Pathology

commercial Columbia agar supplemented by sheep's blood (5%), polymyxin B (5000 U/l), amphotericin B (2 mg/l), trimethoprim (10 mg/l) and vancomycin (15 mg/l) (Biomedics SL, Madrid). Plates were incubated for 7 days at 37°C under an microaerophilic environment (5% O_2, 10% CO_2 and 85% N_2) (Campylobacter Microaerophylic System, Difco, Detroit). *H. pylori* was identified by the typical morphology (S-curved bacilli) and positive reactions for oxidase, catalase and urease (rapid). A patient was considered to be infected with *H. pylori* when at least one sample was positive by the Gram stain and/or culture. In the Gram stain examination a patient was considered to be positive for *H. pylori* when microscopic structures consistent with *H. pylori* were observed on at least one of the four slides. Similarly, a patient was considered to be culture positive when at least one of the four plates yielded this organism.

Susceptibility Testing. The antibiotic susceptibility of daily isolates was determined by the disk-diffusion method. Well-isolated colonies of *H. pylori* were transferred with a wire loop to a tube containing saline solution. Turbidity was adjusted to the 1 standard of the McFarland scale. A swab soaked in this suspension was rubbed, according to the National Committee for Clinical Laboratory Standards, (NCCLS guidelines; USA) [6] onto the surface of plates containing Columbia agar supplemented with sheep's blood (5%). Penicillin (10 μg), amoxicillin/clavulanic acid (20/10 μg), nalidixic acid (30 μg), and ciprofloxacin (5 μg) disks were purchased (Oxoid España SA Madrid, Spain; Diagnostic Pasteur Marnes la Coquette, France). The plates were incubated at 37°C for 7 days under an atmosphere containing 5% O_2, 10% CO_2 and 85% N_2. An *H. pylori* strain was considered sensitive when the inhibition zone was greater than 30 mm. A zone of no inhibition represented a resistant strain.

Therapy. Among 13 patients with gastric mucosa colonised with a fluorquinolone-sensitive *H. pylori* strain, six were randomised to receive ciprofloxacin (Baycip, Química Farmacéutica Bayer) (500 mg q 12 h for 21 days) and ranitidine (Zantac, Glaxo) (150 mg q 12 h for 30 days); and seven ciprofloxacin (500 mg q 12 h) plus amoxicillin/clavulanic acid (Augmentine, Beecham Laboratories) (500 mg q 8 h) for 21 days and ranitidine (150 mg q 12 h) for 30 days. The microbiological study was repeated at the end of treatment and 3 months there after.

Results

The results of the *H. pylori* susceptibility test from patients not previously treated are shown in Fig. 1. All strains were uniformly sensitive to penicillin and amoxicillin/clavulanic acid. Although all strains tested were resistant to nalidixic acid, the quinolones such as norfloxacin and ciprofloxacin were active. Their percentages of sensitivity were 69.3% and 98.6%, respectively.

The *H. pylori* strains isolated from patients before therapy with ciprofloxacin or ciprofloxacin plus amoxicillin/calvulanic acid were always susceptible to norfloxacin and ciprofloxacin (Tables 1, 2). None of the *H. pylori* isolates from

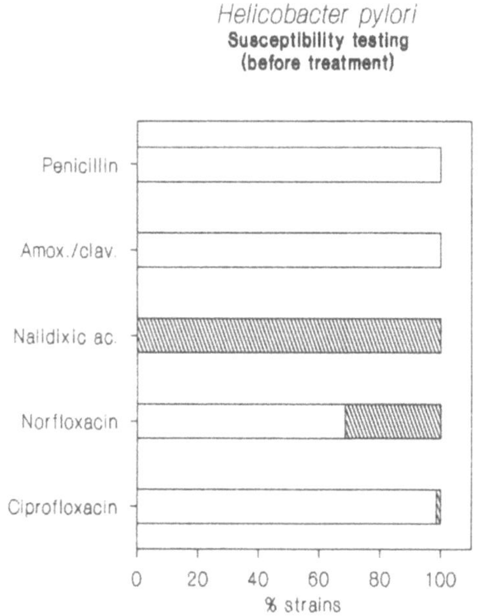

Fig. 1. *Helicobacter pylori* susceptibility testing (before treatment). *Shaded columns*, resistant; *Open columns*, sensitive

the six patients treated with ciprofloxacin and ranitidine was eradicated, even after 30 days of therapy. In four of six patients (66.6%), quinolone-resistant *H. pylori* strains were recovered after treatment.

In three out of the seven patients treated with ciprofloxacin plus amoxicillin/calvulanic acid and ranitidine, *H. pylori* was not recovered by culture from the gastric mucosa after 30 days of therapy. In the four remaining patients only scattered colonies of *H. pylori* were obtained after 30 days of therapy. Nevertheless, none of the isolates was resistant to norfloxacin and/or ciprofloxacin. A new biopsy for culture was obtained 3 months after the end of therapy and in all cases *H. pylori* was recovered again.

Discussion

The lack of a "gold standard" in the therapy of *H. pylori* associated infection is due to the development of resistance to antibacterial agents and failure to eradicate this organism from the gastric mucosa. Although several antimicrobials, such as erythromycin, metronidazole, tinidazole, ofloxacin and ciprofloxacin, and bismuth-containing compounds have good in vitro activity [1–3], their effectiveness in vivo is disappointing [7–9].

The results of in vitro studies show that ciprofloxacin and other fluorquinolones are highly effective against *H. pylori* [1–3]. Ciprofloxacin with minimal inhibiting concentrations ranging from 0.06 to 2 μ/ml was more active than ampicillin or amoxicillin (MIC values 0.01–1 μg/ml. As previously reported [10], ciprofloxacin maintains a moderate activity even at acid pH levels, and the

Table 1. Culture followup of patients treated with ciprofloxacin (500 mg q 12 h) for 21 days plus ranitidine (150 mg q 12 h) for 30 days

		Susceptibilities					Duodenal and gastric biopsies				
Patient	Day	P	Aug	Nal	Nor	Cip	Bulb	Antrum	Body	Fundus	Ulcus
1	0	S	S	R	S	NT	–	+	+	+	+
	30	S	S	R	R	R	–	+	+	+	+
	120	NT	NT	NT	NT	NT	NT	NT	NT	NT	NT
2	0	S	S	R	S	NT	–	+	+	+	
	30	S	S	R	R	S	–	–	+	+	
	120	S	S	R	R	R	+	+	+	+	
3	0	S	S	R	S	S	–	+	+	+	+
	30	S	S	R	R	R	+	+	+	+	+
	120	S	S	R	R	R	–	+	+	+	+
4	0	S	S	R	S	S	–	+	+	+	
	30	S	S	R	S	S	–	+	+	+	
	120	S	S	R	S	S	+	+	+	+	
5	0	S	S	R	S	S	–	+	+	+	
	30	S	S	R	S	S	–	+	+	+	
	120	NT	NT	NT	NT	NT	NT	NT	NT	NT	
6	0	S	S	R	S	S	–	+	+	+	
	30	S	S	R	R	R	–	+	+	+	
	120	S	S	R	R	R	+	+	+	+	

[a] 0, Before treatment; 30, end of treatment; 120, 3 months after end of treatment. P, penicillin, Aug, amoxicillin/clavulanate; Nal, nalidixic acid; NOR, norfloxacin; CIP, ciprofloxacin; S, susceptible; R, resistant; NT, not tested.

Table 2. Culture follow up of patients treated with ciprofloxacin (500 mg q 12 h) plus amoxicillin/calvulanate (500 mg q 8 h) for 21 days plus ranitidine (150 mg q 12 h) for 30 days.

		Susceptibilities					Duodenal and gastric biopsies				
Patient	Day[a]	P	Aug	Nal	Nor	Cip	Bulb	Antrum	Body	Fundus	Ulcus
1	0	S	S	R	S	S	–	+	+	+	–
	30						–	–	–	–	–
	120	S	S	R	S	S	+	+	+	+	+
2	0	S	S	R	S	S	+	+	+	+	+
	30	S	S	R	S	S	+[b]	+[b]	+[b]	+[b]	+[b]
	120	S	S	R	S	S	+	+	+	+	+
3	0	S	S	R	S	S	–	+	+	+	+
	30						–	–	–	–	–
	120	S	S	R	S	S	–	+	+	+	+
4	0	S	S	R	S	NT	–	+	+	+	
	30						–	–	–	–	
	120	S	S	R	S	S	+	+	+	+	
5	0	S	S	R	S	S	–	+	+	+	
	30	S	S	R	S	S	–	+[b]	+[b]	+[b]	
	120	NT	NT	NT	NT	NT	NT	NT	NT	NT	
6	0	S	S	R	S	S	–	+	+	+	
	30	S	S	R	S	S	–	+[b]	+[b]	+[b]	
	120	S	S	R	S	S	–	+	+	+	
7	0	S	S	R	S	S	–	+	+	+	
	30	NG	NG	NG	NG	NG	–	+[b]	+[b]	+[b]	
	120	S	S	R	S	S	–	+	+	+	

[a] 0, Before treatment; 30, end of treatment; 120, 3 months after end of treatment.
[b] Scattered colonies; P, penicillin, Aug, amoxicillin/clavulanate; Nal, nalidixic acid; NOR, norfloxacin; CIP, ciprofloxacin; NT, not tested; NG, no growth on antibiotic sensitivity test plates.

concentrations obtained in the gastric mucosa 6 h after an oral dose (500 mg) were 50–100 times greater than the MIC range and decreased beyond this period [11].

Although this study seemed promising, none of our patients treated for 3 weeks with ciprofloxacin was free from *H. pylori*, and resistance to ciprofloxacin with cross-resistance to other fluorquinolones was demonstrated in 66.6% of our patients. These findings are in agreement with those reported by Glupczynski et al. [8] and Martens et al. [12].

In a recent in vitro study Haas et al. [13] found the emergence of fluorquinolone-resistant *H. pylori* strains. These were exposed to subinhibitory concentrations of ciprofloxacin. The acquired resistance was also stable over the period studied. In vivo this circumstance could be due to prolonged intervals of subinhibitory concentrations at the site of infection and should lead to the selection of resistant strains.

To our knowledge, data on the emergence of amoxicillin-resistant *H. pylori* strains have not yet been reported. With serial passages, an in vitro selection method of resistant strains, *H. pylori* strains were selected with amoxicillin MIC values higher (0.06–0.2 μg/ml than that of the original isolate (0.004–0.03 μg/ml, but the organism was still considered sensitive to amoxicillin. Moreover, the spontaneous resistance rate was low and did not correlate well with the ability to increase the MIC for amoxicillin [13].

Interestingly, clavulanic acid is also active against *H. pylori* in vitro [1] and resistance has not yet been reported.

As ciprofloxacin did not eradicate *H. pylori* from the gastric mucosa, it was associated with amoxicillin/clavulanic acid to enhance their antibacterial effect and to decrease the development of resistant strains. The association of ciprofloxacin plus amoxicillin/clavulanic acid achieved the eradication of *H. pylori* in 43% of our patients after 30 days of therapy; in the remaining patients only scattered colonies were obtained. Unfortunately, this eradication should be considered a clearance because *H. pylori* was recovered again 3 months after the end of therapy. However, amoxicillin/clavulanic acid seems to prevent the emergence of fluorquinolone-resistant variants during a combined therapy.

References

1. Lambert T, Mégraud F, Gerbaud G, Courvalin P (1986) Susceptibility of Campylobacter pyloridis to 20 antimicrobial agents. Antimicrob Agents Chemother 30: 510–511
2. Goodwin CS, Blake P, Blincow E (1986) The minimum inhibitory and bacterial concentrations of antibiotics and anti-ulcer agents against Campylobacter pyloridis. J Antimicrob Chemother 17: 309–314
3. Simor AE, Ferro S, Low DE (1989) Comparative in vitro activities of six new fluorquinolones and other oral antimicrobial agent against Campylobacter pylori. Antimicrob Agents Chemother 33: 108–109
4. Rauws EAJ Langenberg W, Houthoff HJ, Zanen HC, Tytgat GNJ (1988) Campylobacter pyloridis-associated chronic active antral gastritis. A prospective study of its prevalence and the effects of antibacterial and antiulcer treatment. Gastroenterology 94: 33–40
5. Goodwing CS, Armstrong JA, Marshall BJ (1986) Campylobacter pyloridis gastritis and peptic ulceration. J Clin Pathol 39: 353–365

6. NCCLS (1990) Performance standards for antimicrobial disk susceptibility test, 4th edn. Approved standard. Document M2–A4
7. Sachdeva M, Lee BL, Sande MA (1989) Ineffectiveness of ciprofloxacin in the eradication of Campylobacter pylori in ulcer disease. Program and Abstract of the 29th interscience conference on antimicrobial agents and chemotherapy. (abstr no 41)
8. Glupczynski Y, Cabke M, Burette A, Delmee M, Avesani V, Bruck C (1987) Treatment failure of ofloxacin in Campylobacter pylori infection (letter). Lancet 1:1096
9. Goodwing CS, Marshall BJ, Blincow ED, Wilson DM, Blackbourn S, Phillips M (1988) Prevention of nitroimidazol resistance in Campylobacter pylori by coadministration of colloidal bismuth subcitrate: clinical and in vitro studies. J Clin Pathol 41:207–210
10. Smith JT, Ratcliffe NT (1986) Effect of pH and magnesium on the in vitro activity of ciprofloxacin. In: Neu HC, Weuta H (eds) Proceedings of the 1st international ciprofloxacin workshop series. Excerpta Medica, New York, pp 12–16 (Current clinical practice series, vol 34)
11. McNulty CAM, Dent JC, Ford GA, Wilkinson SP (1988) Inhibitory antimicrobial concentrations against Campylobacter pylori in gastric mucosa. J Antimicrob Chemother 22:729–738
12. Martens JCC, Dekker W, Ligtvoet EEJ, Blok P (1989) Treatment failure of norfloxacin against Campylobacter pylori and chronic gastritis in patients with nonulcerative dispepsia. Antimicrob Agents Chemother 33:256–257
13. Haas CE, Nix DE, Schentag JJ (1990) In vitro selection of resistant Helicobacter pylori. Antimicrob Agents Chemother 34:1637–1641

Treatment of *Helicobacter pylori*-Positive Lesions: ^{13}C-Urea Breath Test Monitoring and Morphology

F.E. Bauer[1], G. Lepsien[2], I. Meineke[1], J.V. Geletneky[1], A. Schauer[3], and F.E. Lüdtke[2]

Introduction

Helicobacter pylori (HP) is now generally accepted as a major pathogenetic factor in antral gastritis and ulcer disease [1–5]. The evidence which links HP infection to peptic ulceration is strongest in duodenal ulcer (DU) and less powerful in gastric ulcer (GU). As far as DU is concerned, however, eradication of infection markedly alters the natural history of the disease [6, 7]. Therefore, very recently the Sidney Working Party on Gastroenterology recommended HP eradication in patients suffering from DU [4].

At present a variety of therapeutic regimens is under investigation, but only limited information is available. Even though the triple therapy (bismuth compounds, amoxycillin/tetracycline, metronidazole) achieves most effective eradication rates, treatment of HP is considered to be unsatisfactory regarding the number of drugs which have to be used and the duration of treatment.

The lack of a simple, non-invasive, but predictive test to evaluate different therapeutic regimens is a major obstacle in monitoring the effect of such a therapy.

The diagnostic methods used up to now (microbiology, urease tests, histology) are all based on endoscopy and therefore unpleasant to the patient, time consuming and expensive. Therefore none of them is ideally suited for monitoring eradication of HP.

Since the first description of a ^{13}C-urea breath test (^{13}C-UBT) [8] this method has been further optimized [9]. The optimized ^{13}C-UBT is completed within 30 min and requires as little as 75 mg of urea making it much more convenient to patients and staff and considerably cheaper [9]. Now an ideal method is available to monitor therapeutic efficacy in patients suffering from HP positive lesions.

The decision whether a patient has an HP-positive or -negative status is based on the recovery of $^{13}CO_2$ in breath (expressed as a percentage of dose).

[1] Department of Clinical Pharmacology, Georg August University of Göttingen, Robert-Koch-Str. 40, 3400 Göttingen, FRG
[2] Department of Surgery, Georg August University of Göttingen, Robert-Koch-Str. 40, 3400 Göttingen, FRG
[3] Department of Pathology, Georg August University of Göttingen, Robert-Koch-Str. 40, 3400 Göttingen, FRG

J.M. Pajares et al. (Eds.)
H. pylori and Gastroduodenal Pathology

Therefore, this method also opens the possibility to examine a possible quantitative relationship between the $^{13}CO_2$ recovery (a measure of HP mass) and the extent of the mucosal lesion.

The aim of the present study was to monitor HP eradication with the ^{13}C-UBT in treated patients suffering from gastroduodenal lesions. In addition, a possible quantitative relationship between $^{13}CO_2$ recovery and the extent of the lesion was investigated.

Patients and Methods

Patients

Since March 1989, 94 patients aged 19–79 years, referred for routine upper endoscopy because of ulcer-like symptoms, were enrolled into this study. Inclusion criteria were HP-positive DU, GU and non-ulcer dyspepsia (NUD), the latter being defined as mucosa without a macroscopically visuable lesion, histologically proven chronic gastritis and at least one of the following symptoms: epigastric or retrosternal pain, discomfort, heartburn, nausea, vomiting or other symptoms considered to be referable to the proximal alimentary tract and lasting for at least 4 weeks [10]. Patients were excluded from the study if they had an HP-negative status, neoplasm, erosive or ulcerative oesophagitis, recently consumed non-steroidal anti-inflammatory drugs (NSAIDs), partial gastrectomy, pregnancy, debilitating diseases or were unable to cooperate. Each subject gave written informed consent. The protocol had been approved by the local ethics committee.

Endoscopy and Biopsy

An upper endoscopy was performed using an Olympus GIF Q fibreoptic panendoscope at the following time intervals: before treatment (0), immediately after treatment (1 month), and after 3, 6 and 12 months. At every occasion two biopsy specimens were taken from the antrum for determination of the gastritis score.

Histological Assessment of Gastritis Score

Quantitative assessment of chronic gastritis was determined as previously described [11]. Briefly, a scoring system ranging from 0 to 10 was used and the following parameters assessed: density of the inflammatory infiltrate in the lamina propria (0–2); density of polymorphonuclear leukocytes in the lamina propria (0–3); presence of intraepithelial polymorphonuclear leukocytes (0–3); and superficial erosions (0–2).

Identification of HP Infection

HP status was assessed using the optimized ^{13}C-UBT as described previously [9]. Briefly, this breath test has been validated against culture, CLO test and fuchsin stain, showing high sensitivity (96%) and specificity (78%). The results were comparable with the quality parameters of others [12, 13]. The ^{13}C-UBT was performed as follows: after fasting for at least 6 h and 30 mins after endoscopy, 150 ml 0.1 *N* citric acid (E. Merck, Darmstadt, FRG) was administered to patients to inhibit gastric emptying; 1 min later patients received 20 ml 0.1 *N* citric acid containing 75 mg ^{13}C-urea (99% ^{13}C, Tracer Technologies Inc., Somerville, MA, USA) followed by 30 ml 0.1 *N* citric acid ("wash"). Breath samples were taken before, and 10, 20 and 30 min after ingestion of urea. Breath samples were collected in 20 ml siliconized vacutainers (Becton-Dickinson, Lincoln, Park, NJ, USA). The $^{13}CO_2/^{12}CO_2$ ratio (δ-value) was measured by isotope ratio mass spectrometry (IRMS). The results were expressed as % dose kg/mmol CO_2 (recovery). HP-negative status was defined as $< 0.1\%$ dose kg/mmol CO_2, HP clearance as an HP-negative status immediately after cessation of treatment, and eradication as an HP-negative status for at least 1 month after cessation of treatment.

Treatment design

Patients were prospectively randomized into groups as shown in Table 1.

Calculations

Gastritis scores at the follow-up visits were compared with the score prior to treatment and assigned (+) when the score was improved or (−) when no improvement was seen. On the basis of this transformation a sign test was used for the statistical evaluation. The critical values were derived from the corresponding binomial distribution. In order to describe the relationship of $^{13}CO_2$

Table 1. Treatment design ($n = 94$)

Group	Drug(s)	Dose (mg/day)	Duration (days)
1	Famotidine	1 × 40	28
2	BSS	3 × 600	28
3	Amoxycillin	3 × 500	28
4	Cefalexin	3 × 500	28
5	Triple therapy		
	BSS	3 × 600	28
	Amoxycillin	3 × 500	28
	Metronidazole	3 × 250	10
6	Famotidine + metronidazole	1 × 40	28
		3 × 250	10
7	BSS + metronidazole	3 × 600	28
		3 × 250	10

Patients of groups 1–4 and 6–7 with primary therapeutic failure or HP relapse turned into group 5.
BSS: bismuth subsalicylate

recovery with the gastritis score, the coefficient of association (Kendall's Tau) was calculated. A p value less than 0.05 was considered to be significant.

Results

A total of 94 patients were enrolled into the study. Seven dropped out. Of the remaining 87 patients, 15 had DU, 22 GU and 50 NUD. The drugs were well tolerated. No serious side effects were observed. Six patients who received antibiotics suffered from moderate diarrhoea without the need of a dose reduction. Four patients suffered from constipation. One patient developed an allergy to amoxycillin.

In Table 2 the therapeutic efficacy of seven regimens with regard to HP clearance and eradication is shown. Famotidine and cefalexin had no effect on HP status. The monotherapy with bismuth subsalicylate (BSS) or amoxycillin led to HP eradication rates of 8% and 20%, respectively. Double therapy applying famotidine and metronidazole revealed an eradication rate of 8%. Double therapy using BSS and metronidazole and triple therapy achieved eradication rates of 67% and 80%, respectively (Fig. 1).

The effect of treatment on the improvement of the gastritis score is shown in Figs. 2–4. HP eradication revealed a significant improvement of the histological lesion in both double (BSS + ME; $p < 0.0006$) and triple treated patients ($p < 0.03$). A persistent HP status was accompanied by failure of the gastritis score to improve (Fig. 3).

The association between $^{13}CO_2$ recovery and the extent of the mucosal lesion is demonstrated in treated patients (Table 3 and Fig. 5). A predictive relationship exists between $^{13}CO_2$ recovery and gastritis score. This "recovery–lesion link" applies for the time immediately after cessation of treatment and in the long term (1 month, $p < 0.0001$; 6 months, $p < 0.003$).

Discussion

The present study provides information on three topics: (a) drug efficacy on HP eradication; (b) relationship between HP status and the extent of the mucosal lesion; and (c) application and options of the ^{13}C-UBT in the management of patients under therapy.

Drug Efficacy on HP Eradication

The results presented above show that neither monotherapy with BSS nor monotherapy with antibiotics nor the combination of famotidine with metronidazole led to an acceptable eradication rate of HP. In the present study the eradication rate of HP after BSS therapy was only 8%. This is similar to the results of other investigators, who report rates between 13% and 33% [11,

Table 2. Therapeutic efficacy, assessed by the ^{13}C-urea breath test, in patients suffering from HP positive lesions.

Treatment	*n*	Diagnosis before treatment			Diagnosis immediately after treatment			HP clearance[b]		HP eradication[c]	
		DU	GU	NUD[a]	DU	GU	NUD[a]	n	%	n	%
Famotidine	11	3	5	3	1	2	3	0	0	0	0
BSS	12	4	2	6	0	1	4	8	67	1	8
Amoxycillin	10	0	1	9	0	0	3	7	70	2	20
Cefalexin	7	1	0	6	0	0	3	0	0	0	0
Triple	23	1	4	18	0	2	3	20	87	16[d]	80
FAM + ME	13	4	6	3	1	3	1	1	8	1	8
BSS + ME	11	2	4	5	0	1	1	10	91	6[e]	67

FAM, famotidine; BSS, bismuth subsalicylate; Me, metronidazole; triple, BSS + AM + ME; DU, duodenal ulcer; GU, gastric ulcer; NUD, non-ulcer dyspepsia.

[a] Diagnosis NUD was rejected if at that time point the patient was free of symptoms; the additional improvement or dissapearance of gastritis was facultative.

[b] HP negative status immediately after cessation of treatment.

[c] HP negative status at least 1 month after cessation of treatment.

[d] Out of 20 patients.

[e] Out of nine patients.

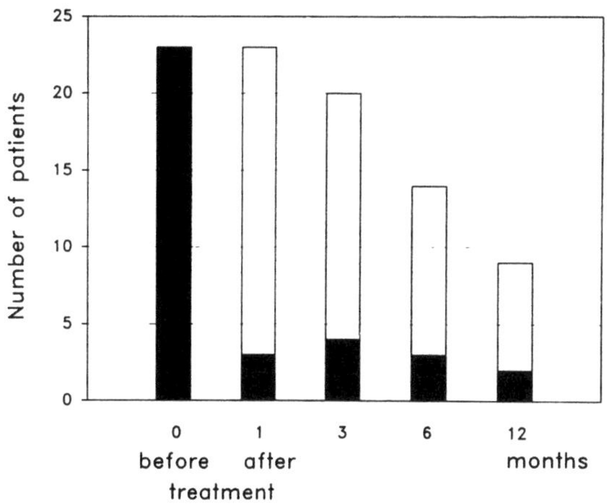

Fig 1. HP status in patients treated with triple therapy (bismuth subsalicylate, amoxycillin, metronidazole) : primary therapeutic failure 13%; eradication rate 80%. After HP eradication no relapse appeared within 1 year. *Black*, HP positive; *white*, HP negative

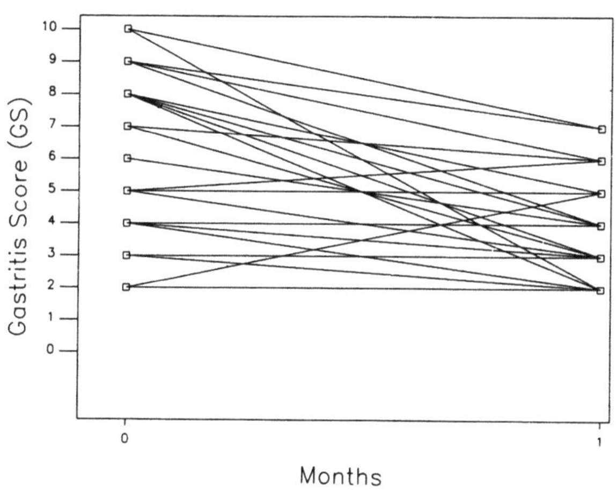

Fig 2. Gastritis score in patients successfully treated with triple therapy. There was a significant improvement of the gastritis score immediately after the end of treatment ($p < 0.03$). For Double therapy (bismuth subsalicylate, metronidazole) : $p < 0.0006$ (not shown)

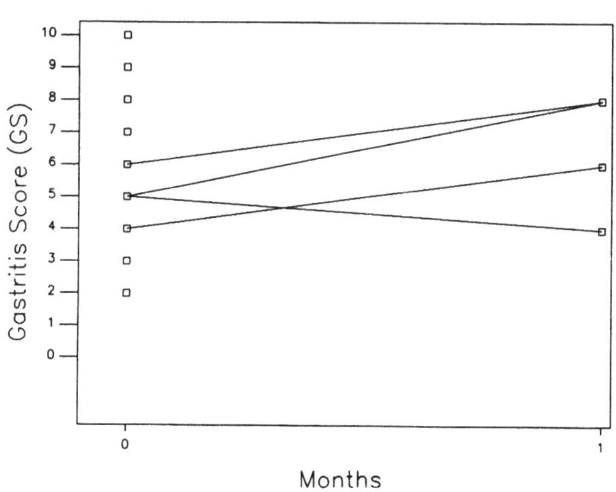

Fig 3. Gastritis score in patients unsuccessfully treated with triple therapy: no improvement of gastritis score immediately after the end of treatment

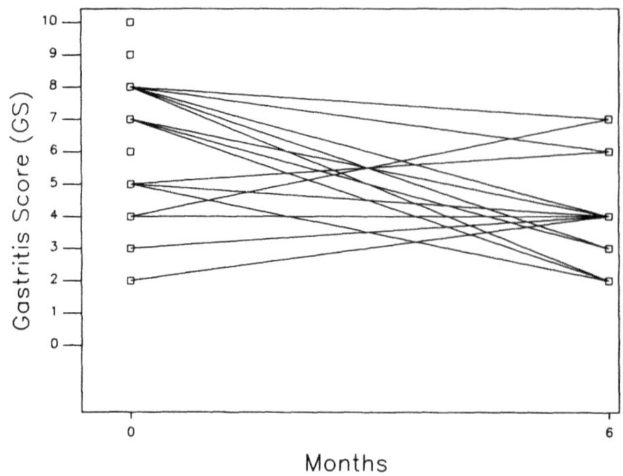

Fig 4. Gastritis score in patients after successful treatment with triple therapy (6 months). Improvement of gastritis score in 10/15 patients (present status)

Table 3. Association (Kendall's tau) between $^{13}CO_2$ recovery (a measure of HP mass) and gastritis score (0–10) in patients before and after different treatment regimens (see Table 1, nos. 1–7). A "recovery–lesion link" is shown after treatment, opening the possibility of non-invasive prediction of the extent of the mucosal lesion by the ^{13}C-UBT.

Time (months)	0	1	6	12
No. of patients	85	85	27	19
Condition	No treatment	End of treatment	Post treatment	Post treatment
Kendall's tau	0.12	0.41	0.44	0.25
Significance	NS	$p < 0.0001$	$p < 0.003$	NS

NS, not significant.

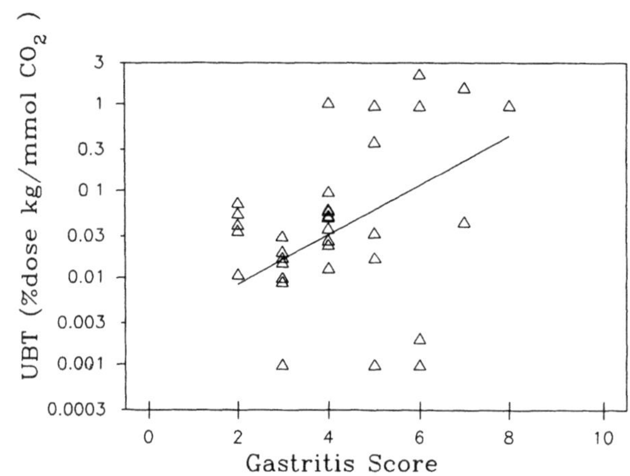

Fig 5. Association between $^{13}CO_2$ recovery (a measure of HP mass) (y axis) and the extent of the mucosal lesion (x axis) in patients immediately after cessation of therapy (1 month, successfully and unsuccessfully treated patients, $n = 46$). A recovery–lesion link is demonstrated (Kendall's tau = 0.41; $p < 0.001$). This provides evidence that the non-invasive ^{13}C-UBT reveals predictive information on the extent of the mucosal lesion

14–17]. With amoxycillin poor HP eradication (20%) was also observed, again in agreement with the results in the literature (0%–26%) [11, 18–20]. When amoxycillin was given together with omeprazole (20 mg/day), eradication increased to 53% [21].

Cefalexin did not influence HP status at all. This is in accordance with the findings of Bayerdörffer and Ottenjann [18]. All antibiotics used for monotherapy were ineffective (< 30%) [22]. A poor eradication rate of only 8% was also seen with a combination of famotidine and metronidazole; this is similar to the findings of other authors (0%—5%) [1, 23].

In contrast to the therapeutic regimens discussed so far, the present data indicate that long-term HP eradication can be achieved if bismuth treatment is combined with metronidazole, or if this therapy is expanded by amoxycillin (triple therapy). The eradication rates in the present study are comparable with the limited information available in the literature. For double therapy (BSS/CBS + tinidazole/metronidazole), eradication rates ranged between 73% and 85% [1, 15, 24, 25], compared with 67% in the present study. Under triple therapy, eradication rates ranged between 65% and 94% [7, 25–29]. In the present study an 80% eradication rate was achieved.

The key drugs in the treatment of HP infection appear to be bismuth compounds and metronidazole. Ultrastructure studies of biopsies of patients treated with CBS have shown different effects on HP within 1–2 h: (a) loosing of adherence properties of HP on the epithelium; (b) destruction of bacterial structure by vacuolization; (c) deposits of bismuth complexes beneath the cell wall of the bacterium [30, 31]. The poor effect of a combination of bismuth with H2 blockers may be based on the interference of H2 blockers with the efficacy of bismuth, which depends in part on the presence of acid in the stomach [32]. Metronidazole has some inherent advantages compared with other antibiotics as its activity is not dependent on pH and it is secreted into the gastric juice where it is able to gain access to bacteria deep in the gastric pits [33]. Other factors for the poor therapeutic effect, especially of antibiotics, on HP may be: poor penetration, local inactivation, inappropriate formulation of the drug, inadequate duration of treatment and poor compliance. A special problem of metronidazole is the development of resistance.

Pretreatment resistance depends on geography, age and ethnic origin. In Belgium the rate of metronidazole resistance is 17% in native-born patients while in immigrants from mediterranean countries resistance ranged from 32% (< 30 years) to 59%. In Zaire a resistance rate up to 87% is found [34]. In this country many patients have received metronidazole for amoebiasis or giardiasis. To predict therapeutic efficacy of metronidazole, a pretreatment assessment of sensitivity is mandatory. Combinations of metronidazole with bismuth or amoxycillin will reduce the risk of resistance to metronidazole during treatment.

Relationship Between HP Status and Extent of Mucosal Lesion

Quantitative Linkage of $^{13}CO_2$ Recovery on the Extent of Lesion

As already mentioned, the recovery of $^{13}CO_2$ in breath was expressed as % dose kg/mmol CO_2 and was used as a index for the mass of HP. On theoretical grounds it can be expected that a high number of HP organisms will metabolise a high amount of urea, leading to an increased recovery of $^{13}CO_2$ in breath. Because of the known pathogenetic role of HP in peptic ulcer disease, it is likely that such a recovery–lesion link exists. Indeed, such a link has been revealed immediately after cessation of treatment and in the long term up to 6 months. The low number of patients after 12 months (ongoing study) accounts for the significance level at this time point. The association before treatment is not as strong as in the follow up. At present it is not clear why this is the case. The post-treatment associations show that recovery of $^{13}CO_2$ (an index of HP mass) predicts the gastritis score. Therefore, apart from diagnosis of HP status, the non-invasive ^{13}C-UBT is also suitable to give predictive information on the extent of the mucosal lesion. In addition, this result contributes to defining the casual role of HP in ulcer disease. Information on the recovery–lesion link has been described here for the first time and further studies are warranted to provide more information on this topic. Using conventional histological methods there is some information available regarding HP colonisation and the extent of acute inflammation. A significant correlation between HP colonisation and the degree (infiltration with lymphocytes and plasma cells) and the activity (infiltration with polymorphonuclear neutrophils) of gastritis has been demonstrated in the antrum and the body of the stomach [35–37].

Therapeutic Linkage of HP Eradication on Extent of Lesion

The second approach to the HP colonisation–inflammation relationship was the comparison of the gastritis score in the untreated (HP positive) and in the treated patients. The study revealed a significant reduction of the gastritis score in successfully treated patients. In cases where HP eradication failed no improvement in the gastritis score was seen. This result corroborates to the pathogenetic role of HP in ulcer disease. Similar results were achieved before and immediately after cessation of treatment by others [11, 38–42] and also in the long term up to 1 year [11].

Application and Options of ^{13}C-UBT

As shown in the present study, the ^{13}C-UBT is a noninvasive, predictive, safe, rapid and easy method and therefore ideally suited to monitor HP status in patients suffering from HP positive lesions. Since the first description in 1987 [8] it has been more widely used [9, 12, 13, 43]. The ^{14}C-UBT is even more frequently applied because of the general availability of scintillation counters, but this test has the major disadvantage of radiation [44–50]. In contrast to the ^{14}C tracer, the ^{13}C-UBT opens the possibility of a broad application in any hospital because the measurement can be done in a few centres where the technical equipment is available. Samples are stable for several months (F.E. Bauer, unpublished observation) and can therefore be mailed. On this basis, our own group has 3 years' experience cooperating with 20 hospitals all over Germany. Furthermore, the method is so easy to perform that patients themselves could carry out the test at home, as some of the patients in this study have done (home testing).

In summary, HP is now accepted as a major pathogenetic factor in ulcer disease. At least one highly effective and reasonably safe therapy is available (triple therapy) which is recommended for routine use in HP-positive DU. For GU and NUD, triple therapy should still be restricted to clinical trials. Double therapy (i.e. BSS plus metronidazole) is presently under clinical investigation. Preliminary data indicate eradication rates comparable with those of the triple therapy. The eradication of HP is a new and highly effective therapeutic concept in the successful treatment of DU because it cures the disease. The complete interruption of ulcer relapse by HP eradication makes this therapeutic concept superior to the H2 blockers which are not able to do so, showing relapse rates of up to 100% [51].

References

1. Marshall BJ, Warren JR, Blincow ED, Phillips M, Goodwin CS, Murray R, Blackbourn SJ, Waters TE, Sanderson CR (1988) Prospective double-blind trial of duodenal ulcer relapse after eradication of Campylobacter pylori. Lancet II: 1437–1441
2. Goodwin CS (1988) Duodenal ulcer, Campylobacter pylori, and the "leaking roof" concept. Lancet II: 1467–1469
3. Graham DY (1989) Campylobacter pylori and peptic ulcer disease. Gastroenterology 96: 615–625
4. Tytgat GNJ, Axon ATR, Dixon MF, Graham DY, Lee A, Marshall BJ (1990) Helicobacter pylori: causal agent in peptic ulcer disease? In : Falk H (ed) Helicobacter pylori. Literature Review no 9, 3rd quarter. Falk, Freiburg, pp 360–369
5. Drumm B (1990) Helicobacter pylori. Arch Dis Child 65: 1278–1282
6. Rauws EAJ, Tytgat GNJ (1990) Cure of duodenal ulcer associated with eradication of Helicobacter pylori. Lancet I: 1233–1235
7. George LL, Borody TJ, Andrews P, Devine M, Moore-Jones D, Walton M, Brandl S (1990) Cure of duodenal ulcer after eradication of Helicobacter pylori. Med J Aust 153: 145–149
8. Graham DY, Klein PD, Evans DJ, Evans DG, Alpert LC, Opekun AR, Boutton TW (1987) Campylobacter pylori detected noninvasively by the ^{13}C-urea breath test. Lancet I: 1174–1178
9. Eggers RH, Kulp A, Tegeler R, Lüdtke FE, Lepsien G, Meyer B, Bauer FE (1990) A methodological analysis of the ^{13}C-urea breath test for detection of Helicobacter pylori infections: high sensitivity and specificity within 30 min using 75 mg of ^{13}C-urea. Eur J Gastroenterol Hepatol 2: 437–444
10. Colin-Jones DG (1988) Management of dyspepsia: report of a working party. Lancet I: 576–579
11. Rauws EAJ, Langenberg W, Houthoff HJ, Zanen HC, Tytgat GNJ (1988) Campylobacter pyloridis-associated chronic active antral gastritis. Gastroenterology 94: 33–40
12. Cooreman M, Hengels KJ, Krausgrill P, Strohmeyer G (1990) ^{13}C-Harnstoff-Atemtest als nicht invasive Methode zum Nachweis von Helicobacter (Campylobacter) pylori. Dtsch Med Wochenschr 115: 367–371
13. Dill S, Payne-James JJ, Misiewicz JJ, Grimble GK, McSwiggan D, Pathak K, Wood AJ, Scrimgeour CM, Rennie MJ (1990) Evaluation of ^{13}C-urea breath test in the detection of Helicobacter pylori and in monitoring the effect of tripotassium dicitratobismuthate in non-ulcer dyspepsia. Gut 31: 1237–1241
14. Mannes GA, Bayerdörffer E, Höchter W, Weingart J, Heldwein W. Müller-Lissner S, Oertel H, Blendinger Ch, Kuntzen O, Bornschein W, Malfertheiner P, Wilkening J, Ruckdeschel G, Pfaller P, van Wulffen H, Köpcke W, Stolte M (1990) Relapse rate of H. pylori (HP) positive duodenal ulcers (DU) following antibacterial therapy. In: Abstracts of the world congress of gastroenterology, Sydney, Australia. Medicine Group, Abingdon, p 955
15. Goodwin CS, Marshall BJ, Warren JR, Blackbourn S, Blincow ED (1987) Clearance of Campylobacter pyloridis and reduced duodenal ulcer relapse with bismuth and tinidazole compared

to cimetidine. In: Kaijser B, Falsen E (eds) Campylobacter, vol 4. University of Göteborg, Göteborg, pp 368–369
16. Lambert JR, Borromeo M, Korman MG, Hansky J, Eaves ER (1987) Effect of colloidal bismuth (De-Nol) on healing and relapse of duodenal ulcers-role of Campylobacter pyloridis. In: Kaijser B, Falsen E (eds) Campylobacter, vol 4. University of Göteborg, Göteborg, p 383
17. Börsch G, Mai U, Müller KM (1988) Monotherapy or polychemotherapy in the treatment of Campylobacter pylori-related gastroduodenal disease. Scand J Gastroenterol 23 [Suppl 142]: 101–106
18. Bayerdörffer E, Ottenjann R (1988) The role of antibiotics in Campylobacter pylori associated peptic ulcer disease. Scand J Gastroenterol 23 [Suppl 142]: 93–100
19. Glupczynski Y, Burette A, Labbe M, Deprez C, De Reuck M, Deltenre M (1988) Campylobacter pylori-associated gastritis: a double-blind placebo-controlled trial with amoxycillin. Am J Gastroenterol 83: 365–372
20. Glupczynski Y, Burette A, Nyst JF (1988) Campylobacter pylori-associated gastritis: attempts to eradicate the bacteria by various antibiotics and anti-ulcer regimens. Acta Gastroenterol Belg 5: 329–337
21. Lamouliatte H, de Mascarel A, Megraud FM, Barberis C, Bernard PH, Cayla R, Quinton A (1990) Omeprazole improves amoxycillin therapy directed towards Helicobacter pylori associated chronic gastritis. Gastroenterology 98: A75
22. Börsch G (1991) Treatment regimens to eradicate Helicobacter pylori. In: Menge H, Gregor M, Tytgat GNJ, Marshall BJ, McNulty CAM (eds) Helicobacter pylori 1990. Springer, Berlin Heidelberg New York, pp 209–215
23. Hirschl AM, Hentschel E, Schütze K, Nemec H, Pötzi R, Gangl A, Weiss W, Pletschette M, Stanek G, Rotter ML (1988) The efficacy of antimicrobial treatment in Campylobacter pylori associated gastritis and duodenal ulcer. Scand J Gastroenterol 23 [Suppl 142]: 76–81
24. O'Morain C (1990) How to eradicate Helicobacter pylori and prevent reinfection. In: Malfertheiner P, Ditschuneit H (eds) Helicobacter pylori, gastritis and peptic ulcer. Springer, Berlin Heidelberg New York, pp 388–394
25. O'Riordan T, Mathai E, Tobin E, McKenna D, Keane C, Swenney E, O'Morain C (1990) Adjuvant antibiotic therapy in duodenal ulcers treated with colloidal bismuth subcitrate. Gut 31: 999–1002
26. Graham DY, Lew GM, Opekun AR, Klein PD, Evans DG, Evans DJ Jr, Alpert LC, Michaletz PA (1990) Controlled evaluation of the effectiveness of triple therapy in eradication of Helicobacter pylori infection. Gastroenterology 98: A 452
27. Lambert JR, Lin SK, Borromeo M, Nicholson L, Korman MG, Hansky J, Monash (1990) Eradication of Helicobacter pylori with colloidal bismuth subcitrate/antibiotic combinations. Gastroenterology 98: A74
28. McNulty CAM, Eyre-Brook IA, Uff JS, Dent JC, Wilkinson SP (1989) Triple therapy is not always 95% effective. 5th international workshop on Campylobacter infections, Puerto Vallarata, Mexico
29. Borody TJ, Cole P, Noonan S, Morgan A, Lenne J, Hyland L, Brandl S, Borody EG, George LL (1989) Recurrence of duodenal ulcer and Campylobacter pylori infection after eradication. Med J Aust 151: 431–435
30. Tytgat GNJ, Rauws EAJ, De Koster E (1988) Campylobacter pylori. Scand J Gastroenterol Suppl 155: 68–79
31. Eberhardt R, Kasper G (1990) Effect of oral bismuth subsalicylate on Campylobacter pylori and on healing and relapse rate of peptic ulcer. Rev Infect Dis 12: 115–119
32. Goodwin CS, Armstrong JA, Wilson DH (1988) Differences between in vitro and in vivo sensibility of Campylobacter pylori to antibacterials. In: Menge H, Gregor M, Tytgat GNJ, Marshall BJ (eds) Campylobacter pylori. Springer, Berlin Heidelberg New York, pp 29–36
33. Hollingsworth JA, Goldie J, Silette LY, Richardson H, Hunt RH (1987) Gastric secretion of antibiotics used for Campylobacter pyloridis. Gut 28: A 1409
34. Glupczynski Y, Burette A, De Koster E, Nyst JF, Deltenre M, Cadranel S, Bourdeaux L, De Vos D (1990) Metronidazole resistance in Helicobacter pylori. Lancet I: 976–977
35. Stolte M, Eidt S, Ohnsmann A (1990) Differences in Helicobacter pylori associated gastritis in the antrum and body of the stomach. Z Gastroenterol 28: 229–233
36. Stolte M, Eidt S, Ritter M, Bethke B (1989) Campylobacter pylori und Gastritis. Pathologe 10: 21–26

37. Gad A, Hradsky M, Furugard K, Malmodin B (1989) Campylobacter pylori and gastroduodenal ulcer disease. Scand J Gastroenterol 24 [Suppl 167]: 81–85
38. Rokkas T, Pursey C, Uzoechina E, Dorrington L, Simmons NA, Filipe MI, Sladen GE (1988) Non-ulcer dyspepsia and short term De-Nol therapy: a placebo controlled trial with particular reference to the role of Campylobacter pylori. Gut 29: 1386–1391
39. Drumm B, Sherman P, Chiasson D, Karmali M, Cutz E (1988) Treatment of Campylobacter pylori-associated antral gastritis in children with bismuth subsalicylate and ampicillin. J Pediatr 113: 908–912
40. Dooley CP, McKenna D, Humphreys H, Bourke S, Keane CT, Sweeney E, O'Morain C (1988) Histological gastritis in duodenal ulcer: relationship to Campylobacter pylori and effect of ulcer therapy. Am J Gastroenterol 83: 278–282
41. Gad A, Hradsky M, Furugard K, Malmodin B, Nyberg O (1989) Campylobacter pylori and non-ulcer dyspepsia. Scand J Gastroenterol 24 [Suppl 167]: 44–48
42. Hirschl AM, Hentschel E, Schütze K, Nemec H, Pötzi R, Gangl A, Weiss W, Pletschette M, Stanek G, Rotter ML (1988) The efficacy of antimicrobial treatment in campylobacter pylori-associated gastritis and duodenal ulcer. Scand J Gastroenterol 23 [Suppl 142]: 76–81
43. Cadranel S, Keppens E, Koslowski M, Verhas M, Glupczynski Y, De Prez C (1990) Detection of HP infection in children by means of a standardised ^{13}C-urea-breath test. Rev Esp Enferm Dig 78 [Suppl 1]: A 84
44. Bell GD, Weil J, Harrison G, Morden A, Johnes PH, Gant PW, Trowell JE, Yoong AK, Daneshmed TK, Logan RF (1987) ^{14}C-Urea breath analysis as non-invasive test for Campylobacter pylori in the stomach. Lancet I: 1367–1368
45. Rauws EAJ, Royen EAV, Langenberg W, Woensel JV, Vrij AA, Tytgat GN (1989) ^{14}C-Urea breath test in Campylobacter pylori gastritis. Gut 30: 798–803
46. Marshall BJ, Surveyor I (1988) Carbon-14 urea breath test for the diagnosis of Campylobacter pylori associated gastritis. J Nucl Med 29: 11–16
47. Henze E, Malfertheiner P, Clausen M, Glasbrenner B, Lietzenmayer R, Schoetensack C, Burkhardt H, Adam WE (1989) Der C-14 Urea Test – Ein neuer nuklearmedizinischer Test für die Campylobacter-pylori-Diagnostik. Nucl Med 28: A34
48. Husebye E, O'Leary D, Skar V, Melby K (1990) How reliable are the ^{14}C-urea breath test and specific serology for the detection of gastric campylobacter? Scand J Gastroenterol 25: 725–730
49. Ormand JE, Talley NJ, Carpenter HA, Shorter RG, Conley CR, Wilson WR, DiMagno EP, Zinsmeister AR, Phillips SF (1990) ^{14}C urea breath test for diagnosis of Helicobacter pylori. Dig Dis Sci 35: 879–884
50. Coelho LGV, Chausson Y, Passos MCF, Sadala RU, Costa EL, Sabino CVS, Queiroz DMM, Mendes EN, Rocha GA, Oliveira CA, Lima GF, Fernandes MLM, Castro LP (1990) Test respiratoire á l'urée marquée au carbone-14 pour le diagnostic de la colonisation gastrique par Helicobacter pylori. Gastroenterol Clin Biol 14: 801–805
51. Bardhan K, Cole DS, Hawkins BW, Franks CR (1982) Does treatment with cimetidine extended beyond initial healing of duodenal ulcer reduce the subsequent relapse rate? Br Med J 284: 621–623

Colloidal Bismuth Subcitrate Plus Amoxycillin Versus Colloidal Bismuth Subcitrate Plus Ofloxacin in the Treatment of Chronic Antral Gastritis Positive for *Helicobacter pylori*

M. Vincenzi[1], F. Benini[2], M. Martinelli[2], P. Tessaro[3], and D. Vincenzi[4]

Introduction

Helicobacter pylori is indicated as a bacterium responsible for human chronic gastritis and is probably one of many major pathogenic factors for peptic disease.

Following the first description by Warren and Marshall [1–3] a number of studies have shown a close link between colonization and gastritis. This is particularly so in the case of active gastritis. Positivity for *H. pylori* in these cases ranges from 70% to 92% compared with 13% for nonactive gastritis.

The link between *H. pylori* and gastritis, as well as the association of the latter with dyspepsia, prompted therapeutic efforts to eradicate the microorganism and to improve dyspeptic symptoms [4].

A double therapeutic approach with colloidal bismuth subcitrate (CBS) and various antibiotics has been used in different trials to implement long-term *H. pylori* eradication [5–8]. Success may be explained by the local bactericide effect of CBS together with the systemic effect of the antibiotic. Another assumption is that CBS can make the microorganism sensitive (or more sensitive) to antibiotics by damaging its cytoplasmic membrane.

However, these combinations could have serious side effects, e.g., hypersensitivity to the drug, colitis associated with the use of antibiotics, as well as a possible onset of bacterial resistance.

The best combinations used to date with few side effects are CBS plus amoxicillin or CBS plus tinidazole. A triple therapeutic approach is more effective in eradicating *H. pylori*, but it is worsened by a number of side effects [7].

Materials and Methods

Sixteen patients (nine women and seven men) had endoscopy of the upper digestive tract because of dyspeptic symptoms. Chronic antral gastritis in an

[1]Divisione Medica II, Unita Sanitaria Locale 37, 48018 Faenza, Italy
[2]Laboratorio Analisi Sez. Microbiologia, Unita Sanitaria Locale 37, 48018 Faenza, Italy
[3]Cattedra Malattie Apparato Digerente, Universita di Padova, Italy
[4]Servizio Trasfusionale, Unita Sanitaria Locale 37, 48018 Faenza, Italy

J.M. Pajares et al. (Eds.)
H. pylori and Gastroduodenal Pathology

active stage and positive for *H. pylori* was shown in all patients. Gastritis was found to be superficial in eight patients and atrophic in the remaining eight (according to the Whitehead score). The average age was 48 years (mean 33–72); the sex-specific average was 41.9 years (mean 37–67) in women and 46.6 years (mean 33–72) in men. At least two antral biopsies were obtained from each patient with subsequent hematoxylin and eosin stain, Giemsa's stain, and the Warthin-Starry method was used to detect *H. pylori*.

Eight randomized patients were treated with CBS (240 mg b.d.) for 4 weeks plus amoxycillin (2 g per day) for 2 weeks (group 1). The remaining eight patients were treated with CBS (240 mg b.d.) for 4 weeks plus ofloxacin (600 mg per day) for 2 weeks (group 2).

Three weeks after the treatment was discontinued, endoscopy and a histologic examination of the antral mucosa were performed to assess *H. pylori* eradication. A serologic test for *H. pylori* was carried out in six patients (three in group 1 and three in group 2) along with endoscopy.

Results

Helicobacter pylori eradication was successful in six of 16 patients (37.5%) (three in group 1 and three in group 2). Inflammation was histologically improved in the microorganism-free patients, whereas it did not change in the others. The serologic test for *H. pylori* showed a decrease of antibody titers in the microorganism-free patients, whereas the values were unchanged in the remaining ones.

Conclusions

There is an established association of *H. pylori* with gastritis, ulcer disease, and possibly non-ulcer dyspepsia. Yet some gastroenterologists are not convinced of the pathogenic role of *H. pylori*, especially because the microorganism has been found in asymptomatic individuals with histologic evidence of gastritis.

In addition, *H. pylori* is often absent and the gastric mucosa is normal in patients complaining of dyspeptic symptoms. At present, the priority is to identify those dyspeptic patients whose symptoms are clearly associated with *H. pylori* and to carry out double-blind studies aimed at evaluating and accounting for a combined therapeutic approach with CBS and antimicrobial agents. This can also be supported by the hypothesis that "active" *H. pylori* gastritis might be seen as a condition promoting the onset of gastric neoplasm (even though this is a hazardous assumption).

Some gastroenterologists are thus persuaded by experience of the benefits obtained from a therapeutic approach to *H. pylori* eradication.

Our data do not seem to suggest a supremacy of CBS plus amoxicillin compared with CBS plus ofloxacin in the treatment of chronic antral gastritis positive for *H. pylori*. Moreover, neither combination seems to be better than CBS alone when we compare the proportions of success reported in literature.

Only a triple therapy with CBS plus amoxicillin (or tetracycline) plus metronidazole is appropriate to obtain a higher proportion of *H. pylori* eradication and to improve gastritis histologically during the long-term follow up. However, this regimen is certainly not recommended for all patients with non-ulcer dyspepsia colonized with *H. pylori*. Only patients with long-standing severe dyspeptic symptoms inadequately responding to conventional treatment could benefit from a triple therapy.

References

1. Warren JR, Marshall BJ (1983) Unidentified curved bacilli on gastric epithelium in active chronic gastritis. Lancet i: 1273–1275
2. Marshall BJ, Warren JR (1984) Unidentified bacilli in the stomach of patients with gastritis and peptic ulceration. Lancet i: 1311–1315
3. Marshall BJ, Hislop I, Glancy R, Armstrong J (1984) Histological improvement of active gastritis in patients treated with De-Nol. Aust N Z J Med 14 [Suppl 4]: 907
4. McNulty CAM, Gearty JC, Crump B et al. (1986) Campylobacter pyloridis and associated gastritis: investigator blind, placebo controlled trial of bismuth salicylate and erythromycin ethylsuccinate. Br Med J 293: 645–649
5. Shungu DL, Nalin DR, Gilman RH et al. (1987) Comparative susceptibilities of Campylobacter pylori to norfloxacin and other agents. Antimicrob Agents Chemother 31: 949–950
6. Hirschl AM, Hentschl E, Schütze K et al. (1988) The efficacy of antimicrobial treatment in Campylobacter pylori associated gastritis and duodenal ulcer. Scand J Gastroenterol 23 [Suppl 142]: 76–81
7. Börsch G, Mai U, Opferkuch W (1988) Oral triple therapy may effectively eradicate Campylobacter pylori in man: a pilot study. Gastroenterology 94: A44
8. Rokkas T, Pursey C, Simmons NA, Filipe MI, Sladen GE (1987) Non-ulcer dyspepsia and colloidal bismuth subcitrate therapy: the role of Campylobacter pyloridis. Gastroenterology 92: 1599
9. Borody T, Hennesey W, Daskalopoulos G, Carrick J, Hazell S (1987) Double-blind trial of De-Nol in non-ulcer dyspepsia associated with Camplyobacter pyloridis gastritis. Gastroenterology 92: 1324
10. Lambert JR, Borromeo M, Eaves EA, Hansky J, Korman M (1988) Efficacy of different dosage regimes of bismuth in eradicating Campylobacter pylori. Gastroenterology 94: A248

Factors Affecting Eradication of *Helicobacter pylori* with Triple Therapy

D.Y. Graham[1,4], G.M. Lew[1], H.M. Malaty[1], D.G. Evans[1], D.J. Evans, Jr.[1], P.D. Klein[1,2,5], L.C. Alpert[3], and R.M. Genta[1,3]

Introduction

Eradication of *Helicobacter pylori* infections has proven to be difficult [2–4, 6–8, 13]. Therapy with a single or with two antimicrobials has proven ineffective as evidenced by eradication rates much below 50% [3]. Recently, combinations of three antimicrobial drugs have been shown to eradicate more than 50% of *H. pylori* infections. Despite these apparent successes, the results have been variable and the factors influencing effectiveness of therapy remain unclear [3, 7, 8]. In this study we evaluated the effectiveness of a combination of three antimicrobial drugs (triple therapy) for eradication of *H. pylori* infection with special emphasis on the factors that might predict good versus poor responses.

Methods

The patients in this study had received antimicrobial triple therapy consisting of tetracycline hydrochloride (500 mg q.i.d.), metronidazole (250 mg t.i.d.) and bismuth subsalicylate tablets (150 mg bismuth per tablet; Pepto-Bismol, Procter and Gamble, Cincinnati, OH) (Table 1). Bismuth subsalicylate tablets were administered at a dosage of one with each meal and two at bedtime (70 patients) or two with each meal and two at bedtime (30 patients). For those with active peptic ulcers, antiulcer therapy (ranitidine, 300 mg after the evening meal) was administered for the initial 14 days of triple therapy. Assessment of compliance with antimicrobial therapy was done by pill count.

Pretreatment *H. pylori* status was confirmed by the ^{13}C-urea breath test [10, 11], and a sensitive, specific enzyme-linked immunosorbent assay (ELISA) [5] in all and by culture and histology in most. Eradication was defined as inability to demonstrate *H. pylori* one or more months after ending therapy.

[1] Department of Medicine,
[2] Department of Pediatrics,
[3] Department of Pathology,
[4] Division of Molecular Virology,
[5] USDA/ARS Children's Nutrition Research Center, Baylor College of Medicine, and Veterans Affairs Medical Center, Houston, Texas, USA

J.M. Pajares et al. (Eds.)
H. pylori and Gastroduodenal Pathology

Table 1. Summary of the therapies used in an attempt to eradicate *H. pylori* infections

Drugs	Dosage	Duration (days)
Therapy 1		
Metronidazole	250 mg t.i.d.	14
Tetracycline	500 mg q.i.d.	14
Bismuth subsalicylate	5 tablets daily	14
Therapy 2		
Metronidazole	250 mg q.i.d.	14
Tetracycline	500 mg q.i.d.	14
Bismuth subsalicylate	8 tablets daily	14
Therapy 3		
Metronidazole	250 mg q.i.d.	14
Tetracycline	500 mg q.i.d.	10
Bismuth subsalicylate	5 tablets daily	10
Therapy 4		
Metronidazole	250 mg q.i.d.	14
Tetracycline	500 mg q.i.d.	28
Bismuth subsalicylate	5 tablets daily	28

Statistical Methods

The data were analyzed by chi-square and stepwise regression analysis using the SAS program (SAS Institute, Cray, NC). We categorized patients into groups: three groups according to the diagnosis (duodenal ulcer, gastric ulcer, or simple gastritis), three groups according to the duration of therapy (10 days, 14 days, and 28 days), two groups according to the number of bismuth subsalicylate tablets taken per day (five or eight tablets), and two groups according to the percentage of medication taken. The stepwise regression was also repeated using only the patients in whom accurate assessment of compliance was available by actual pill count. The Mantel-Haenszel chi-square distribution was used to examine the relationship between the eradication rate and the type of ulcer, duration of therapy, number of bismuth tablets, and the percentage of prescribed medications taken.

Results

The evaluable patients included 70 with duodenal ulcer, 17 with gastric ulcer, and six with *H. pylori* gastritis not associated with ulcer disease. The median follow up for those in whom the infection was eradicated was 29 weeks (mean 33 weeks, range 4–79 weeks). The overall eradication rate was 87%. The factors evaluated for their effect on predicting eradication included age, gender, type of disease, duration of therapy, amount of bismuth subsalicylate (five or eight tablets daily), and compliance with the prescribed medications. Stepwise regression showed that compliance was the most important factor predicting success; the success rate was 96% for those who took > 60% of the prescribed medications and 69% for those who took less. For those taking more than 60% of the prescribed therapy the eradication rates were similar for (a) those receiving therapy for 14 days or when tetracycline and bismuth subsalicylate were continued for an additional 14 days; (b) patients with duodenal ulcer, gastric ulcer, and simple *H. pylori* gastritis; and (c) whether five or eight bismuth subsalicylate tablets were taken.

Complications to triple therapy were experienced by 20 patients (18.5%), 19 of whom were patients in whom follow up was obtained. Nausea was the most common side effect (45%); the majority of complications were minor.

Discussion

Helicobacter pylori infections have proven to be difficult to eradicate. The best eradication rates have been reported with combinations of antimicrobials with most combinations employing a bismuth salt, metronidazole or tinidazole, and amoxicillin or tetracycline. Our results are consistent with these observations as our overall eradication rate was 87%. Our study was the first to examine the factors responsible for a poor rate of eradication of *H. pylori*; we found that failure to take the prescribed medications was the variable that best predicted a poor result.

It is not clear whether other triple therapies using a bismuth and two other antimicrobials would achieve the same excellent results because a number of other variables were not examined. One possibly important variable that we did not test was the timing of medications in relation to meals. This may be important, especially in relation to bismuth administration, as administration with meals would tend to allow a longer contact time and possibly better intragastric distribution than administration fasting [9]. This hypothesis is also consistent with data that tablet formulations of antacids have a longer duration of action than do antacid liquids [1]. A previous study [12] has shown that administration of the swallow tablet of bismuth subcitrate was more effective if given four times daily than when given twice daily; in that study bismuth subcitrate was administered before meals. Additional studies will be required to examine whether liquid formulations of bismuth subsalicylate are equally effective to tablet formulations.

We conclude that triple therapy is effective for eradication of *H. pylori* and that future studies will need to take compliance into account for comparisons between regimens.

Acknowledgements. This work was supported by the Department of Veterans Affairs, by grant DK 39919 from the National Institute of Diabetes and Digestive and Kidney Diseases, by the US Department of Agriculture/Agricultural Research Service Children's Nutrition Research Center, and by the generous support of Hilda Schwartz. Pepto-Bismol tablets were generously supplied by Procter and Gamble.

References

1. Barnett CC, Richardson CT (1985) In vivo and in vitro evaluation of magnesium-aluminum hydroxide antacid tablets and liquid. Dig Dis Sci 30:1049–1052
2. Bayerdorffer E, Ottenjann R (1988) The role of antibiotics in Campylobacter pylori associated peptic ulcer disease. Scand J Gastroenterol 23 [Suppl 142]:93–100
3. Borsch GMA, Graham DY (1991) Helicobacter pylori. In: Collen MJ, Benjamin SB (eds) Pharmacology of peptic ulcer disease. Springer, Berlin Heidelberg New York (Handbook of experimental pharmacology, vol 99)
4. Borsch G, Mai U, Muller KM (1988) Monotherapy or polychemotherapy in the treatment of Campylobacter pylori-related gastroduodenal disease. Scand J Gastroenterol 23 [Suppl 142]:101–106
5. Evans DJ Jr, Evans DG, Graham DY, Klein PD (1989) A sensitive and specific serologic test for detection of Campylobacter pylori infection. Gastroenterology 96:1004–1008
6. Gastrointestinal Physiology Working Group, Morgan DR, Kraft W, Bender M, Pearson A (1988) Nitrofurans in the treatment of gastritis associated with Campylobacter pylori. Gastroenterology 95:1178–1184
7. Glupczynski Y, Burette A (1990) Drug therapy of Helicobacter pylori infection: problems and pitfalls. Am J Gastroenterol 85:1545–1551
8. Graham DY, Börsch GMA (1990) The who's and when's of therapy for Helicobacter pylori. Am J Gastroenterol 85:1552–1555
9. Graham DY, Evans DG (1990) Prevention of diarrhea by enterotoxigenic Escherichi coli: lessons learned with volunteers. Rev Infect Dis 12:S68–S72
10. Graham DY, Klein PD, Evans DJ Jr, Evans DG, Alpert LC, Opekun AR, Boutton TW (1987) Campylobacter pylori detected noninvasively by the 13C-urea breath test. Lancet i:1174–1177
11. Klein PD, Graham DY (1989) Campylobacter pylori detection by the ^{13}C-urea breath test. In: Rathbone B, Heatley V (eds) Campylobacter pylori and gastroduodenal disease. Blackwell, Oxford, pp 94–106
12. Lambert JR, Borromeo M, Eaves ER, Hansky J, Korman M (1988) Efficacy of different dosage regimes of bismuth in eradicating Campylobacter pylori. Gastroenterology 94:A248
13. Rauws EAJ, Langenberg W, Houthoff HJ, Zanen HC, Tytgat GNJ (1988) Campylobacter pyloridis-associated chronic active antral gastritis. A prospective study of its prevalence and the effects of antibacterial and antiulcer treatment. Gastroenterology 94:33–40

Comparison of Three Triple Therapies in the Eradication of *Helicobacter pylori*

G. Daskalopoulos

Introduction

The eradication of *Helicobacter pylori* infection has proved difficult. As with other chronic infections, evasion of the immune response may be a major factor contributing towards the persistence of infection and the difficulty in eradication with antimicrobials. Although *H. pylori* can be suppressed with single antibiotic therapy, recrudesence of the infection following the cessation of treatment is the rule [1, 2]. This occurs despite a high in vitro sensitivity of *H. pylori* to various single antibiotics [3] and apparent therapeutic mucosal antibiotic levels [4, 5]. The resistence to antibiotic therapy may be due to (a) diminished antibiotic activity in the acid environment of the stomach; (b) the development of resistent forms of *H. pylori*; and (c) the antibiotics not reaching the bacteria in its protected niche under the mucous layer. It is obviously difficult to determine if an antibiotic has reached a precise niche [4]; however, adequate therapeutic mucosal antibiotic levels appear to have been achieved. This would seem to suggest that lack of assistance from an immune response may be a significant factor in failure of single antibiotic therapy.

Empirically, triple antibiotic therapy has been shown to be the most efficacious in eradicating *H. pylori* [6] with eradication rates of up to 90% being reported. Historically, triple therapy has consisted of colloidal bismuth and metronidazole in combination with amoxicillin or tetracycline. Our experience has been that both are equally effective (unpublished data). Amoxicillin has two disadvantages: (a) it is theoretically less active in the acid environment of the stomach; (b) a group of patients will not be able to take this medication because of penicillin allergy. Tetracycline has the theoretical disadvantages of (a) chelating with bismuth (McNulty [4] proposed that this could be viewed as an advantage since it ensures delivery of the antibiotic to the bacteria to which the bismuth irreversibly binds); (b) it cannot be used in children because of staining of teeth. Its activity, however, does not appear to be altered by gastric acidity.

Aim/Methods

We have assessed three different triple therapy regimens in the eradication of *H. pylori* and assessed the efficacy of norfloxacin as a component of a triple

Marrickville Endoscopy Centre, 311 Marrickville Road, Marrickville 2204, Australia

J.M. Pajares et al. (Eds.)
H. pylori and Gastroduodenal Pathology

Table 1. Triple therapy to eradicate *H. pylori*

Patients (n)	Regimen	
	Abbreviation	Description
32	$B_2\ T_2\ M_2$	Colloidal bismuth 1 tablet q.i.d. for 2 weeks Tetracycline 250 mg q.i.d. for 2 weeks Metronidazole 200 mg t.d.s. for 2 weeks
30	$B_4\ T_2\ M_2$	Colloidal bismuth 1 tablet q.i.d. for 4 weeks Tetracycline 250 mg q.i.d. for 2 weeks Metronidazole 200 mg t.d.s. for 2 weeks
29	$B_2\ T_2\ N_2$	Colloidal bismuth 1 tablet q.i.d. for 2 weeks Tetracycline 250 mg q.i.d. for 2 weeks Norfloxacin 400 mg b.i.d. for 2 weeks

therapy. A total of 91 patients with confirmed *H. pylori* both at rapid urease testing and histopathology were treated with one of three different treatment regimens (Table 1).

All patients were re-endoscoped 2 weeks after completion of treatment, and biopsies were again taken for rapid urease testing and histopathology. Giemsa staining was used to assess the presence and grade (0–3) of *H. pylori.*

Results

The treatment regimens were compared for their ability to eradicate *H. pylori* (Fig. 1). Colloidal bismuth given for 4 weeks in combination with tetracycline and metronidazole (B4 T2 M2) was the most effective in eradicating *H. pylori,* achieving an 87% eradication rate. This, however, was not significantly different from the group given colloidal bismuth for only 2 weeks in combination with tetracycline and metronidazole (B2 T2 M2) which achieved a 75% eradication rate. Both of the above groups were superior to the combination of colloidal bismuth, teracycline and norfloxacin (B2 T2 N2) which achieved a 24% eradication rate.

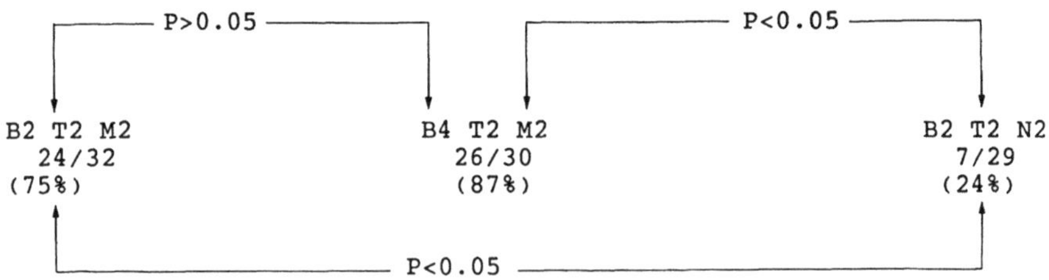

Fig. 1. The effectiveness of three different triple therapies in eradicating *H. pylori* as a percentage of the number of patients treated (p values shown). *B*, colloidal bismuth *T*, tetracycline *M*, metronidazole; *N*, norfloxacin

The grading of *H. pylori* was compared before and after treatment among the three groups (Fig. 2). Before treatment there was no difference in the grading of *H. pylori* among the three groups. After treatment there was improvement in the grading of *H. pylori* in the colloidal bismuth-, tetracycline- and metronidazole-treated groups (B2 T2 M2 and B4 T2 M2) but not in the colloidal bismuth-, tetracycline- and norfloxacin-treated (B2 T2 N2) group.

Twenty-two patients who were treated with B2 T2 N2 and eight patients who were treated with B2 T2 M2 failed therapy. These patients were offered second-line therapy with the alternate treatment combination. Five patients completed second-line therapy with B2 T2 N2 and none of the five had eradication on follow up (Fig. 3). Thirteen patients completed second-line therapy with B2 T2 M2 and 11 (84%) patients had successful eradication on follow up.

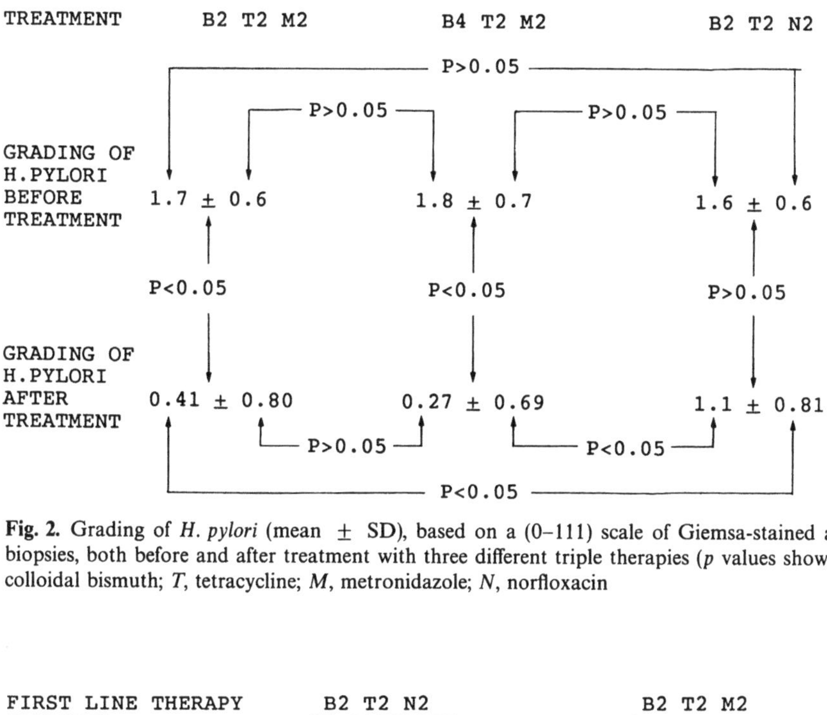

Fig. 2. Grading of *H. pylori* (mean ± SD), based on a (0–111) scale of Giemsa-stained antral biopsies, both before and after treatment with three different triple therapies (*p* values shown). *B*, colloidal bismuth; *T*, tetracycline; *M*, metronidazole; *N*, norfloxacin

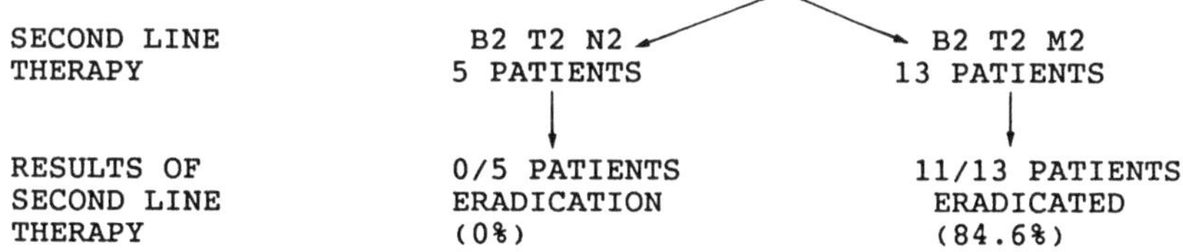

Fig. 3. Patients who failed first-line therapy were offered second-line treatment with the alternative triple therapy. The success of eradication with second-line therapy is shown. *B*, colloidal bismuth; *T*, tetracycline; *M*, metronidazole; *N*, norfloxacin. Both triple therapy regimens were given for 2 weeks

Discussion

These data show that not all triple therapies are equally effective. Despite in vitro evidence of the effectiveness of norfloxacin [3, 7] and the development of high mucosal concentrations of quinolone antibiotics [4], the eradication rate of *H. pylori* with norfloxacin as part of a triple therapy was low. In fact, the eradication rate with norfloxacin as part of a triple therapy was lower than we would have predicted. This may reflect the relative importance of metronidazole in the triple therapy, rather than the failure of norfloxacin. It has, however, been shown in vivo that norfloxacin given as a single agent is ineffective in the eradication of *H. pylori* [8]. As with any prolonged therapy, particularly requiring multiple medication, compliance does become important in treatment failure [9]. By crossing over our treatment failures we feel that we can exclude non-compliance in the norfloxacin triple therapy group as a cause of the low eradication rate as with cross-over these patients achieve an 84% eradication rate.

We chose to change the metronidazole in the triple therapy with norfloxacin because of reported resistance of *H. pylori* to metronidazole [10–12]. In some countries very high incidences of metronidazole resistance have been reported and seem to relate to previous exposure to nitroimidazoles [10, 11]. However, despite in vitro metronidazole resistance, there is debate as to whether this is of clinical significance in *H. pylori* eradication with triple therapy. Replacing the metronidazole in the triple therapy with norfloxacin does not improve the eradication rate and is not a useful solution to the problem of metronidazole resistance. It is likely that changing metronidazole with any other antibiotic, excluding another nitroimidazole, would produce similar results.

In this study we confirmed eradication at 2 weeks following treatment. The ideal time for confirmation of eradication is not known, although a working party at the World Congress of Gastroenterology (1990) recommended 4 weeks. Therefore, to substantiate our data further, we have reviewed our long-term follow up of patients successfully treated with triple therapy consisting of a combination of colloidal bismuth, tetracycline and metronidazole for 2 weeks at the dosage previously shown. "Successful" treatment was deemed as a negative urease test and a negative Giemsa stain for *H. pylori* on antral biopsies. The group consisted of 98 patients re-examined at 2–56 months (mean 38 months) after initial eradication. Twelve (12.2%) patients were found to have recurrence of *H. pylori* infection. Of the 98 patients, 74 (75.5%) had returned for follow up because of recurrence of dyspeptic symptoms and 11 (14.9%) had a recurrence of *H. pylori* compared with one of 24 (4.2%) asymptomatic patients. The difference between the two groups is not statistically significant ($p = 0.17$). It can be seen from these data, even in a selected group of patients to which many patients had returned because of symptomatic recurrence, that eradication at 2 weeks represented "true" eradication in 88% of cases. This figure is the same as that noted by Weil et al. [13] in a group of 32 patients who were *H. pylori* negative 1 month after *H. pylori* therapy, 28 (88%) patients remained negative at 3 months. Similar data have been obtained by Rauws et al. [14]. Logan et al. [15] showed recrudescence of *H. pylori* following treatment in 18 of 19 patients within 7 days by a ^{13}C-urea breath test and this was confirmed endoscopically at 10 days post therapy. One of 19 patients remained eradicated of *H. pylori*. In this instance colloidal bismuth was given alone. Whether a similar rapid recrudescence would be obtained after triple therapy has not been shown. We believe that our long-term recurrence data demonstrate that our triple therapy results would not have been significantly altered by reassessing eradication at 4 weeks as opposed to 2 weeks following completion of therapy. However, for uniformity and to allow for comparisons with other studies we have since changed to a routine 4-week reassessment after completion of therapy.

In vitro antibiotic sensitivity testing has proved to be of limited value in determining appropriate therapies for *H. pylori*. In my view there are, however, three approaches which may help in developing a better treatment for *H. pylori*. Firstly, there are several animal models [16–21] which may be appropriate for assessing antibiotic therapies for *H. pylori*. In the case of the ferret model [16, 17] the organism which naturally infects this animal is Helicobacter mustelae, which is closely related to *H. pylori*. Data obtained from this model with regard to antimicrobial therapy may be applicable to *H. pylori*. The rhesus monkey [21] has naturally occurring *H. pylori* and can also be experimentally infected. This model is, however, expensive and not readily available. The mouse model described by Hegedus Dick and Lee (in press) is of particular interest because of the ease of storage and handling of mice and the similarity to the human infected with *H. pylori* with regard to the difficulty of eradication. Secondly, follow-up assessments after completion of treatment can be markedly assisted by a ^{14}C- or ^{13}C- urea breath test [22–24]. Breath tests are less invasive diagnostic tools and give a result representative of the whole stomach, whereas single gastric biopsies may yield false-negative results. Lastly, biopsies may be taken and stored under liquid nitrogen [25]. These samples can later be thawed and the *H. pylori* cultured. This could be of value in assessing treatment failures and determining the role of antibiotic resistance in eradication failure.

I conclude with a word of caution that although we look towards developing a totally effective triple therapy or, even more desirably, an effective monotherapy, the difficulty we are facing in eradicating *H. pylori* may not be simply due to a failure of the antibiotics being used, but rather, as in the case of *Mycobacterium tuberculosis*, from a lack of assistance from an immune response.

References

1. Langenberg W, Rauws EAJ, Houthoff HJ, Oudbier JH, Widjojokosumo A, Tytgat GJN, Zanen HC (1987) Follow up of Campylobacter pyloridis associated gastritis after treatment with amoxacillin and/or colloidal bismuth subcitrate. In: Kaijser B, Falsen E (eds) Campylobacter, Vol 4. University of Göteborg, Göteborg, pp 366–367
2. Hirschl AM, Pletschette M (1989) Antibiotic treatment of Campylobacter pylori infection. In: Rathbone BJ, Heatly RV (eds) Campylobacter pylori and gastroduodenal disease. Balckwell, Oxford, pp 217–224
3. McNulty CAM, Dent JC (1988) Susceptibility of Campylobacter pylori to twenty-one antimicrobial agents. Eur J Clin Microbiol Infect Dis 7: 566–569

4. McNulty CAM, Dent JC, Ford GA, Wilkinson SP (1988) Inhibitory antimicrobial concentrations against Campylobacter pylori in gastric mucosa. J. Antimicrob Chemother 22:729–738
5. Westblom TU, Duriex DE, Madan E, Belshe RB (1990) Guinea pig model for antibiotic transport across gastric mucosa: inhibitory tissue concentrations of clindamycin against Helicobacter pylori (Campylobacter pylori) following two separate dose regimens. Antimicrob Agents Chemother 34/1:25–28
6. Borody T, Cole P, Noonan S, Morgan A, Ossip G, Maysey J, Brandl S (1988) Long term Campylobacter recurrence post-eradication. Gastroenterology 94:A43.
7. Shungu DL, Nalin DR, Gilman RH, Gadebusch HH, Cerami AT, Gill C, Weissberger (1987) Comparative susceptibilities of Campylobacter pylori to norfloxacin and other agents. Antimicrob Agents Chemother 31:949–950
8. Mertens JCC, Dekker W, Ligtvoet EEJ, Blok P (1989) Treatment failure of norfloxacin against Campylobacter pylori in patients with nonulcerative dyspepsia. Antimicrob Agents Chemother 33:256–257
9. Glupczynski Y, Lubbe M, Vanderlinden MP, Nyst JF (1989) Lack of antibiotic compliance in patients treated for Campylobacter pylori-associated gastritis. Am J Gastroenterol 84/9:1060–1064
10. Becx MCJM, Jansen AJHM, Clusener HAL, DeKoning RW (1990) Metronidazole-resistant Helicobacter pylori. Lancet 335 (8688): 539–540
11. Glupczynski Y, Burette A, De Koster E, Nyst JF, Deltenre M, Cadranel S, Bourdeaux L, De Vos D (1990) Metronidazole resistance in Helicobacter pylori. Lancet 335 (8695): 976–977
12. Goodwin CS, Marshall BJ, Blincow ED, Wilson DH, Blackbourn S, Phillips M (1988) Prevention of nitriomidazole resistance in Campylobacter pylori by coadministration of bismuth subcitrate: clinical and in vitro studies. J. Clin Pathol 41:207–210
13. Weil J, Bell GD, Jones PH, Gant P, Trowell JE, Harrison G (1988) Eradication of Campylobacter pylori: are we being misled? Lancet 2:1245
14. Rauws EAJ, Langenberg W, Houthoff HJ, Zanen HC, Tytgat GNJ (1988) Campylobacter pyloridis-associated chronic active gastritis. Gastroenterology 94:33–40
15. Logan RPH, Gummett PA, Polson RJ, Walker MM, Baron JH, Misiewicz J (1990) The recurrence of *H. pylori* in relation to duodenal ulcer. Gastroduodenal pathology and Helicobacter pylori. Rev Esp Enferm Dig 78 [Supp 1]: 72
16. Otto G, Fox JG, Wu PY, Taylor NS (1990) Eradication of Helicobacter mustelae from the ferret stomach: an animal model of Helicobacter (Campylobacter) pylori chemotherapy. Antimicrob. Agents Chemother 34/6:1232–1236
17. Gottfried MR, Washington K, Harrell LJ (1990) Helicobacer pylori-like microorganisms and chronic active gastritis in ferrets. Am J Gastroenterol 85/7:813–818
18. Lambert JR, Borromeo M, Pinkard KJ, Turner H, Chapman CB, Smith ML (1987) Colonization of gnotobiotic piglets with Campylobacter pyloridis- an animal model? J Infect Dis 155:1344
19. Radin MJ, Eaton KA, Krakowka S, Morgan DR, Lee A, Otto G, Fox J (1990) Helicobacter pylori gastric infection in gnotobiotic beagle dogs. Infect Immun 58/8:2606–2612
20. Engstrand L, Gustavsson S, Jorgensen A, Schwan A, Scheynius A (1990) Inoculation of barrier-born pigs with Helicobacter pylori: a useful animal model for gastritis type B. Infect Immun 58/6:1763–1768
21. Baskerville A, Newell DG (1988) Naturally occurring chronic gastritis and C. pylori infection in the rhesus monkey: a potential model for gastritis in man Gut 29:465–472
22. Bell GD, Weil J, Harrison G et al. (1987) ^{14}C-Urea breath analysis, a non-invasive test for Campylobacter pylori in the stomach. Lancet i:1367–1368
23. Marshall BJ, Surveyor I (1988) Carbon-14 urea breath test for the diagnosis of Campylobacter pylori associated gastritis. J. Nucl Med 29:11–16
24. Graham DY, Klein PD, Evans DJ, Evans DG, Alpert LC, Opekun AR, Boutton TW (1987) Campylobacter pyloridis detected by the ^{13}C-urea test. Lancet i:1174–1177
25. Ribeiro CD, Gray SJ (1987) Long term freeze storage of Campylobacter pyloridis. J. Clin Pathol 40:1265

Ciprofloxacin–Omeprazole Treatment for Eradication of *Helicobacter pylori*

J. Labenz[1], E. Gyenes[2], U. Peitz[1], and G. Börsch[1]

Introduction

Helicobacter pylori is gaining increasing importance in practical gastroenterology because of its causal role in the pathogenesis of chronic type B gastritis, in addition to emerging as a co-causal factor in the etiology of idiopathic ulcer disease. It has been shown that a drug-induced bacterial eradication is accompanied by an improvement or healing of gastritis and that the frequency of ulcer recurrences can be lowered substantially and significantly compared to the spontaneous course of the disease [1, 9]. Elaborate triple therapy schedules are appropriate to eradicate *H. pylori* in a high percentage of cases [2, 4]. However, a desire exists for a simple short-term therapy which is both low in complications and of adequate efficacy. Omeprazole and the gyrase inhibitor ciprofloxacin, which is highly effective in vitro against *H. pylori*, eliminate the bacteria in vivo only in occasional cases when taken one at a time [5–7, 10, 11]. Unge et al. [11] were able to show in a small-scale study that the combination of amoxicillin and omeprazole led to an eradication of bacteria in 62.5% of the cases, whereas amoxicillin and omeprazole on their own were ineffective in this regard.

The objective of the present pilot study was to investigate whether eradication of *H. pylori* can be attained to a noteworthy extent by an anacidity-modified antibiosis with the exquisitely acid-sensitive quinolone ciprofloxacin, which is highly effective in vitro.

Materials and Methods

Patients and Study Design. In an open pilot study, 20 *H. pylori*-positive patients, either outpatients or hospitalized, with an ulcer condition requiring treatment or a severe non-ulcer dyspepsia were treated for 1 week with 40 mg omeprazole (two Antra capsules) before breakfast and 2 × 500 mg ciprofloxacin (Ciprobay 500 film-coated tablets) 1 h after breakfast and after the evening meal. At the end of this therapy phase, ulcer patients were given 20 mg omeprazole in the

[1] Medical Division, Elisabeth Hospital, Academic Teaching Hospital of the University of Essen (GHS), Moltkestr. 61, 4300 Essen 1, FRG
[2] Department of Laboratory Medicine, Elisabeth Hospital, Academic Teaching Hospital of the University of Essen (GHS), Moltkestr. 61, 4300 Essen 1, FRG

J.M. Pajares et al. (Eds.)
H. pylori and Gastroduodenal Pathology

Table 1. Exclusion criteria

– Age under 18 years
– Pregnancy or lactation
– Simultaneous treatment with another antibiotic or bismuth preparation
– Severe concomitant diseases
– Putative lack of compliance
– Known hypersensitivity to quinolones
– Manifest disorder of clotting
– Resected stomach
– Lack of consent to participate in the study

morning preprandially or a full dose of an H2 receptor blocker at night up to the final follow-up examination. Dyspepsia patients did not receive further medication. The exclusion criteria are shown in Table 1. The inadmissible concomitant medication comprised bismuth preparations, antibiotics, pirenzepine, sucralfate and misoprostol. The *H. pylori* status and the symptom score were determined in hospitalized patients in the first 3 days after the end of therapy and, if the *H. pylori* result was negative, again after a further 4 weeks. In outpatients, the post-therapeutic *H. pylori* status was determined, at the earliest, 4 weeks after discontinuation of the medication. A patient was regarded as post-therapeutically *H. pylori*-negative when bacteria were not detected either in the biopsy urease test, microscopically after Gram staining or in culture.

Methods. At the beginning and at follow-up investigations, there was a clinical evaluation and an esophagogastroduodenoscopy in which two biopsies were taken from the antrum and two biopsies from the corpus. An antrum and a corpus biopsy were analyzed together in the biopsy urease test (Telen-Quick) and the two other samples were examined microscopically and cultured. The methods used have been described in detail elsewhere [3]. If necessary, focal findings were biopsied in addition. The data analysis comprised only a descriptive evaluation.

Results

The study group comprised 20 patients (12 women and eight men) with an average age of 56 years (range: 26–82 years). Four patients had a duodenal ulcer, 12 patients a gastric ulcer (one ulcer was ultimately shown to be a carcinoma by biopsy), and four patients had severe non-ulcer dyspepsia without endoscopically visible focal lesions (Table 2).

Helicobacter pylori could be identified in all patients before the beginning of therapy by means of urease test. The bacteria were microscopically visible in the Gram preparation in 18 cases and could be cultured in 16 cases.

The rate of eradication of the bacterium in the study group was 15% (three out of 20 patients). In two of these three *H. pylori*-negative ulcer patients, the macroscopically visible lesions had healed completely. In one patient, a residual defect 1 cm in size at the gastric angle was still found, as compared to an

Table 2. Ciprofloxacin/omeprazole combination therapy for eradication of *H. pylori*: overview of results

				Pretherapeutic			Posttherapeutic				
No.	Sex	Age (years)	Indication	BUT	Micro	Cult.	BUT	Micro	Cult.	Lesion	Eradication
1	F	43	DU	+	+	+	+	+	+	H	No
2	F	76	GU	+	+	–	+	–	–	H	No
3	F	58	GU	+	+	+	–	+	+	H	No
4	M	53	GU	+	+	+	+	+	+	H	No
5	M	68	GU	+	+	+	+	+	+	H	No
6	F	82	GU	+	–	–	–	–	–	H	Yes
7	F	50	NUD	+	+	+	+	+	+	–	No
8	M	26	NUD	+	+	+	+	+	+	–	No
9	F	48	GU	+	+	+	+	+	+	H	No
10	M	49	GU	+	+	–	–	–	–	NH	Yes
11	M	75	GU	+	+	+	+	–	–	H	No
12	M	39	GU	+	+	+	+	+	+	H	No
13	F	48	GU[1]	+	+	+	+	–	+	NH	No
14	F	39	NUD	+	+	+	+	–	+	–	No
15	F	67	GU[a, b]	+	+	+	–	+	+	NH	No
16	F	54	DU	+	–	+	+	+	+	NH	No
17	F	65	DU	+	+	+	+	+	+	H	No
18	F	73	DU	+	+	+	–	–	–	H	Yes
19	M	53	NUD	+	+	–	+	+	+	–	No
20	M	53	GU	+	+	+	+	+	+	NH	No

BUT, biopsy urease test; Micro, Gram-stained, mucosal smear preparation; Cult, specific *H. pylori* culture; *DU*, duodenal ulcer, *GU*, gastric ulcer, NUD, non-ulcer dyspepsia; H, healed; NH, not healed.
[a]Followup 1–3 days after the end of therapy.
[b]Carcinoma confirmed by the histology

ulcer size of 5 cm at the beginning of therapy. Two patients were not followed up 4 weeks after the end of therapy. One of these patients had a histologically confirmed ulcerating stomach carcinoma which was treated surgically. In this case, a persistence of positive bacterial findings had already been demonstrated immediately after therapy. A further female patient with a positive *H. pylori* status immediately after therapy received another schedule of antibiotic therapy. In two of the 11 other ulcer patients who remained *H. pylori*-positive after therapy, the lesions had not yet healed in the final follow up (Table 2).

The medication after the end of therapy up to the final control consisted of 20 mg omeprazole for 11 of the 14 ulcer patients, and three patients received a full dose of H2 receptor blocker at night. The corresponding rates of ulcer healing after 5 weeks were 90.9% (ten out of 11 patients) under omeprazole and 66.7% (two out of three patients) under H2 receptor blocker.

In a partial study group of ten inpatients who were followed up on days 1–3 after the end of therapy, we obtained negative *H. pylori* findings in six cases. Four out of these six patients were once more *H. pylori* positive 4 weeks later.

As possible side effects of the study medication, which did not have to be terminated prematurely by any patient, "subjective feeling of unrest" was reported in a dyspepsia patient with marked improvement of his symptoms of acid type, and "paresthesias in the right leg" were reported by an ulcer patient. The correlation with the medication appeared to be rather doubtful in both cases. Further adverse concomitant effects were not registered.

Discussion

Unge et al. [11] were able to show that the antibiotic efficiency (*H. pylori* eradication) of amoxicillin could evidently be substantially enhanced by a medication-induced anacidity. In this pilot study we therefore investigated whether a similar effect could also be attained with the quinolone ciprofloxacin, which has good efficacy in vitro, but an inadequate efficacy in vivo [5, 7, 10]. Ciprofloxacin acts bactericidally by impeding the bacterial enzyme "gyrase." Omeprazole specifically and selectively inhibits the enzyme H^+/K^+ ATPase (proton pump) of the parietal cells of the stomach. This suppresses the H^+ secretion in the gastric lumen independently of the nature of the acid stimulus. It has proved to be an effective agent for ulcer therapy which is probably superior to H2 receptor blockers in intermittent high-dose therapy. According to the data available, neither agent is able to eradicate *H. pylori* to a noteworthy extent on its own [6, 7, 10].

The antibacterial efficacy of the quinolones is pH dependent and declines with increasing acidity [8]. For this reason, the working hypothesis was formulated for the present pilot study that omeprazole-induced hypoacidity or at least almost complete anacidity would improve the inadequate antibacterial efficacy of ciprofloxacin against *H. pylori* in the acid stomach. Such a drug combination would almost be the ideal therapy of chronic ulcer in an acute phase. Whereas omeprazole-induced hypochlorhydria or achlorhydria would heal practically any acute ulcer, substantial prophylaxis of recurrences would be expected from the eradication of *H. pylori*.

However, the low rate of *H. pylori* eradication of 15% is to be emphasized as a main result of the present study, so that an adjuvant ciprofloxacin medication over 1 week with the objective of eliminating the bacteria and thereby reducing the rate of recurrence cannot be recommended in the concept for treatment of acute ulcers. Moreover, our data also support the previous view that omeprazole on its own does not eliminate *H. pylori* to a clinically relevant extent and that the *H. pylori* status (biopsy urease test, microscopic detection in Gram-stained mucosal smears, specific culture) can only be reliably appraised 4 weeks after the end of a bacterial eradication treatment.

References

1. Börsch G, Mai U, Opferkuch W (1989) Short- and medium-term results of oral triple therapy to eradicate C. pylori. Gastroenterology 96: A53
2. Börsch G, Wegener M, Mai U, Opferkuch W (1989) Efficiency of oral triple therapy to eradicate Campylobacter pylori. In: Megraud F, Lamouliatte H (eds) Gastroduodenal pathology and Campylobacter pylori. Elsevier, Amsterdam, pp 595–598
3. Börsch G, Labenz J, Rehner M, Rühl GH, Gyenes E (1990) Nachweis von Helicobacter pylori. Muench Med Wochenschr 132: 391–394
4. Borody T, Cole P, Noonan S, Morgan A, Ossip G, Maysay J, Brandl S (1988) Long-term Campylobacter recurrence post-eradication. Gastroenterology 94: A43
5. Glupczynski Y, Labbe M, Burette A, Delmee M, Avesani V, Bruck C (1987) Treatment failure of ofloxacin in Campylobacter pylori infections. Lancet i: 1096
6. Graham DY (1989) Campylobacter pylori and peptic ulcer disease. Gastroenterology 96: 615–625
7. Hirschl AM, Stanek G, Rotter M, Hentschel E, Schütze K (1987) Ulcus duodeni und Antibiotika-Therapie. Dtsch Med Wochenschr 112: 781
8. McNulty CAM (1989) Bacteriological and pharmacological basis for the treatment of Campylobacter pylori infection. In: Rathbone BJ, Heatley RV (eds) Campylobacter pylori and gastroduodenal disease. Blackwell, Oxford, pp 209–216
9. Marshall BJ, Goodwin CS, Warren JR, Murray R, Blincow ED, Blackbourn SJ, Phillips M, Waters TE, Sanderson CR (1988) Prospective double-blind trial of duodenal ulcer relapse after eradication of Campylobacter pylori. Lancet ii: 1437–1441
10. Stone JW, Wise R, Donovan IA, Gearty J (1988) Failure of ciprofloxacin to eradicate Campylobacter pylori from the stomach. J Antimicrob Chemother 22: 92–93
11. Unge P, Olsson J, Gad A, Gnarpe H (1989) Does omeprazole (40 mg O.M.) improve antimicrobial therapy directed towards gastric Campylobacter pylori in patients with antral gastritis? A pilot study. In: Megraud F, Lamouliatte H (eds) Gastroduodenal pathology and Campylobacter pylori. Elsevier, Amsterdam, pp 641–645

Amoxicillin–Omeprazole Treatment for Eradication of *Helicobacter pylori*

J. Labenz[1], E. Gyenes[2], G.H. Rühl[3], and G. Börsch[1]

Introduction

The spiral bacterium *Helicobacter pylori* is regarded as the causal factor in chronic type B gastritis and is said to play a co-causal role in the etiology of peptic ulcer disease. This bacterium is gaining increasing importance in practical gastroenterology since it could be shown that an eradication of the bacterium is accompanied by an improvement or healing of gastritis [9]. Also, the frequency of ulcer recurrences can be reduced substantially and significantly [1, 6, 10]. Up to now, eradication of *H. pylori* in a high percentage of cases (80%–94%) required elaborate triple therapy schedules consisting of a bismuth salt and a double antibiotic combination [2, 4]. A simple schedule of short-term therapy with low complications and adequate efficiency is still not available. Amoxicillin and omeprazole on their own bring about at best an elimination of the bacteria in individual cases [5]. In this connection, Unge et al. [12] were able to show in a small-scale study that the combination of omeprazole and amoxicillin led to an eradication in five out of eight patients, whereas omeprazole and amoxicillin alone were ineffective in the control groups.

The objective in the present study was to check in a large patient population whether amoxicillin therapy after induction of anacidity by omeprazole does indeed lead to an eradication of bacteria in more than 50% of the treated patients.

Materials and Methods

In an open study, 35 outpatient or hospitalized *H. pylori*-positive patients with ulcer disease or severe functional dsypepsia requiring treatment were treated for 1 week with 40 mg omeprazole before breakfast and 4×500 mg amoxicillin suspension 1 h before meals and before going to bed. As a rule, there followed a therapy with 20 mg omeprazole in the morning preprandially or an H2 receptor

[1]Medical Division, Elisabeth Hospital, Academic Teaching Hospital of the University of Essen (GHS), Moltkestr. 61, 4300 Essen 1, FRG
[2]Department of Laboratory Medicine, Elisabeth Hospital, Academic Teaching Hospital of the University of Essen (GHS), Moltkestr. 61, 4300 Essen 1, FRG
[3]Institute of Pathology, Ruhr University of Bochum, Universitätsstr. 150, 4630 Bochum, FRG

J M. Pajares et al. (Eds.)
H. pylori and Gastroduodenal Pathology

antagonist in full dose at night in ulcer patients up to the final follow-up examination. After completion of the study medication, dyspepsia patients did not receive any further specific pharmacotherapy. Exclusion criteria were a patient age under 18 years, pregnancy or lactation, simultaneous treatment with a bismuth preparation or another antibiotic, severe concomitant diseases, putative lack of compliance, known penicillin allergy, manifest disorder of clotting, stomach resection, and lack of consent to participate in the study.

Clinical evaluation (including symptom score) and a proximal intestinoscopy with taking of four antrum and four corpus biopsies were carried out before the beginning and not before 4 weeks after discontinuation of the study medication and in inpatients also on days 1–3 after the end of therapy. The antrum and corpus biopsies were analyzed with a biopsy urease test (Telen-Quick), specific culture, and histology after Giemsa staining. The respective methods have already been described in detail elsewhere [3]. In addition, focal findings were biopsied as required. The patient was regarded as post-therapeutically *H. pylori* negative when the final follow-up examination did not reveal bacteria either in the urease test, culturally, or histologically. The analysis of data was restricted to a descriptive evaluation.

Results

Of the 35 patients participating in the study, 31 completed the study without contravening the protocol. Three patients did not attend the envisaged follow-up examinations. One patient received intercurrent antibiotic treatment. The study group comprised 20 women and 11 men with an average age of 58.1 years (range: 27–86 years). Sixteen patients had a duodenal ulcer, eight a gastric ulcer, three a gastroduodenal double ulcer, and four had dyspepsia (three with presently inactive, but previously confirmed ulcer disease and one with functional dyspepsia). *H. pylori* could be demonstrated in all patients pretherapeutically by means of the biopsy urease test and histologically. It could be also detected in culture in 30/31 cases.

The rate of eradication of the bacteria at least 4 weeks after cessation of therapy was 61.3% (19 out of 31 patients) in the study group (Table 1). In four out of 11 patients who showed negative *H. pylori* findings on days 1–3 after therapy the bacterium could be detected again 4 weeks later. In two patients (one with omeprazole pretreatment), no bacteria of typical morphology could be detected pretherapeutically in preparations from antrum biopsies processed histologically, whereas *H. pylori* could be detected in the corpus biopsies. This constellation was found in four further patients after 5 weeks of omeprazole therapy.

Nineteen patients with ulcers received 20 mg omeprazole subsequent to the study medication. In all cases the peptic lesions had completely healed at the final follow-up examination. An eradication was attained in 14 cases, and five patients remained *H. pylori* positive. The ulcers healed in four patients with eradication of the bacterium under 300 mg ranitidine at bedtime. In a 54-year-old woman with persistent *H. pylori* colonization, a duodenal ulcer did not heal

Table 1. Results of omeprazole/amoxicillin combination therapy for eradication of *H. pylori*

	Patients	
	(*n*)	(%)
Eradication rates		
Total	19/31	61.3
Duodenal ulcer	10/16	62.5
Gastric ulcer	5/8	62.5
Gastroduodenal double ulcer	2/3	66.7
Functional dyspepsia	2/4	50.0
Ulcer healing		
H. pylori eradication	17/17	100
H. pylori persistance	8/10	80

under ranitidine medication. In two further patients who were treated with pirenzepine or an antacid, the peptic lesions had healed at the final follow-up examination. *H. pylori* could be detected in both cases. An 80-year-old female patient with gastric ulcer did not take any anti-ulcer medication after completion of the study medication. Both the ulcer and the *H. pylori* colonization of the gastric mucosa could be demonstrated in the follow-up investigation.

Eleven patients who were followed up on days 1–3 after therapy were already free of symptoms at this time with regard to the principal signs of the ulcer disease or the functional dyspepsia. At the final examination, 26 of the 31 patients did not show any symptoms of their underlying disease. Five patients complained of persistent symptoms. Four patients had a duodenal ulcer with persistent *H. pylori* colonization. Three of the ulcers had healed at the endoscopic follow-up examination, and one ulcer could still be demonstrated. In addition, there was the 80-year-old female patient with gastric ulcer who did not receive any specific medication.

A burning or furry feeling in the pharynx was complained of by four patients as a side effect during the therapy phase. This disappeared again spontaneously and did not lead to discontinuation of the study medication in any patient. Such paresthesias were not observed under therapy with omeprazole alone. In a 74-year-old female patient with duodenal ulcer, the therapy had to be terminated after 6 days because of diarrhea which stopped spontaneously after discontinuation of the medication. *H. pylori* could nevertheless be eradicated.

Discussion

On the basis of a study by Unge et al. [12], who were able to show in a small patient population that the antibiotic efficiency of amoxicillin with regard to eradication of *H. pylori* could be substantially enhanced after induction of luminal anacidity or hypoacidity, we checked the reproducibility of the results in a large group of patients.

Omeprazole specifically and selectively inhibits the enzyme H^+/K^+-ATPase (proton pump) of the parietal cells of the stomach. The H^+ secretion into the gastric lumen is thereby inhibited independently of the nature of the acid stimulus. It has proved to be an effective therapeutic agent which is probably superior to H2 receptor antagonists in intermittent high-dose therapy of peptic ulcer [7]. Major side effects have not been reported up to now in humans with a limited period of administration. Neither omeprazole nor the bactericidal amoxicillin on their own result in a noteworthy eradication of *H. pylori*. After oral administration of 500 mg amoxicillin, the drug concentration in the gastric mucosa attains values which are several times higher than the minimal inhibitory concentration of this substance for *H. pylori* [8]. The minimal inhibitory concentration (MIC) for *H. pylori* are appreciably lower at neutral pH than for example at pH 5.5 for a large number of antibiotics [8]. An induction of resistance by amoxicillin has not become known up to now in *H. pylori*.

The main result of the present study is a rate of eradication of *H. pylori* of 61.3% by a combined amoxicillin/omeprazole treatment for 1 week. This is a rate of success which has not been attained before with antibiotic monotherapy. The therapeutic schedule presented here thus comes close to the ideal conceptions of ulcer pharmacotherapy. Practically any peptic ulcer can be expected to heal with an omeprazole-induced anacidity or hypoacidity. Moreover, a substantial prophylaxis of recurrences is to be expected from the eradication of *H. pylori*. The concept of this therapy is simple and inexpensive. Apart from clinically insignificant paresthesias in the region of the pharynx, we only observed one significant side effect (self-limiting diarrhea) in our series. This compelled discontinuation of treatment after 6 days. Because of the hypothesis that the acid suppression attained with 40 mg omeprazole might not be sufficient in some patients to allow an optimal action of amoxicillin, we are presently carrying out a further study using a still higher dose of omeprazole.

As a supplementary finding, we found *H. pylori* histologically on the corpus mucosa but not on the antrum mucosa in six patients. Five of these patients were under omeprazole therapy. Similar observations have already been reported elsewhere [11]. They underscore that false-negative *H. pylori* findings are to be expected under omeprazole medication, especially when only antrum biopsies are analyzed.

References

1. Börsch G, Mai U, Opferkuch W (1989) Short- and medium-term results of oral triple therapy to eradicate C. pylori. Gastroenterology 96: A53
2. Börsch G, Wegener M, Mai U, Opferkuch W (1989) Efficiency of oral triple therapy to eradicate Campylobacter pylori. In: Megraud F, Lamouliatte H (eds) Gastroduodenal pathology and Campylobacter pylori. Elsevier, Amsterdam, pp 595–598
3. Börsch G, Labenz J, Rehner M, Rühl GH, Gyenes E (1990) Nachweis von Helicobacter pylori. Muench Med Wochenschr 132: 391–394
4. Borody T, Cole P, Noonan S, Morgan A, Ossip G, Maysay J, Brandl S (1988) Long-term Campylobacter recurrence posteradication. Gastroenterology 94: A43

5. Burette A, Glupczynski Y, Dereuck M, Labbe M, Deltenre M (1987) Campylobacter pyloridis and associated gastritis: investigator blind, placebo controlled trial with amoxicillin. 4th international workshop on Campylobacter infections, Göteborg, Sweden, p A28
6. Marshall BJ, Goodwin CS, Warren JR, Murray R, Blincow ED, Blackbourn SJ, Phillips M, Waters TE, Sanderson CR (1988) Prospective double-blind trial of duodenal ulcer relapse after eradication of Campylobacter pylori. Lancet ii: 1437-1441
7. McFarland RJ, Bateston MC, Green JRB, O'Donoghue DP, Dronfield MW, Keeling PWN, Burke GH, Dickinson RJ, Shreeve DR, Peers EM, Richardson PDI (1990) Omeprazole provides quicker symptom relief and duodenal ulcer healing than ranitidine. Gastroenterology 98: 278
8. McNulty CAM (1989) Bacteriological and pharmacological basis for the treatment of Campylobacter pylori infection. In: Rathbone BJ, Heatley RV (eds) Campylobacter pylori and gastroduodenal disease. Blackwell, Oxford, pp 209-216
9. Rauws EAJ, Langenberg W, Houthoff HJ, Zanen HC, Tytgat GNJ (1988) Campylobacter pyloridis-associated chronic active antral gastritis. A prospective study of its prevalence and the effects of antibacterial and antiulcer treatment. Gastroenterology 94: 33-40
10. Rauws EAJ, Tytgat GNJ (1990) Cure of duodenal ulcer associated with eradication of Helicobacter pylori. Lancet i: 1233-1235
11. Stolte M, Bethke B (1990) Elimination of Helicobacter pylori under treatment with omeprazole. Z Gastroenterol 28: 271-274
12. Unge P, Olsson J, Gad A, Gnarpe H (1989) Does omeprazole (40 mg O.M.) improve antimicrobial therapy directed towards gastric Campylobacter pylori in patients with antral gastritis? A pilot study. In: Megraud F, Lamouliatte H (eds) Gastroduodenal pathology and Campylobacter pylori. Elsevier, Amsterdam, pp 641-645

Eradication of *Helicobacter pylori* Cures Duodenal Ulcer Disease

E.A.J. Rauws

Introduction

The integrity of the gastroduodenal mucosa is often said to depend on the balance between aggressive and defensive factors. Gastric acid and pepsin are generally regarded as essential factors in ulcer disease but recently *Helicobacter pylori* has been identified as an important factor as well. Many studies have appeared since *H. pylori* proved to be the cause of active chronic gastritis. Duodenal ulcer has been known to be associated with antral gastritis, and in all patients with duodenal ulcer there is concomitant antral gastritis. As *H. pylori* is the most common cause of antral gastritis, it is not surprising that *H. pylori* infection is found in almost all patients with duodenal ulcer disease. However, the strong association between gastritis and duodenal ulceration does not prove a causal role for *H. pylori.*

Fewer studies have looked at the occurrence of *H. pylori* in the duodenal bulb. *H. pylori* does not colonize normal duodenal mucosa, but can be found on islands of gastric type epithelium in the duodenum. It has been suggested that *H. pylori* colonization of these islands of gastric metaplasia, probably induced by acid injury, causes duodenitis and probably duodenal ulceration [1].

So far no study has shown that *H. pylori* infection leads to peptic ulceration. However, several studies have now been carried out in duodenal ulcer patients evaluating the efficacy of bismuth compounds and antibiotics on ulcer healing and ulcer relapse.

Duodenal Ulcer and Ulcer Relapse

The reasons for the lower incidence of duodenal ulcer relapse after healing with colloidal bismuth subcitrate (CBS) compared to H2 receptor blockers are not clear [2–5]. CBS is a combination of various complex bismuth salts of citric acid and binds to protein in the ulcer crater, inhibits pepsin activity, stimulates the endogenous prostaglandin synthesis, but has also antimicrobial effects against *H. pylori.* In contrast, all other convential ulcer healing agents (H2 receptor antagonists, pirenzepine, sucralfate) and also omeprazole lack antimicrobial

Department of Gastroenterology, Academic Medical Centre, Meibergdreef 9, 1105 AZ Amsterdam, The Netherlands

J.M. Pajares et al. (Eds.)
H. pylori and Gastroduodenal Pathology

effects against *H. pylori* [6]. This difference might explain the lower or retarded ulcer relapse rate known for more than a decade after initial ulcer healing with CBS [3, 4].

Bismuth Salts

Treatment of duodenal ulcer with various forms of bismuth results in healing rates after 4 weeks of treatment that are similar (78%–84%) to those obtained with H2 receptor antagonists. Of more clinical interest is the observation that the period of remission following initial ulcer healing with CBS is longer than after treatment with H2 receptor antagonists. The lower ulcer relapse rate following treatment with CBS is now explained by either the eradication of *H. pylori* or by one of the many other poorly understood effects of CBS [7]. In 1983 it was suggested that *H. pylori* was the cause of antral gastritis and was also strongly involved in the etiology of duodenal ulcers. Since that time, many trials have been performed comparing the ulcer healing and ulcer relapse rate after treatment aimed at eradicating *H. pylori*. Coghlan et al. [8] allocated 66 consecutive patients with endoscopically diagnosed duodenal ulceration randomly to 6 weeks of CBS ($n = 32$), 5 ml four times a day, or cimetidine ($n = 34$) 400 mg twice a day. In each group 93% were initially positive for *H. pylori*. Post-treatment endoscopy showed that 23 patients in each group had healed ulcers. All 46 healed patients entered the follow-up study to evaluate ulcer relapse rates in relation to *H. pylori* status. They were allowed to use antacids, but not to take specific anti-ulcer therapy. Endoscopy was repeated after 1 year or if the patient suffered from recurrent ulcer-like symptoms. During the follow-up year, six patients were lost and one patient refused follow-up endoscopy, leaving 39 patients for whom full data were available.

Of these 39 patients, 24 remained *H. pylori* positive after treatment, and relapse occurred in 19 (79%), compared with only four of the 15 patients (27%) in whom *H. pylori* could not be isolated at the end of the treatment ($p < 0.01$). Of the ten patients who remained *H. pylori* negative in the follow-up year, only one patient (10%) had a recurrence of duodenal ulceration; this contrasted with 22 such recurrences in the 29 in whom *H. pylori* was still present or recurred after apparent eradication ($p < 0.001$).

Bismuth Salts and Antimicrobial Combinations

Marshall et al. [9] treated 100 consecutive patients with both duodenal ulcer and *H. pylori* infection. The patients were followed up after ulcer healing to see if eradication of *H. pylori* affected ulcer relapse. They were randomly assigned to 8 weeks of treatment with cimetidine or CBS, with tinidazole or placebo given concurrently from day 1 to 10 inclusively. Endoscopy with biopsies taken for culture were performed at entry, in weeks 10, 22, 34, and 62 and whenever symptoms recurred. No maintenance therapy was given. *H. pylori* persisted in all the cimetidine-treated patients ($n = 22$) and in 95% of those treated with cimetidine and tinidazole ($n = 29$).

Helicobacter pylori was eradicated in six of the 22 CBS/placebo treated group (27%), and in 19 of the 27 CBS/tinidazole group (70%). When *H. pylori* persisted, the duodenal ulcer healed in 61% and relapsed in 37/44 (84%) within 12 months. When *H. pylori* was cleared after treatment, the ulcer healed in 92% ($p < 0.001$). Only five of 24 (21%) of the ulcers relapsed in the following 12 months ($p < 0.0001$).

The authors concluded that the relapse rate was identical to that in the cimetidine-treated group if *H. pylori* was not eradicated. The presence of *H. pylori* and not the type of therapy was the factor that determined duodenal ulcer relapse.

The highest success rates for *H. pylori* eradication has been reported after using triple therapy (two antibiotics in combination with a bismuth compound) [10–12]. George et al. [10] treated 82 patients whose duodenal ulcers were recurrent or resistant to H2 receptor antagonist therapy. All *H. pylori*-positive patients were treated in a treatment protocol of ranitidine full dose over 4 weeks, followed by a 4-week triple therapy course consisting of CBS (one tablet q.d.s.) and tetracycline hydrochloride (500 mg q.d.s.) for 4 weeks, with metronidazole (200 mg q.d.s.) for the first 10 days. No further maintenance therapy was given. Repeat endoscopy and rebiopsy to determine *H. pylori* status was performed 1 month after stopping treatment, during further follow up if symptoms recurred, and once every following year. Side effects were common during triple therapy, especially during the first 10 days of therapy, probably caused by metronidazole. Nausea (25%), diarrhea (11%), drowsiness or dizziness (5%), and a burning sensation in the mouth (2%) were reported. No cases of antibiotic-induced colitis occurred. Duodenal ulcer healed in all 78 patients available for reendoscopy 4 weeks after cessation of triple therapy. *H. pylori* was eradicated in 75 of these 78 patients (96%). At 1 year 73 patients were reevaluated. Of those, 71 remained *H. pylori* negative and without duodenal ulcer relapse. The remaining two patients were *H. pylori* positive again with endoscopic duodenitis. Both these two patients were using H2 receptor antagonists again. After 2 years, 57 patients could be reevaluated. All 57 were still *H. pylori* negative and without duodenal ulcer relapse; they were not taking H2 receptor antagonists. At 3 years, 34 patients underwent reendoscopy and biopsy; 33 patients remained *H. pylori* negative. The remaining patient was *H. pylori* positive again, but no ulcer was found during endoscopy. Four years after stopping triple therapy, 15 patients were again reendoscoped. All were still *H. pylori* negative and ulcer free.

All the remaining 14 patients who did not return for endoscopy after 1–4 years were interviewed by telephone. All were symptom free and without medication.

Remarkably, during this prolonged *H. pylori*-negative follow up the sometimes heavily distorted anatomy of the duodenal bulb was restored.

The authors concluded that long-term eradication of *H. pylori* is possible and that in these patients the ulcer has not relapsed during a follow-up time of 4 years. Rauws and Tytgat [12] studied 50 patients with intractable duodenal ulcer disease to examine the relationship between ulcer recurrence and *H. pylori* infection. Intractability was defined as breakthrough ulcers on maintenance H2 receptor blockers or the recurrence of two or more ulcers in 1 year. Some patients did not comply with or had refused maintenance therapy. The patients were randomly assigned to treatment with CBS monotherapy (26 patients) for 4 weeks or CBS with amoxicillin (375 mg t.i.d.) for 4 weeks and metronidazole (500 mg t.i.d.) for the last 10 days of this 4-week period of therapy. After these regimens were taken, all patients started taking ranitidine, 150 mg, until the first control endoscopy in week 8.

Five patients, all on triple therapy, withdrew because of side effects (nausea, diarrhea, rash). Thus, 45 patients completed their treatment. Of these 45 patients, 38 had ulcer healing at reendoscopy 4 weeks after stopping treatment (21 of 26 patients after CBS followed by ranitidine and 17 of 19 patients after triple therapy followed by ranitidine). These 38 patients were followed up and underwent reendoscopy and biopsy 3, 6, and 12 months after treatment was stopped or earlier if symptoms recurred. No antacids were prescribed and no further ranitidine maintenance therapy was given after the endoscopy in week 8.

At endoscopy 1 month after the study treatment period (week 8) histological and bacteriological examinations failed to detect *H. pylori* in 17 patients, 15 of whom had received triple therapy. There was no endoscopic ulcer relapse in any of these 17 patients who remained *H. pylori* negative over the following 12 months, whereas 17 of the 21 patients who were still *H. pylori* positive 1 month after stopping treatment had endoscopically confirmed duodenal ulcer relapse during that time. The difference in duodenal ulcer relapse rates between the *H. pylori*-positive and -negative groups was highly significant ($p < 0.001$). Nine patients who remained positive for *H. pylori* and again had ulcer relapse were subsequently given triple therapy. Seven of these nine patients were eradicated of *H. pylori* and had no further ulcer relapse within the next 12 months.

In this study, all patients received the same dose of bismuth. The rate of *H. pylori* eradication was, however, lower in those patients treated with CBS monotherapy than in those treated with triple therapy. Eradication of *H. pylori* and not the prolonged action of CBS after absorption explains the lower ulcer relapse rate in these patients.

Conclusion

There is no proof that *H. pylori* infection causes duodenal ulceration. However, the markedly reduced relapse rate after *H. pylori* eradication and concomitant reduction of inflammation support a dominant role for *H. pylori* in ulcer diathesis. It has now been established that eradication of *H. pylori* changes the natural history of duodenal ulcer disease. After *H. pylori* eradication, no duodenal ulcer relapses were reported unless recrudescence of *H. pylori* occurred.

The Working Party that met during the World Congress of Gastroenterology 1990 discuss "*Helicobacter pylori*: causal agent in peptic ulcer disease?" advises that anti-*H. pylori* therapy should be considered in patients requiring either maintenance therapy or surgery, or where complications such as bleeding or perforation have occurred. The Working Party recommends a triple therapy of either CBS or bismuth subsalicylate, one tablet q.d.s., tetracycline hydrochloride 500 mg q.d.s., and metronidazole 400 mg t.d.s. over 14 days.

References

1. Wyatt JL, Rathbone BJ, Dixon MF, Heatly RV (1987) Campylobacter pyloridis and acid induced gastric metaplasia in the pathogenesis of duodenitis. J Clin Pathol 40: 841–884
2. Bardhan K, Cole DS, Hawkins BW, Franks CR (1982) Does treatment with cimetidine extended beyond initial healing of duodenal ulcer reduce the subsequent relapse rate? Br Med J 284: 621–623
3. Martin DF, Way SJ, Tweedle DEF, Hollanders D, Ravenscroft MM, Miller JP (1981) Difference in relapse rates of duodenal ulcer after healing with cimetidine or tripotassium dicitrate bismuthate. Lancet i: 7–10
4. Hamilton I, O'Connor HJ, Wood NC, Bradbury I, Axon ATR (1986) Healing and recurrence of duodenal ulcer after treatment with tripotassium citrato bismuthate (TBD) tablets or cimetidine. Gut 27: 106–110
5. Lane MR, Lee SP (1988) Recurrence of duodenal ulcer after medical treatment. Lancet i: 1147–1149
6. Rauws EAJ, Langenberg W, Houthoff HJ, Zanen HC, Tytgat GNJ (1988) Campylobacter-pyloridis-associated chronic active gastritis. Gastroenterology 94: 33–40
7. Konturek SJ, Radecki T, Piastucki I et al. (1987) Studies on the gastroprotective and ulcer-healing effects of colloidal bismuth subcitrate. Digestion 37 [Suppl] : 8–15
8. Coghlan JG, Gilligan D, Humphries H et al. (1987) Campylobacter pylori and recurrence of duodenal ulcers - a 12 months follow-up study. Lancet 2: 1109–1111
9. Marshall BJ, Goodwin CS, Warren JR et al. (1988) Prospective double-blind trial of duodenal ulcer relapse after eradication of Campylobacter pylori. Lancet 2: 1437–1442
10. George LL, Borody TJ, Andrews P et al. (1990) Cure of duodenal ulcer after eradication of Helicobacter pylori. Med J Aust 153: 145–149
11. Borody TJ, Cole P, Noonan S et al. (1989) Recurrence of duodenal ulcer and Campylobacter pylori infection after eradication. Med J Aust 151: 431–435
12. Rauws EAJ, Tytgat GNJ (1990) Eradication of Helicobacter pylori cures duodenal ulcer. Lancet i: 1233–1235

Two-Month Assessment of Duodenal Ulcer Healing and *Helicobacter pylori* Eradication Rates in Patients Treated for One Month with Omeprazole Alone or Combined with Antibiotics

V. Lamy[1], B. Ramdani[1], A. Floriani[1], D. Famerée[2], J. Cappelli[1], and R. Moisse[1]

Introduction

Helicobacter pylori is strongly associated with chronic gastritis and duodenal ulcer. Some authors have claimed that omeprazole is effective against *H. pylori* [1–8], and certainly omeprazole has been described as being one of the best treatments for duodenal ulcer disease.

Material and Methods

Eighteen patients (16 men and two women) with endoscopically proven duodenal ulcer were included in a prospective study. They received omeprazole 20 mg daily for 1 month, either as monotherapy in nine patients, combined with tinidazole 2 × 500 mg for 10 days in seven patients, combined with amoxicillin 4 × 500 mg for 10 days in one patient, and combined with amoxicillin 4 × 500 mg plus metronidazole 3 × 500 mg for 10 days in one patient.

The control examination was done 1 month after the end of these treatments. The patients were observed for duodenal ulcer healing and *H. pylori* eradication. Eradication of *H. pylori* was defined as negative culture, urease test, cytology and histology with six antral biopsies 1 month after the end of treatment.

Results

Table 1 summarizes the results of this study.

- All the treatments of this study are able to heal duodenal ulcer. However, the results are mediocre with only ten of 18 patients being cured.
- The patients treated with omeprazole alone remained *H. pylori* positive (nine out of nine patients) and were healed only in three out of nine cases.

[1]Gastroenterological Unit, Hôpital Civil de Jumet, 73 rue de Gosselies, 6040 Charleroi, Belgium
[2]Microbiological Unit, Hôpital Civil de Jumet, 73 rue de Gosselies, 6040 Charleroi, Belgium

J.M. Pajares et al. (Eds.)
H. pylori and Gastroduodenal Pathology

Table 1. Results

Patient No.	Sex	Day 0 HP	Day 0 DU	Treatment (28 days)	Day 56 control HP	Day 56 control DU
1	M	+	+	2	+	–
2	M	+	+	2	+	B
3	M	+	+	1	+	–
4	M	+	+	1	+	+
5	M	+	+	1	+	+
6	M	+	+	2	+	–
7	M	+	+	1	+	+
8	F	+	+	1	+	–
9	M	+	+	1	+	–
10	M	+	+	4	–	–
11	M	+	+	1	+	+
12	M	+	+	1	+	+
13	M	+	+	2	+	–
14	M	+	+	1	+	–
15	M	+	+	2	+	B
16	M	+	+	2	+	B
17	M	+	+	3	+	–
18	F	+	+	2	+	–

HP, *Helicobacter pylori* status; DU, duodenal ulcer; 1, omeprazole; 2, omeprazole plus tinidazole; 3, omeprazole plus amoxicillin; 4, omeprazole plus metronidazole plus amoxicillin; B, bulbitis.

- All five patients with an active duodenal ulcer at the control 1 month after the end of treatment had received omeprazole alone.
- All three patients with active bulbitis were treated with omeprazole plus tinidazole.
- Four tinidazole-sensitive patients became resistant to tinidazole after treatment, two had received omeprazole alone and two omeprazole plus tinidazole.
- Most of the patients remained *H. pylori* positive after these treatments.
- The one patient treated with omeprazole plus two antibiotics became *H. pylori* negative and was cured of his duodenal ulcer.

Discussion

It seems that omeprazole, one of the most potent anti-ulcer drugs currently available, can clear, but cannot really eradicate *H. pylori* [2, 4]. There has been confusion in the recent literature regarding this topic [3, 5–8]. Our study seems to confirm that omeprazole is effective in healing duodenal ulcer but cannot be used alone for *H. pylori* eradication. The real explanation as to the clearance effect of omeprazole on *H. pylori* is not known. It probably acts by pH changes induced in the stomach. The use of omeprazole combined with appropriate antibiotics will be a good therapeutic challenge in the future for both the healing of duodenal ulcer and *H. pylori* eradication.

Conclusion

- It seems interesting to continue the study of the clearance mechanism induced on *H. pylori* using omeprazole.
- A large clinical controlled study on duodenal ulcer healing and *H. pylori* eradication after omeprazole combined with appropriate antibiotics treatment is also very attractive.

References

1. Unge P, Gad A, Gnarpe H, Olsson J (1989) Does Omeprazole improve antimicrobial therapy directed towards gastric Campylobacter pylori in patients with antral gastritis? Scand J Gastroenterol 24 [Suppl 167]: 49–54
2. Mainguet P, Delmée M, Debongnie JC (1989) Omeprazole, Campylobacter pylori, and duodenal ulcer. Lancet 12: 389
3. Biasco G, Migliori M, Barbara L, Corinaldesi R, di Febo G (1989) Omeprazole, Helicobacter pylori, gastritis, and duodenal ulcer. Lancet 9: 1403
4. De Koster E, Nyst JF, Glupczunski Y, Deprez C, Van Gossum M, Deltenre M (1990) Treatment of Helicobacter pylori: one week CBS + Amoxicillin + Minocyclin + Omeprazole. In: Abstracts of the world congress of Gastroenterology, Sydney, 1990. Medicine Group Abington, UK, PD 94
5. Pretolani S, Bonvicini F, Careddu, Cilla D, Acompara P, Gasbarrini A, Gasbarrini G (1990) Effect of short term therapy with Omeprazole in patients with resistant ulcers and Helicobacter pylori gastritis. In: Abstracts of the world congress of Gastroenterology, Sydney, 1990. Medicine Group, Abington, UK, PD 100
6. Tessaro P, Di Mario F, Rugge M, Baffa R, Pasqualetti P, Vianello F, Glorioso S, Naccarato R (1990) Efficacy of Omeprazole in eradicating Helicobacter pylori from gastric mucosa. In: Abstracts of the world congress of gastroenterology, Sydney, 1990. Medicine Group, Abington, UK, PD 109
7. Catalano F, Mangiameli A, Toscano MA, Inserra G, Monello S, Brogna A, Ayoubi Khajekini MT, Rizzo G, Blasi A (1990) Helicobacter pylori and non-ulcer dyspepsia: efficacy of Omeprazole treatment. In: Abstracts of the world congress of gastroenterology, Sydney, 1990. Medicine Group, Abington, UK, PD 116
8. Vigneri S, Termini R, Scialabba A, Pisciotta G, Scarpignato E, Tessaro P, Di Mario F, Naccarrato R (1990) Efficacy of Omeprazole in healing duodenal ulcer and eradicating Helicobacter pylori from gastric mucosa. In: Abstracts of the world congress of Gastroenterology, Sydney, 1990. Medicine Group, Abington, UK, PP 928

Helicobacter pylori and Healing of Refractory Duodenal Ulcer: Preliminary Results of a Randomised Trial

O. Sangaletti, M. Lazzaroni, H. Zhu, and G. Bianchi Porro

Introduction

Helicobacter pylori (HP) infection is very frequently observed in patients suffering from gastric and, in particular, duodenal ulcer with a mean incidence of 60% and 86%, respectively, as opposed to a 24% occurrence in asymptomatic volunteers [1]. HP infection of the gastric antrum is considered potentially capable of influencing ulcer healing rates and may well play a role in duodenal ulcer resistance to H2 antagonists. To explore the possibility of HP involvement, we undertook a randomised study of refractory duodenal ulcers comparing the efficacy of colloidal bismuth subcitrate (CBS) plus antibiotics active on HP with that of sucralfate, a drug with similar anti-ulcer properties but devoid of anti-HP activity.

Patients and Methods

Eleven patients five men, six women; mean age 47.8 $\pm$ 8.07 years) with HP-positive duodenal ulcers resistant to 8 weeks of therapy with full-dose H2 antagonists were entered into the study. They were randomly allocated to treatment with CBS 120 mg q.i.d for 4 weeks with the addition of amoxicillin 1 g t.i.d. during week 1 and 500 mg b.i.d. during week 2 or with only sucralfate 1 g q.i.d. for 4 weeks. Ulcers endoscopically unhealed after 4 weeks were openly treated with the other form of treatment for a further 4 weeks. Endoscopy was performed at the end of treatment and 4 weeks after cessation of therapy. HP infection was demonstrated on gastric biopsies by histology (Giemsa staining) and rapid urease test. Eradication of HP was defined as negative histology and rapid urease test of antral biopsies 1 month after stopping therapy. Immunoglobulin G (IgG) levels were quantified by enzyme-linked immunosorbent assay (ELISA) (G.A.P. Test; BIO-RAD Laboratories, Segrate, Milan). All the samples, diluted 1:200, were assayed simultaneously, and the absorbance of the test was read at 405 nm. The sensitivity and the specificity of the test were 80% and 100%, respectively. The antibody activity was expressed as optical density (OD), and samples with an OD above 0.150 were considered as positive.

Gastrointestinal Unit, L.Sacco Hospital, Via G.B. Grassi 74, 20157 Milan, Italy

J.M. Pajares et al. (Eds.)
H. pylori and Gastroduodenal Pathology

Statistical analysis was performed where appropriate using the Fisher exact test and signed rank test.

Results

The results relative to ulcer healing rate and HP infection are summarised in Table 1. Sucralfate therapy was associated with a healing rate of 13% (1/8) of refractory duodenal ulcers without HP clearance from the gastric antrum, whereas CBS plus antibiotics healed 100%. So far, only three patients of the CBS group have been endoscopically controlled at 4 weeks after stopping therapy; they are still healed and HP negative.

Antibody levels were measured in eight patients of the CBS group and in six patients of the sucralfate group. Figure 1 shows the antibody profiles for each patient at 4 and 8 weeks. The median values before and after therapy were 1.116 and 1.090 (p not significant) for the sucralfate group and 1.090 and 0.955 ($p < 0.05$) for the CBS plus antibiotics group. In the CBS group, three patients underwent control 4 weeks after treatment cessation; a further reduction of antibody titres was observed (Fig. 1).

Table 1. Preliminary results of a controlled comparison between CBS plus antibiotics and sucralfate in the treatment of resistant duodenal ulcers

	CBS + antibiotics (n = 8)		Sucralfate (n = 8)		
	(n)	(%)	(n)	(%)	P
Healing at 4 weeks	8/8	100	1/8	13	< 0.05
Healing at 8 weeks	3/3	100			
HP clearance	8/8	100	0/8		
HP eradication (1 month after healing)	3/3	100			

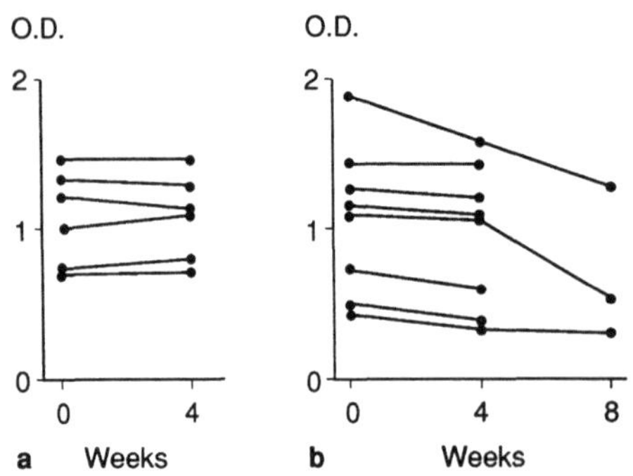

Fig. 1 a, b. Antibody profile before and after therapy with sucralfate (**a**) and CBS plus antibiotics (**b**)

Discussion

The efficacy of CBS in healing duodenal ulcer [2] and in preventing relapse [3–9] has been widely shown. The activity of CBS against HP is also well established, and several studies suggest that eradication of HP is the key to preventing ulcer relapse [3, 7, 10–12]. To date few studies have explored the possibility that HP might be involved in the development of refractoriness to H2 blockers. Two controlled clinical trials have shown that these ulcers respond well to short-term treatment with CBS (with healing rates of 80% and 90% after 4 and 8 weeks of treatment, respectively) [13–14], and a pilot study showed that the combination of H2 antagonists with an antibiotic (200 mg b.i.d. ofloxacin) is capable of 100% healing of HP-positive resistant duodenal ulcers [15]. Our preliminary results confirm the possibility of HP infection playing an important role in the development of refractoriness to H2 antagonists; the healing rate in the CBS plus antibiotic group was significantly higher than that in the sucralfate group. After therapy, all patients in the CBS group were HP negative, and a significant decrease of IgG antibodies was also observed. The eradication of HP, only observed in three patients, was associated with a further decrease of IgG levels thus confirming the usefulness of IgG assay in evaluating the efficacy of an antibacterial therapy.

References

1. Arakawa T, Kobayashi K (1990) Association of Helicobacter pylori with gastritis, duodenitis and peptic ulcer disease. Drug Invest 2 [Suppl 1]:46–51
2. Wagstaff AJ, Benfield P, Monk JP (1988) Colloidal bismuth subcitrate: a review of its pharmacodynamic and pharmacokinetic properties, and its therapeutic use in peptic ulcer disease. Drugs 36:132–157
3. Smith AC, Price AB, Borrielli P, Levi AJ (1988) A comparison of ranitidine and tripotassium dicitrate bismuthate (TBD) in relapse rates of duodenal ulcer: the role of Campylobacter pylori (CP) (abstract). Gastroenterology 94:431
4. Martin DF, Hollanders D, May SJ, Ravenscroft MM, Tweedle DEF, Miller JP (1981) Difference in relapse rates of duodenal ulcer after healing with cimetidine or tripotassium dicitrato bismuthate. Lancet i:7–10
5. Hamilton I, O'Connor HJ, Wood NC, Bradbury I, Axon ATR (1986) Healing and recurrence of duodenal ulcer after treatment with tripotassium dicitrato bismuthate (TDB) tablets or cimetidine. Gut 27:106–110
6. Lee FI, Samlof IM, Hardman M (1985) Comparison of tripotassium dicitrato bismuthate tablets with ranitidine in healing and relapse of duodenal ulcers. Lancet i:1299–1301
7. Coghlan JG, Gillgan D, Humphries H, McKenna D, Dooley C, Sweeney E, Keane C, O'Morain C (1987) Campylobacter pylori and recurrence of duodenal ulcers - 12 month follow-up study. Lancet ii:1109-1111
8. Kang JY, Piper DW (1982) Cimetidine and colloidal bismuth subcitrate in the treatment of chronic duodenal ulcer, comparison of initial healing and recurrence after healing. Digestion 23:73–79
9. Schreeve DR, Klass HJ, Jones PE (1983) Comparison of cimetidine and tripotassium dicitrato bismuthate in healing and relapse of duodenal ulcers. Digestion 38:96–101
10. Lambert JR, Borromeo M, Korman MC, Hansky J, Eaves ER (1987) Effect of colloidal bismuth subcitrate (De-Nol) on healing and relapse of duodenal ulcers—role of Campylobacter pyloridis. Gastroenterology 92:1489 (abstr)

11. Borody T, Cole P, Noonan A, Morgan A, Ossip G, Mesey J, Brand L (1988) Long-term Campylobacter pylori recurrence post-eradication. Gastroenterology 94:43
12. Goodwin CS, Marshall BJ, Blincow ED, Wilson DH, Blackbourn S, Phillips M (1988) Prevention of notomidazole resistance in Campylobacter pylori by co-administration of colloidal bismuth subcitrate: clinical and in vitro studies. J Clin Pathol 41:207–210
13. Lam SK, Lee NW, Koo J, Hui WM, Fok KH, N M (1984) Randomised cross-over trial of tripotassium dicitrato bismuthate vs high dose cimetidine for duodenal ulcers resistant to standard dose of cimetidine. Gut 25:703–706
14. Bianchi Porro G, Parente F, Lazzaroni M (1987) Tripotassium dicitrato bismuthate (TDB) vs two different dosages of cimetidine in the treatment of resistant duodenal ulcers. Gut 28:907–911
15. Bayerdorffer E, Pirlet TH, Sommer A, Kasper G, Ottenjan R (1988) Ofloxacin in der Therapie "resistenter ulcera duodeni". Eine Pilotstudie. Z Gastroenterol 26:155–159

Treatment with Amoxicillin and Colloidal Bismuth Subcitrate in Children with *Helicobacter pylori*-Associated Peptic Disease

C. De Giacomo[1], L. Villani[2], and R. Fiocca[2]

Helicobacter pylori Infection in Children

In children *Helicobacter pylori* infection is strikingly associated with peptic disease [1]. Family history of peptic disease, previous digestive procedures, and nonspecific epigastric pain are the most frequently encountered clinical features [2]. Isolated vomiting and more severe symptoms such as hematemesis are rarely described. At endoscopy antral nodularity, and gastric and duodenal ulcers are the typical lesions [2]. The infection is almost invariably associated with histologic evidence of chronic gastritis [2, 3]. Serum immune response against the bacterium with production of specific immunoglobulin G (IgG) and IgA is found in the majority of infected children [4, 5]. An increase in the level of serum pepsinogen I usually occurs [5].

Treatment in Children

Long-term eradication of the infection is associated with symptomatic improvement, healing of peptic ulcers, associated with a reduced relapse rate, and histologic improvement [2, 5, 6]. In spite of these results, treatment is not well established. In fact, as in adults, antibiotic monotherapy demonstrated poor efficacy in eradicating *H. pylori* and improving gastritis [7, 8]. The association of amoxicillin with tinidazole [5] or bismuth salts [2, 9] allowed better results, corresponding to a similar experience in adults. In addition, the number of pediatric patients is low, and long-term follow up to confirm effective eradication are not always available [9].

Open Pilot Study on Amoxicillin and Colloidal Bismuth Subcitrate in Children: Patients and Methods

Twenty-nine children and adolescents (mean age $\pm$ SD 12.08 $\pm$ 4.54) with *H. pylori* infection were studied. The clinical findings at diagnosis were: recurrent epigastric pain in 21, associated with recurrent vomiting in three, and with

[1]Clinica Pediatrica, IRCCS Policlinico S. Matteo, Università di Pavia, 27100 Pavia, Italy
[2]Istituto di Anatomia Patologica, IRCCS Policlinico S. Matteo, Università di Pavia, 27100 Pavia, Italy

J.M. Pajares et al. (Eds.)
H. pylori and Gastroduodenal Pathology

hematemesis in one. At the time of endoscopy, eight children had no symptoms related to gastroduodenal disease; in these eight, endoscopy was carried out for indications other than dyspepsia (follow up of previous diseases, evaluation of esophageal varices, investigation of failure to thrive).

After diagnosis of *H. pylori* infection, an oral treatment was started as follows: amoxicillin (70 mg/kg per day) in two doses for 2 weeks and colloidal bismuth subcitrate (CBS) (De-Nol, Gist-Brocades, The Netherlands, 480 mg Bi_2O_3/1.73 per square meter of body surface per day) in three or four divided doses for 4 weeks. In seven younger children (mean age of 8.4 years, range of 5–12 years), the blood bismuth level was monitored by atomic mass spectrometry. All but one of the patients completed the course of therapy. The one patient who did not complete the course was a 17-year-old girl affected by erosive antral gastritis who discontinued the treatment because of vomiting. After 2–13 months from cessation of therapy endoscopy, with antral biopsies, was repeated in 22 patients, while eight refused further endoscopy. Parental informed consent was requested to start the treatment.

Endoscopy

Endoscopic examination of the upper digestive tract was carried out by an Olympus GIF P3 endoscope before starting the treatment. In 22 patients endoscopy and antral biopsy were repeated after a period of 2–13 months after the treatment was discontinued. Endoscopy was performed after i.v. diazepam sedation. Parental informed consent was requested before endoscopy.

Histology

Hematoxylin and eosin- and Giemsa-stained antral biopsies were evaluated by two different pathologists (RF, LV). Classification of chronic gastritis was made following published criteria [10]. Grading of gastritis was based on the criteria of Marshall et al. [11], slightly modified as follows: briefly, the presence and density of mononuclear cells in the lamina propria (score of 0–3), the presence and density of intraepithelial granulocytes (0–2), the extent of micropapillary and microerosive alterations of surface epithelium (0–2), and the presence and severity of mucous depletion (0–2) were each assessed separately. The presence and number of bacteria were scored from 0 to 2, but this score was excluded from the total gastritis score.

Serology

An enzyme-linked immunosorbent assay (ELISA) with a suspension of six isolates of *H. pylori* (whole cells) as antigen was developed. Serum samples for specific IgG and IgA were evaluated before treatment in 22 patients and also after treatment in 17. This method showed 100% of positive predictivity when both IgG and IgA are positive in the same patient [4].

Statistical Analysis

Comparison of gastritis scores before and after therapy was performed using the Student's *t* test for paired samples and the Wilcoxon test. Correlation between gastritis and *H. pylori* scores was obtained with the calculation of the linear correlation coefficient and the Spearman's test. Analysis of anti-*H. pylori* antibody levels was carried out by the Student's *t* test for paired samples after logarithmic transformation of values.

Results

Disappearance or improvement of symptoms was recorded in 14 of 21 symptomatic children. One child, who had not complained of epigastric pain before endoscopy and treatment, developed typical dyspepsia some months after unsuccessful treatment. Eight children had persistence of epigastric pain during follow up: six of these were re-evaluated by endoscopy and antral biopsy showing persistence of *H. pylori* infection in four.

Endoscopy

Endoscopic follow up was performed in 22 patients (Table 1). Healing of gastric ulcer in one and duodenal ulcer in two patients was found. Antral nodularity was still present in four patients (in three of these the infection was cleared).

Table 1. Results of treatment in 22 children with endoscopic and histologic follow up

	Before treatment	*H. pylori* status After treatment	
		Negative	Positive
Children (*n*)	22	16	6
Digestive symptoms (*n*)	16	2	4
Endoscopy			
Normal antrum (*n*)	5	8	2
Nodular antritis (*n*)	11	3	1
Hyperemia (*n*)	6	5	3
Duodenopathy (*n*)	13	6	3
Gastric ulcer (*n*)	1	–	–
Duodenal ulcer (*n*)	2	–	–
Histology			
Normal antrum (*n*)	–	3	–
Chronic gastritis (*n*)	22	12	6
Activity (*n*)	9	–	2
Gastritis score (mean ± SD)	3.9 ± 1.6[ab]	1.2 ± 0.7[a]	3.5 ± 1.3[b]

[a] Significant difference; [b] not significant.

Table 2. Effects of treatment with or without eradication of *H. pylori* on specific antibody titers

		IgG	IgA
Patients cured ($n = 10$)	Pre-treatment	1.19 ± 0.29	1.29 ± 0.42
		$p < 0.001$	$p < 0.001$
	Post-treatment	0.30 ± 0.35	0.72 ± 0.45
Patients still infected ($n = 7$)	Pre-treatment	1.08 ± 0.24	1.27 ± 0.46
		NS	NS
	Post-treatment	0.95 ± 0.35	1.17 ± 0.50

Results are expressed as means ± SD after log transformation of values. NS, not significant.

Histology

The association of amoxicillin and CBS cleared *H. pylori* in 16 of the 22 children with histologic follow-up. The mean gastritis score ± SD for the 22 children was significantly reduced by treatment (means ± SD of the scores before vs. after treatment = 3.90 ± 1.61 vs. 1.86 ± 1.42). In the 16 patients in whom *H. pylori* was eradicated the difference between the scores before and after treatment was highly significant (means ± SD = 4 ± 1.73 vs. 1.20 ± 1.77). In the six persistently infected patients, no significant difference of the scores was found (means ± SD = 3.67 ± 1.37 vs. 3.50 ± 1.38). The calculation of the correlation coefficient ($r = 0.76$, $p < 0.001$) indicated a strong correlation between the gastritis score and the *H. pylori* score.

Serology

Six months after stopping treatment, the titers of *H. pylori*-specific IgG and IgA were significantly lower in children in whom treatment eradicated the infection, but not in children who remained infected (Table 2).

Side Effects and Blood Bismuth Levels

Only one child discontinued treatment because of recurrent vomiting. The other 28 completed the course of treatment without significant side effects. Values of blood bismuth level in the eight youngest children ranged from 5 to 28 μg/l with a mean of 11.6 μg/l.

Conclusions

Early unsuccessful attempts to obtain long-term eradication of *H. pylori* with antibiotic monotherapy [7, 8] prompted us to evaluate oral therapy combining amoxicillin with CBS. We performed a pilot study on 29 children and adolescents with *H. pylori*-associated chronic gastritis (three with peptic ulcers). Owing to the high rate of relapse reported in adults within the first 4 weeks after stopping therapy [12], we preferred to repeat endoscopy over 2 months after discontinuation of therapy. Our results show that combination of amoxicillin and CBS is effective in producing long-term eradication in 75% of children. The clearance of *H. pylori* was strikingly associated with the healing of peptic ulcers and significant improvement of chronic gastritis. The therapy including amoxicillin and CBS was well tolerated in all but one child. Blood bismuth levels were far below the proposed (and debated) threshold for discontinuing therapy [13], confirming previously reported results on the safety of CBS at the pediatric age [14].

References

1. Drumm B, Sherman P, Cutz E, Karmali M (1987) Association of Campylobacter pylori on the gastric mucosa with symptomatic gastritis in children. N Engl J Med 316:1557–1561
2. De Giacomo C, Fiocca R, Villani L, Lisato L, Licardi G, Diegoli N, Donadini A, Maggiore G (1990) Helicobacter pylori infection and chronic gastritis: clinical, serological, and histologic correlations in children treated with amoxicillin and colloidal bismuth subcitrate. J Pediatr Gastroenterol Nutr 11:310–316
3. Czinn SJ (1990) Helicobacter pylori induced gastritis in childhood. In: Malfertheiner P, Ditschuneit H (eds) Helicobacter pylori, gastritis and peptic ulcer. Springer, Berlin Heidelberg New York, pp 181–187
4. De Giacomo C, Lisato L, Negrini R, Licardi G, Maggiore G (1991) Serum immune response to Helicobacter pylori in children: epidemiologic and clinical applications. J Pediatr 119:205–210
5. Oderda G, Vaira D, Holton J, Ainley C, Altare F, Ansaldi N (1989) Amoxycillin plus tinidazole for Campylobacter pylori gastritis in children: assessment by serum IgG antibody, pepsinogen I, and gastrin levels. Lancet 1:690–692
6. Yeung CK, Fu KH, Yuen KY, Ng WF, Tsang TM, Branicki FJ, Saing H (1990) Helicobacter pylori and associated duodenal ulcer. Arch Dis Child 65:1212–1216
7. De Giacomo C, Maggiore G, Licardi G, Scotta MS, Fiocca R (1988) Effects of antibacterial treatment of Campylobacter pylori-associated gastritis in children. Gastroenterology 95:1699
8. Oderda G, Dell' Olio D, Morra I, Ansaldi N (1989) Campylobacter pylori gastritis: long term results of treatment with amoxycillin. Arch Dis Child 64:326–329
9. Drumm B, Sherman P, Chiasson D, Karmali M, Cutz E (1989) Treatment of Campylobacter pylori-associated antral gastritis in children with bismuth subsalicylate and ampicillin. J Pediatr 113:908–912
10. Cheli R, Perasso A, Giacosa A (eds) (1983) Gastritis. A critical review. Springer, Berlin Heidelberg New York
11. Marshall BJ, Armstrong JA, Francis GJ, Nokes NT, Wee SH (1987) Antibacterial action of bismuth in relation to Campylobacter pyloridis colonization and gastritis. Digestion 37 [Suppl 2]:16–30
12. Rauws EAJ, Langemberg W, Houthoff HJ, Zanen HC, Tytgat GNJ (1988) Campylobacter pylori-associated chronic active antral gastritis. A prospective study of its prevalence and the effects of antibacterial and antiulcer treatment. Gastroenterology 94:33–40
13. Benet LZ (1990) Consensus statement on bismuth pharmacokinetics following oral dosing of colloidal bismuth subcitrate. The world congress of gastroenterology, 26–31 August 1990, Sydney
14. Cadranel S, Goyens P, Zeghlache S (1990) Bismuthemia in children treated for Campylo (Helico) bacter pylori (HP) primary gastritis. Third joint meeting ESPGAN/NASPGN, 23–26 May, Amsterdam, p 110

Helicobacter pylori and Duodenal Ulcer Relapse: Effect of Bismuth-Free Short-Term Triple Therapy

L.G.V. Coelho[1], M.C.F. Passos[1], Y. Chausson[2], M.L.M. Fernandes[3], M.L.G. Nogueira[3], and L.P. Castro[1]

Introduction

Recent studies have demonstrated that the eradication of *Helicobacter pylori* is associated with a significant reduction in the rate of duodenal ulcer relapse or even the abolition of the disease [1–5]. The best results have been obtained with the use of a combination of bismuth salts with one or two antimicrobial agents [2–5]. However, bismuth salts are not available in many countries and in some their use has been banned because of past reports of toxicity from France and Australia [6–8].

Before the identification of *H. pylori*, Chinese studies had shown that furazolidone could heal peptic ulcer and lower the relapse rate after treatment [9, 10], and recently in vitro studies have demonstrated that *H. pylori* is very sensitive to furazolidone [11]. In a double-blind, randomized, prospective pilot study comparing furazolidone and cimetidine we demonstrated that furazolidone has an anti-ulcer action similar to cimetidine (Fig. 1) and an 18% *H. pylori* eradication rate 6 months after treatment. A significantly lower relapse rate was also demonstrated in patients treated with furazolidone over a 6-month follow-up period (Fig. 2) [12]. It was later observed that the combination of furazolidone, amoxicillin, and metronidazole was effective in eradicating *H. pylori* in 83% of asymptomatic *H. pylori*-positive volunteers [13].

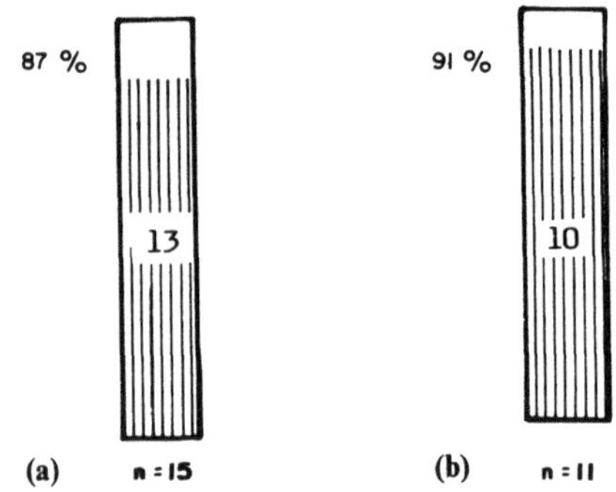

Fig. 1a, b. Duodenal ulcer healing rates. **a** Cimetidine group; **b** furazolidone group

[1] Gastroenterology, Nutrition, and Digestive Surgery Unit, University Hospital, Federal University of Minas Gerais, Av. Alfredo Balena 110, Belo Horizonte, MG 30.130, Brazil
[2] National Committee of Nuclear Energy (CDTN), Pampulha, Belo Horizonte, Brazil
[3] Department of Statistics, ICEX, Federal University of Minas Gerais, Pampulha, Belo Horizonte, MG, Brazil

J.M. Pajares et al. (Eds.)
H. pylori and Gastroduodenal Pathology

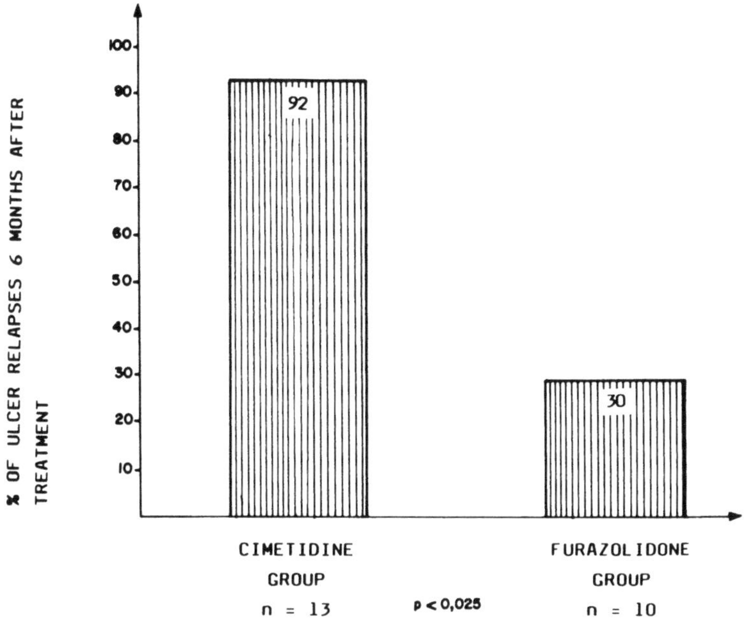

Fig. 2. Comparison of relapse rates of duodenal ulcer 6 months after treatment

This study was carried out to evaluate the use of this treatment regimen for the eradication of *H. pylori* and for the reduction of duodenal ulcer relapses. The detailed results of this investigation have already been published elsewhere [14].

Patients and Methods

After informed consent 61 patients with endoscopically proven duodenal ulcer and *H. pylori* infection detected on ^{14}C-urea breath test as previously described [15] were included in the study. In addition to classical anti-ulcer agents prescribed by their attending physicians, all patients received furazolidone (one 200-mg tablet), metronidazole (one 250-mg tablet) and amoxicillin (one 500-mg capsule) in combination three times a day for 5 days. Patients were advised to avoid alcoholic drinks because of the possible occurrence of Antabuse-like effects triggered by metronidazole and furazolidone.

The breath test was repeated after an interval of at least 2 months after the end of treatment and again 6.5 months after treatment for the detection of a possible recurrence of infection. Endoscopy was also performed 6.5 months, or whenever necessary, after treatment to determine possible ulcer relapse.

Statistical analysis was performed by means of the Student's *t* test, chi-square test, and the exact Fisher test.

Results

A total of 61 patients were entered into the study. Eight patients were withdrawn because they failed to attend follow-up appointments; five others were excluded: three were submitted to surgical treatment due to pyloric stenosis or clinical intractability, and two because of side effects (one had nausea and one a rash).

Thus, 48 patients (26 men and 22 women; mean age, 41.5 years) completed their treatment and were available for check ups as required by the study protocol. Before treatment 22 patients had a healed ulcer, 22 had active ulcers, and in four patients the staging of the duodenal ulcer were not clearly described. Of the 48 patients treated with triple therapy, 33 were using H2 bockers, three were using antacids, and 12 took no medication.

Side effects were frequent and occurred in 39.3% (24/61) of the patients: 30% had nausea, 10% had occasional vomiting, 4% complained of dizziness and of a bitter taste in the mouth, one patient reported diarrheal bowel movements, and one patient reported cephalalgia. The treatment had to be interrupted in only two patients (3, 3%) due to serious side effects.

At 2.5 months (55–138 days; median 69 days) after the end of the treatment, patients were invited to return for a breath test. All the 48 patients were available for study: the test was negative in 60% (29/48) of the patients and 40% (19/40) remained *H. pylori*-positive.

At 6.5 months (156–364 days; median 198 days) after treatment all 48 patients returned for re-endoscopy and a breath test to assess the endoscopic stage of the duodenal ulcer and to determine the rate of delayed eradication.

The *H. pylori*-positive and *H. pylori*-negative patients were homogeneous in relation to age, sex, endoscopic staging of the ulcer and use of anti-ulcer drugs before the antimicrobial treatment as well as in relation to the occurrence of adverse reactions (Table 1).

Table 1. Clinical and endoscopic characteristics of *H. pylori*-positive and *H. pylori*-negative patients as determined by a breath test performed 6.5 months after treatment

Characteristics	Status *H. pylori* positive *n*	 *H. pylori* negative *n*	*p*
Patients	22	26	
Sex (male/female)	14/8	12/14	NS
Healed ulcer upon admission[a]	8/19	14/25	NS
Use of anti-ulcer drugs upon admission	16	20	NS
Side effects of treatment	9	9	NS
Healed ulcers 6.5 months after treatment[a]	16	26	0.012
Symptoms 6.5 months after treatment	9	2	0.01
Use of anti-ulcer drugs 6.5 months after treatment	8	0	0.02

[a]Endoscopic staging was not clearly determined in three *H. pylori*-positive and one *H. pylori*-negative patient on admission.

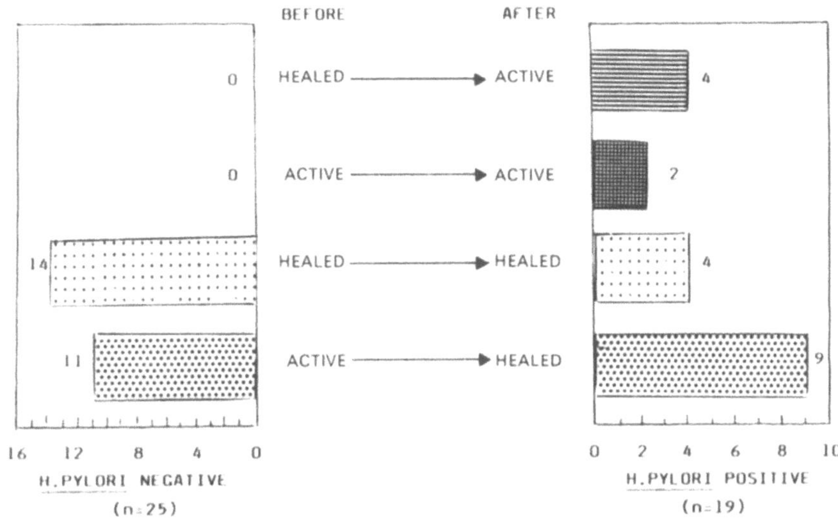

Fig. 3. Endoscopic staging of duodenal ulcer before and after antimicrobial treatment in both groups

Of the 29 *H. pylori*-negative patients at 2.5 months after treatment only four (13.8%) showed a return to positivity at 6.5 months with a eradication rate of 54% ranging from 40% to 60%, with 95% confidence limits. All of the 26 patients in whom the bacterium was eradicated showed healed duodenal ulcers, whereas the 22 patients in whom *H. pylori* persisted six (27%) showed active ulcers, a statistically significant correlation ($p = 0.012$). Of the 26 patients who remained *H. pylori*-negative 6.5 months after treatment only two referred to occasional epigastric burning and none was using anti-ulcers drugs. Otherwise, of the 22 patients who remained *H. pylori*-positive, eight (36%) continued with symptoms and still used anti-ulcer drugs. There was a significant correlation between the presence of *H. pylori*, the persistence in the use of anti-ulcer drugs ($p = 0.002$), and the presence of dyspeptic symptoms ($p < 0.01$) at the end of the 6.5 month follow-up period.

The endoscopic staging of the duodenal ulcers before and after antimicrobial treatment in *H. pylori*-positive and *H. pylori*-negative patients is shown in Fig. 3. The only changes that occurred in the *H. pylori*-negative group were towards healing, whereas in the *H. pylori*-positive group four ulcers recurred and two patients continued to have active ulcers until the end of the study.

Discussion

The etiology of peptic ulcer remains unknown. Aggressive and host defense mechanisms play a role, and an imbalance in these factors is thought to be important. The idea that peptic ulcer disease may have an infective cause is not new. In 1940 Freedberg and Barron [16] suggested that the success of bismuth compounds in peptic ulcer might be due to suppression of gastric spirochetes. The discovery of *H. pylori* and its strong association with duodenal ulcer disease has heightened interest in a possible infective cause for duodenal ulcer disease and has in turn led to investigations to discover whether eradication of this organism would succeed in preventing the frequent ulcer relapses and ultimately cure the disease. The prolonged remission from duodenal ulcer achieved if *H. pylori* is eradicated has intensified efforts to eradicate the organism and prevent reinfection. The easiest possible method to study this causal relationship is, in fact, to eradicate *H. pylori* and monitor the recurrence of duodenal ulcer. In a large Australian series reported by Borody et al. [3], an eradication rate of 94% was obtained in more than 100 patients treated with triple therapy [colloidal bismuth subcitrate (CBS) one tablet daily for 1 month, tetracycline 2 g daily for 1 month, and metronidazole 200 mg four times daily for 10 days]. Similar success has been reported by Graham et al. [17] using bismuth subsalicylate (BSS) as the bismuth salt. In the most recent report from one Australian group [5], the duodenal ulcer relapses and the rate of reinfection were followed over 4 years in a large group of resistant or intractable duodenal ulcer patients rendered *H. pylori* negative after triple therapy with a combination of CBS, tetracycline, and metronidazole. Of the 78 patients available for re-endoscopy 4 weeks after completion of triple therapy, 75 (96%) were *H. pylori* negative and the duodenal ulcer had healed in all 78 patients. In the 75 remaining patients the relapse rates for *H. pylori* infection and duodenal ulcer were studied endoscopically, yearly and at any recurrence of symptoms. At year 1, 71 of 73 patients remained free of *H. pylori* infection and duodenal ulcer. The corresponding figures were subsequently: year 2, 57/57; year 3, 33/34; year 4, 15/15. No duodenal ulcer recurred in *H. pylori*-negative patients who were followed for up to 4 years.

In the present paper we demonstrate that triple therapy combining furazolidone, metronidazole, and amoxicillin for 5 days was effective in eradicating *H. pylori* in 54% of patients with duodenal ulcers.

Even though the eradication rate observed here was lower than those reported in studies employing bismuth in combination with one or more antibiotics for 1 or 2 weeks [1–3, 18, 19], our therapeutic schedule is the shortest-lasting regimen reported so far for the eradication of *H. pylori*, with a consequent increase in patient adherence to treatment, lower financial cost, and a lower possibility of the occurrence of the serious adverse reactions often observed with the prolonged use of antimicrobial agents. The absence of bismuth salts in the combination used here facilitates its use in several countries where these salts are not commercially available. In addition, since the half-life of bismuth salts is estimated to be about 20 days [20], their use several times a year (as in case of recurrent ulcers, for example), in patients with kidney or liver disease, or in pediatric patients may cause problems requiring frequent monitoring of bismuthemia to avoid ocurrence of toxic effects [6–8, 21].

There was a high return of recurrence of the infection (approximately 14% in 4 months or 3.5% per month) in comparison with the figures reported by Rauws et al. [18] – 1% per month – and George et al. [5] – 7.6% in 4 years or 0.15% per month. These findings favor the idea that tests to evaluate the eradication of the micro-organism should be performed 1–3 months after the end of treatment. However, in view of the high prevalence of *H. pylori* in Brazil (more than 85%), even in the asymptomatic population (unpublished data), and in other developing countries [22, 23], additional studies with longer follow-up periods should be performed to learn more about the natural history of patients in whom the infection was eradicated with respect to reinfection with *H. pylori*, especially in those populations with high prevalence of the infection.

The present study also demonstrates that the eradication of *H. pylori* is associated with a significant reduction of active ulcers during a mean follow-up time of 6.5 months. None of the present patients whose breath test was *H. pylori* negative presented active ulcers during the study period, in contrast to 27% of the patients who continued to be infected with the microorganism. In addition, the patients in whom the microorganism was eradicated no longer needed any anti-ulcer medication during follow up, most of them categorically reporting a sensation of well-being they had not experienced in previous treatments, with tolerance of foods previously avoided, alcoholic drinks, and smoking. Although not scientifically quantified, these observations have also been reported by other investigators [2, 3].

Several studies have demonstrated that the eradication of *H. pylori* in patients with ulcer disease is accompanied by histologic resolution of the chronic gastritis almost invariably present and by marked reduction [2, 3] or even abolition [4, 5] of ulcer relapse. This epidemiologic evidence suggests that *H. pylori* is pathogenetically important in ulcer diathesis although the physiopathologic mechanisms involved are still obscure. It has been recently observed [24] that *H. pylori*-positive ulcer patients have significantly higher acid secretion in response to pentagastrin stimulation and significantly greater postprandial gastrinemia than *H. pylori*-negative ulcer patients, and that eradication of the microorganism significantly reduces the postprandial serum gastrin levels.

The encouraging results observed after antimicrobial treatment of duodenal ulcer are still hampered by various obstacles. The difficulty in eradicating *H. pylori* from the gastric mucosa has stimulated the use of several antimicrobial agents [25, 26], which in turn favor the development of bacterial resistance and increase the incidence of side effects [3, 25, 26]. Otherwise, in many parts of the world, where giardiasis and amebiasis are endemic, nitroimidazoles are frequently used, and primary resistance of *H. pylori* to these compounds is common. Even in those communities where primary resistance is relatively low, triple therapy fails in up to 20% and side effects are significant in 30% [27].

In conclusion, we strongly believe that the antimicrobial treatment offers the perspective of a definitive cure of the ulcer process. The combined treatment with furazolidone, amoxicillin, and metronidazole for 5 days represents a well-tolerated, inexpensive, and effective therapeutic regimen for the eradication of *H. pylori* in more than 50% of ulcer patients. Studies must be conducted to obtain better treatment regimens in terms of simplicity, safety, and effectiveness. Finally, two emerging problems, particularly in developing countries, must be pointed out in the management of *H. pylori*-positive duodenal ulcer patients: the high rate of primary metronidazole resistance and the probably higher risk of reinfection after treatment due to the higher prevalence of *H. pylori* infection in the population.

References

1. Coghlan JG, Humphreis H, Dooley C et al. (1987) Campylobacter pylori and recurrence of duodenal ulcers—a 12 month follow-up study. Lancet ii: 1109–1111
2. Marshall BJ, Warren JR, Blincow ED et al. (1988) Prospective double-blind trial of duodenal ulcer relapse after eradication of Campylobacter pylori. Lancet ii: 1437–1441
3. Borody TJ, Cole P, Nooman S et al. (1989) Recurrence of duodenal ulcer and Campylobacter pylori infection after eradication. Med J Aust 151: 431–435
4. Rauws EAJ, Tytgat GNJ (1990) Cure of duodenal ulcer associated with eradication of Helicobacter pylori. Lancet 335: 1233–1235
5. George LL, Borody TJ, Andrews A et al. (1990) Cure of duodenal ulcer after eradication of Helicobacter pylori. Med J Aust 153: 145–149
6. Burns R, Thomas DW, Barron VJ (1974) Reversible encephalopathy possibly associated with bismuth subgallate ingestion. Br Med J 1: 220–223
7. Hillemand P, Polliere M, Laquais B et al. (1977) Traitement bismuthique et bismuthemie. Semin Hop 53: 1663–1669
8. Monseu G, Struelens M, Roland M (1976) Bismuth encephalopathy. Acta Neurol Belg 76: 301–308
9. Zheng ZT, Wang ZY, Chu YA et al. (1985) Double-blind short-term trial of furazolidone in peptic ulcer. Lancet i: 1048
10. Zhao HY, Li G, Guo J et al. (1985) Furazolidone in peptic ulcer. Lancet ii: 276
11. Howden A, Boswell P, Tovey F (1986) In vitro sensitivity of Campylobacter pyloridis to furazolidone. Lancet ii: 1035
12. Coelho LGV, Queiroz DMM, Barbosa AJA et al. (1989) Furazolidone x cimetidine in duodenal ulcer C. pylori-positive patients. Gastroenterology 96/5: A91
13. Coelho LGV, Passos MCF, Queiroz DMM et al. (1990) Five day triple therapy and 15-day double therapy on H. pylori eradication. In: Malfertheiner P, Distschneit H (eds) Helicobacter pylori, gastritis and peptic ulcer. Springer, Berlin Heidelberg New York, pp 438–440
14. Coelho LGV, Passos MCF, Chausson Y, Castro LP (1991) Five day bismuth-free triple therapy for the eradication of Helicobacter pylori and reduction of duodenal ulcer relapse. Am J Gastroenterol 86(8): 971–975
15. Coelho LGV, Chausson Y, Passos MCF et al. (1990) Test repiratoire ā l'urēe marquēe au carbone-14 pour le diagnostic de la colonisation gastrique par le Helicobacter pylori. Gastroenterol Clin Biol 14: 801–805
16. Freedberg AS, Barron LE (1940) The presence of spirochetes in human gastric mucosa. Am J Dig Dis 7: 443–445
17. Graham DY, Lew GM, Michaletz PA (1989) Randomized trial of the effect of eradication of C. pylori on ulcer healing and relapse. Gastroenterology 96 [Suppl]: A181
18. Rauws EAJ, Langenberg W, Houthoff HJ et al. (1989) Campylobacter pyloridis-associated chronic active antral gastritis. Gastroenterology 94: 33-40
19. Börsch G, Mai U, Opjerkuch W (1989) Short and medium-term results of oral triple therapy to eradicate C. pylori. Gastroenterology 96: A53
20. Hirschl AM, Hentschel E, Schütze K et al. (1988) The efficacy of antimicrobial treatment in Campylobacter pylori - associated gastritis and duodenal ulcer. Scand J Gastroenterol 23 [Suppl 142]: 76–87
21. Playford RJ, Matthews CH, Campbell MJ et al. (1990) Bismuth induced encephalopathy caused by tri potassium dicitrato bismuthate in a patient with chronic renal failure. Gut 31: 359–360
22. Glupczynski Y, Bourdeaux L, Verhas M, Balegamire B, Devos D, Devrecker T (1989) Epidemiology of Campylobacter pylori infection in Zaire. Klin Wochenschr 67 [Suppl XVIII]: 23

23. Perez-Perez GI, Taylor DN, Bodhidatta L et al. (1990) Seroprevalence of Helicobacter infections in Thailand. J Infect Dis 161:1237–1241
24. Levi S, Beardshall K, Swift I et al. (1989) Antral Helicobacter pylori, hypergastrinaemia, and duodenal ulcers: effect of eradicating the organism. Br Med J 299:1504–1505
25. Bayerdörffer E, Ottenjann R (1988) The role of antibiotics in Campylobacter pylori associated peptic ulcer disease. Scand J Gastroenterol 23 [Suppl 142]:93–100
26. Graham DY, Klein PD, Opekum AR et al. (1989) In vivo susceptibility of Campylobacter pylori. Am J Gastroenterol 84:233–238
27. Tytgat GNJ, Axon ATR, Dixon MF, Graham DY, Lee A, Marshall BJ (1990) Helicobacter pylori: causal agent in peptic ulcer disease? In: Working party reports. World congress of gastreonterology, Sydney. Blackwell, Oxford, p36–45

Bismuth Subnitrate in Comparison to Ranitidine in Patients with Non-ulcer Dyspepsia: A Randomized Controlled Trial

M. Mateblowski, P. Topfmeier, M. Fischer, and H. Rohde

Introduction

The prevalence of dyspeptic disorders is high and has not changed within the last 40 years [13]. The term non-ulcer dyspepsia (NUD) describes a complex of symptoms, e.g., postprandial feeling of fullness, burping, flatulence, heart burn, nausea, epigastric pain, and lack of appetite [21]. NUD is very heterogenous and hard to define; [5, 32] it has great socioeconomic impact because the average notification of sickness is 26 days annually for NUD patients [19].

Helicobacter pylori (HP) is detectable in 45% [20] to 60% [9] of all NUD patients, and it cannot been excluded that, for a part, of these patients this bacterium is responsible for their symptoms [28].

The aim of this investigation was to compare bismuth subnitrate (BSN; Ulkowis, Temmler Pharma, Marburg, FRG) and ranitidine (Zantic 300; Glaxo, Hamburg, FRG) regarding symptoms and HP eradication, as well as activity and gradation of gastritis.

Methods

By the use of several clinical, ultrasound, blood chemistry, and X-ray tools, further diseases of the stomach and gut were disclosed. Furthermore, the following were also excluded: patients with anamnesis of clinically verified diseases of the liver, kidney, or lung; patients with known incompatibility to bismuth preparations or ranitidine; pregnant or nursing women; women without safe contraception; patients with known malign, hematologic, or cardiac diseases or relevant co-medication.

An endoscopy was performed after premedication of 2–4 mg midazolam (Dormicum, Hoffman-La-Roche, Grenzach Whylen, FRG) using Fujinou endoscopes (UGI-FP3 or UGI-F49). To avoid cross-contamination, all material in putative contact with biopsies and all endoscopes were sterilized directly after the examination using Gigasept 1% and Gigasept FF 6% (Schülke und Meyer, Norderstedt, FRG).

After evaluation of endoscopic findings, biopsies were taken in the last 2 cm of the pylorus. One sample was used for histology, the other one in a biochemi-

Am Hofacker 38, 3551 Lahntal 3, FRG

J.M. Pajares et al. (Eds.)
H. pylori and Gastroduodenal Pathology

cal urease test system (CUT, Temmler Pharma, Marburg, FRG). In the case of a positive result (within 30 min) patients were asked to take part in this study.

Histology was performed according to the new classification and grading of gastritis [11, 26]. No pathologist had knowledge about the treatment. Samples were stained using hematoxyline and eosin, and gradation and activity of gastritis were determined. Mucosa colonization by HP was semiquantitatively evaluated according to the following scale:

0 No bacteria detectable
1 Low density: longer search necessary to find bacteria
2 Moderate density: bacteria often detectable at low magnification, flocks of bacteria promptly visible at higher magnification.
3 High density: partly gross colonization, clearly detectable at low magnification.

All patients were asked to state the severity of their symptoms on a visual – analogue scale (VAS) before and at the end of a 4-week treatment period [12, 17]. Patients were randomized to take either two tablets of BSN three times daily (2.100 mg/day) before meals or 1 tablet of Ranitidine before bedtime (300 mg/day).

To determine bismuth concentration, methemoglobin, serum glutamic oxaloacetic transaminase (SGOT), and creatinine, 10 ml blood was analyzed on the first and last days of treatment. By the use of CRFs which did not contain any information regarding the substances used, the objectivity of the test was guaranteed.

Biometry and Evaluation

The comparison between both medications was based on the decrease of pain, changes in the gradation and activity of gastritis, and decrease of HP colonization. Differences between BSN and ranitidine were analyzed in those patients who revealed a decrease in pain of 20% or more; the statistical significance was determined using Fischer's exact test (double sided).

For the gradation of gastritis, a total score system based on histologic findings was defined: severity of gastritis plus 2 × HP colonization plus 3 × gastritis activity = total score (points). Each parameter was classified as follows: 0 = no; 1 = low; 2 = moderate; 3 = severe status. Differences between the patient groups with regard to a change in the severity of gastritis were tested using the Ulemann test (double sided). Before treatment, alpha error was defined as 0.05 and adjusted to 0.025 corresponding to a number of comparisons. The whole evaluation was performed by the intent-to-treat method.

Results

Over a period of 4.5 months 48 NUD patients (24 in each group) were treated. The demographic and anamnestic data (Table 1) as well as the endoscopic

Table 1. Comparison of demographic and anamnestic data of NUD patients treated with bismuth subnitrate or ranitidine

	NUD patients treated with BSN (n = 24)	Ranitidine (n = 24)
Sex		
Female (n)	15	15
Male (n)	9	9
Median	49	39
Age (years)		
Range	17–81	23–66
Median	168	170
Height (cm)		
Range	150–192	150–182
Median	75	74
Weight (kg)		
Range	46–91	42–95
Smokers [n]	8	13
Consumers of alcohol (n)	18	18
Consumers of coffee (n)	23	21
Anamnesia with		
Epigastric pain (n)	24	23
Peptic ulcer (n)	7	10
Duration of symptoms		
< 30 days (n)	15	13
> 30 days (n)	9	10

diagnosis confirmed that both groups were comparable except that the average age of the patients receiving ranitidine was 10 years less in comparison to those receiving BSN. Controls were performed after 28 days (median) in the bismuth group and 29 days in the ranitidine group. Before treatment all patients were HP positive.

After 4 weeks 71% of all patients treated with BSN, but only 29% of those who received Ranitidine reverted to being urease negative as evaluated by the *Campylobacter* urease test (CUT). The histologically proven eradication rate after BSN was 58% but 0% after Ranitidine.

The grading of HP infection decreased after treatment for the BSN group: low infection 26% to 11%, moderate 58% to 26%, and severe 16% to 5%. In contrast, no remarkable changes were found in patients treated with ranitidine before therapy 27% of patients showed low infection compared to 33% after therapy, and 67% showed moderate infection before and after treatment.

In most cases, gradation of gastritis before therapy was moderate, although there was a higher proportion of severe gastritis among BSN patients. After treatment 53% of BSN patients, but only 29% of Ranitidine patients showed an improvement. Gastritis activity was decreased in 55% of patients in the BSN group and 70% of the Ranitidine group.

On average, the total degree of gastritis showed a decrease for BSN of four points from 8 ± 3 to 4 ± 3 points, whereas for patients treated with Ranitidine a one-point decrease from 7 ± 3 to 6 ± 3 points was noted. The difference

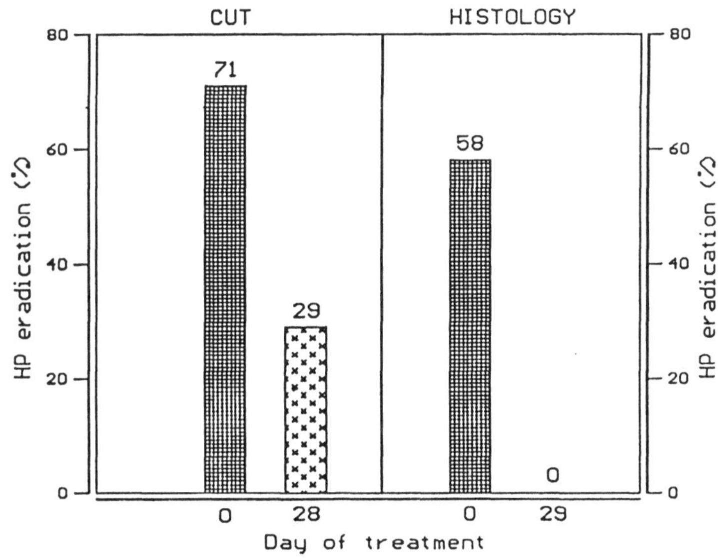

Fig. 1. Eradication of *H. pylori* by bismuth subnitrate and ranitidine in patients with non-ulcer dyspepsia. *Shaded columns*, bismuth subnitrate; *crossed column*, ranitidine

(a)

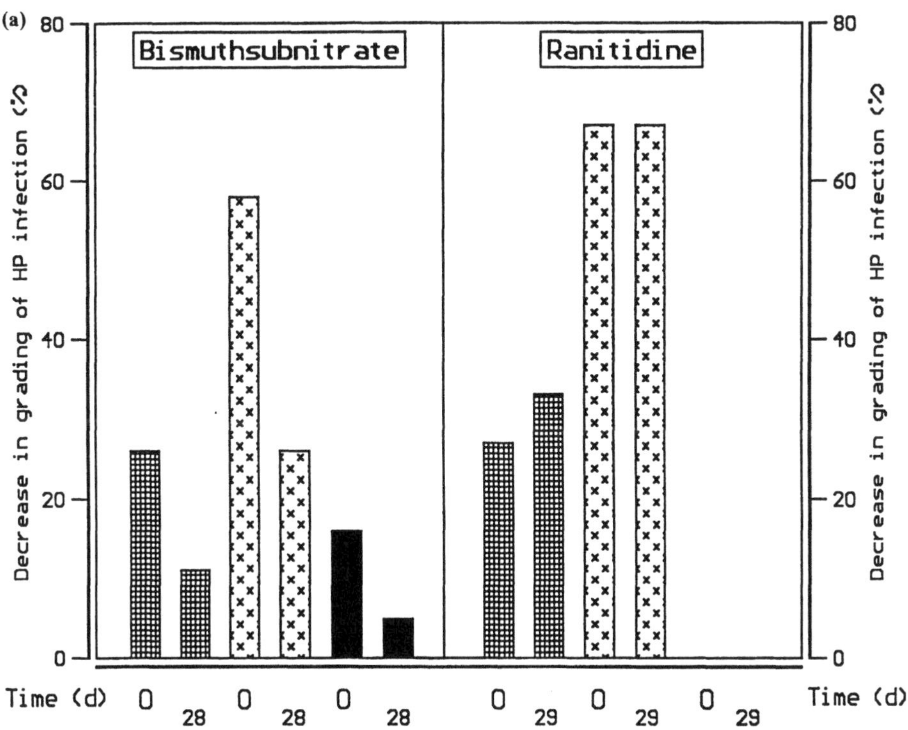

Fig. 2a, b. Grading of HP infection before and after treatment with bismuth subnitrate (**a**) or ranitidine (**b**). *Shaded columns*, low; *crossed columns*, moderate; *solid columns*, severe

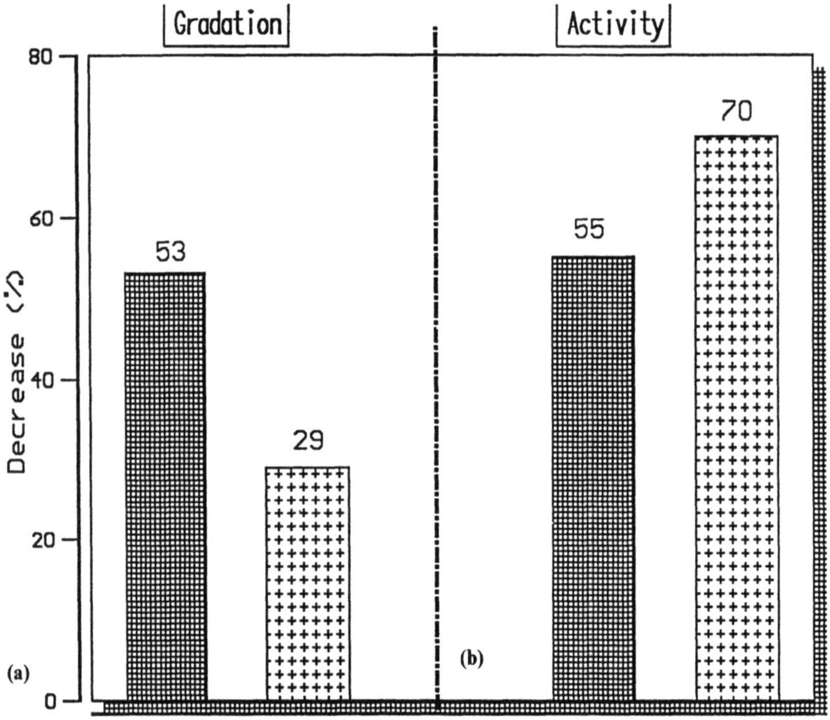

Fig. 3a, b. Gradation (**a**) and activity (**b**) of gastritis in NUD patients after treatment with bismuth subnitrate (*shaded columns*) ranitidine (*crossed columns*)

between both medications was statistically significant: $p = 0.0199$ Ulemann test, double sided.

Symptoms were improved after both treatments with slight differences. A minimal decrease in pain of 20% (VAS) was found for 96% and 88% of all patients treated with BSN and Ranitidine, respectively. No statistically significant changes were found for the blood levels of methemoglogin, SGOT, bismuth, and creatinine in both treatment groups. There were no values beyond the accepted reference ranges.

Discussion

Over the last decade there has been a dramatic increase in the number of NUD patients [33]. There was a change in the attitude of pathologists [25] parallel to the discovery of the pathogenic nature of HP in chronic gastritis [30] and the proof of a statistically significant coherence between the density of HP colonization in antrum mucosa and the grade as well as the activity of gastritis [27].

Also, gastroenterologists had to clarify the "normal" endoscopic findings in a new way. Biopsy monitoring in clinical trials became more significant [25]. This

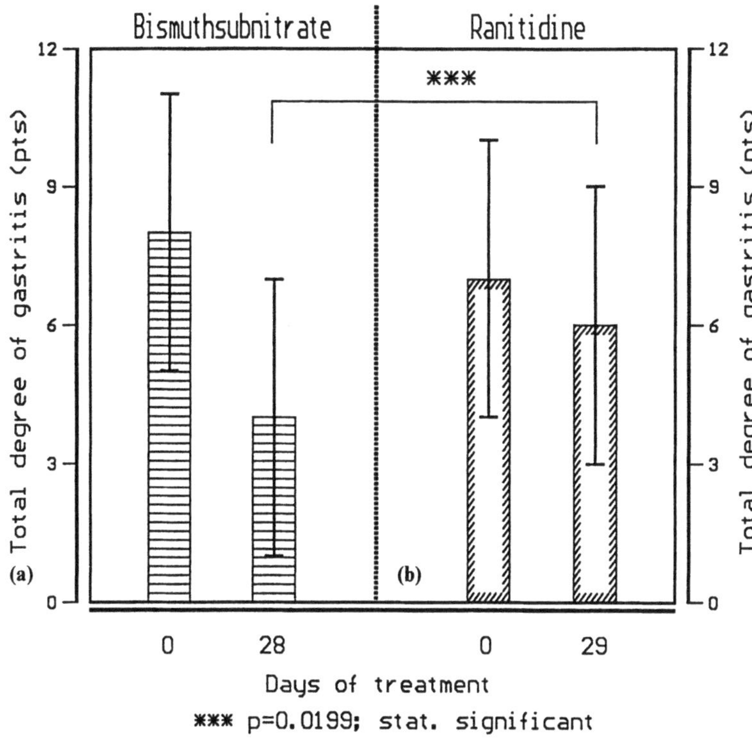

Fig. 4a, b. Total degree of gastritis in NUD patients after treatment with bismuth subnitrate (**a**) or ranitidine (**b**). *Asterisks*, $p = 0.0199$, statistically significant; *pts*, points

is important, especially for NUD where only very few verified endoscopic findings are possible [4, 8, 31] and the clinical symptoms do not correlate with the histological status [2, 16]. The hope that healing [16] and a decrease in the relapse rate [20] would become possible in NUD associated with HP gastritis, as in peptic ulcers, by the use of bismuth preparations seems legitimate because of the cytoprotective effect of bismuth salts. HP is not able to survive in the altered micromilieu of the mucosa [7, 10] and healing of gastritis is induced [1, 4, 24].

In NUD patients colloidal bismuth citrate [22, 23], gallate [29] and nitrate [24, 28] stood the test. Apart from the loss of hydrophilic properties, the behaviour of salicylate and nitrate salts within the acid stomach is comparable [6]. BSN accelerates healing of duodenial [3] and ventricularules [28] and proved to be very good in this investigation in NUD patients.

A total of 95.8% of patients treated with BSN, in contrast to 87.5% treated with ranitidine, showed a decrease in pain of more than 20%, estimated by the use of VAS. This value is comparable to those of other investigations using BSN [24, 28], bismuth subcitrate [23] and bismuth subsalicylate [15].

In contrast to H2 receptor antagonists [19], bismuth preparations showed an improvement of symptoms in placebo-controlled trials [14, 24]. This advantage of bismuth preparations with regard to a decrease in pain and improved histological findings was borne out in our investigation. Furthermore, after ranitidine there was no change in HP colonization after treatment, whereas in patients treated with BSN 58% reverted to being HP negative, and the gradation of HP infection – low, moderate, or severe – was lowered from 26% to 11%, 58% to 26% and 16% to 5% respectively. These results are comparable to those of Archimandreitis et al. [1] and Niedobitek et al. [18].

In contrast to the ranitidine group, the improvement in the grading and activity of gastritis was impressive in patients receiving BSN and so comparable to other investigations [24, 28]. With regard to HP colonization, we found a statistically significant difference ($p = 0.0199$) for BSN. This therapeutic success confirms earlier findings that HP is the main cause of B-gastritis [27]. Overall, this preparation containing bismuth can be considered to be an effective and safe drug for the treatment of non-ulcer dyspepsia.

References

1. Archimandritis AJ, Tjivras M, Davaris P, Alexion A, Bisikas J (1990) Helicobacter associated gastritis in patients with duodenal ulcer: the influence of various drugs. Gut 31:481–482
2. Börsch G (1988) Therapie der Campylobacter-pylori-Infection. Leber Magen Darm 1:38–45
3. Carr-Locke D, Wicks AB (1986) A double-blind, endoscopically assessed evaluation of bismuth subnitrate preparation (Roter) and cimetidine in the treatment of duodenal ulcer. Br J Clin Pract 40:373–375
4. Chodos JE, Dworkin BM, Smith F, van Horn K, Weiss L, Rosenthal WS (1988) Campylobacter pylori and Gastroduodenal Disease: a prospective endoscopic study and comparison of diagnostic tests. Am J Gastroenterol 83:1226–1230
5. Colin-Jones DG (1988) Management of dyspepsia: report of a working party. Lancet I:576–578
6. DuPont HL (1989) Bismuth subsalicylate in the treatment and prevention of diarrheal disease. Drug Intell Clin Pharm 21:687–693
7. Gebbers JO, Altermatt HJ, Altdorfer J (1988) Campylobacter pylori: Ursache von Gastritis und Ulkuskrankheit? Schweiz Med Wochenschr 118:577–583
8. Glupczynski Y, Burette A, Labbe M, Deprez C, De Reuck M, Deltenre M (1988) Campylobacter associated gastritis: a double-blind placebo-controlled trial with amoxycillin. Am J Gastroenterol 83:365–372
9. Graham DY, Klein PD (1987) Campylobacter pylori gastritis. The past, the present, and speculations about the future. Am J Gastroenterol 82:283–287
10. Hall DWR (1988) Review of the modes of action of colloidal bismuth subcitrate. Scand J. Gastroenterol 24 [Suppl 157]: 3–6
11. Heilmann KL, Stolte M, Borchard F, Heine M, Löning TH, Ottenjann W, Remmle W, Rühl G, Schaefer HE, Schlake W, Seib HJ, Stamm B, Steininger H, Wiebecke B (1989) Gastritis Graduierung und Klassifikation. Pathologe 10:194–196
12. Huskission EC (1983) Visual analogue scales. In: Melzack R (ed) Pain measurement and assessment. Raven, New York, p 33
13. Jones R, Lydeard S (1989) Prevalence of symptoms of dyspepsia in the community. Br Med J 298:30–32
14. Lambert JR, Dunn K, Turner H, Korman MG (1986) Effect on histological gastritis following eradication of campylobacter pyloridis. Gastroenterology 90:1509–1512
15. Malfertheiner P, Stanescu A, Baszako K, Vanek E, Bode G, Ditschuneit H (1988) Wismutsubsalicalate-Behandlung bei campylobacter-pylori assoziierter chronischer erosiver Gastritis. Dtsch Med Wochenschr 113:923–929
16. Marshall BJ, Goodwin CS, Warren JR, Murray R, Blincow ED, Blackbourn TJ, Phillips M, Waters ThE, Sanderson ChR (1988) Prospective double-blind trial of duodenal ulcer relapse after eradication of campylobacter pylori. Lancet 24:1437–1442

17. Maxwell C (1978) Sensitivity and accuracy of the visual analogue scale: a psycho-physical classroom experiment. Br J Clin Pharmacol 6: 16–21
18. Niedobitek F, Grosse G, Hammer M, Bonk G, Nehls R, Volkheimer G (1989) Gastritis and bacterial colonization of the gastric mucosa in adolescents. Am J Gastroenterol 84: 239–244
19. Nyren O, Adami HO, Gustavsson S, Lööf L, Nyberg A (1985) Social and economic effects of non-ulcer-dyspepsia. Scand J Gastroenterol 20 [Suppl 109]: 41–49
20. Rauws EAJ, Tytgat GNJ (1990) Cure of duodenal ulcer associated with eradication of helicobacter pylori. Lancet i: 1233–1235
21. Rösch W (1988) Motilitätsstörungen des Magens. Schwerpunkt Dyspepsie. Z. Gastroenterol 26 [Suppl 4]: 18–21
22. Rokkas T, Pursey C, Simmons NA, Filipe MJ, Sladen GE (1987) Non-ulcer dyspepsia an colloidal bismuth subcitrate therapy. The role of campylobacter pyloridis. Gastroenterology 92A: 1599
23. Rokkas T, Pursey C, Uzoechina E, Dorrington L, Simmons NA, Filipe MJ, Sladen GE (1988) Non-ulcer dyspepsia and short term De-Nol therapy: a placebo controlled trial with particular reference to the role of campylobacter pylori. Gut 29: 1386–1391
24. Stanescu A, Malfertheiner P, Mayer D, Baczako K, Ditschuneit H (1989) Wismut-subnitrat – Therapie der campylobacter-pylori-positiven chronischen erosiven Gastritis. Gastroenterohepatologie 7: 3–10
25. Stolte M (1988) Normal endoscopic findings in the gastrointestinal tract – when should a biopsy be taken? Endoscopy 20: 111–113
26. Stolte M, Heilmann KL (1989) Neue Klassifikation und Graduierung der Gastritis. Leber Magen Darm 5: 220–226
27. Stolte M, Eidt S, Ritter M, Bethke B (1989) Campylobacter pylori and Gastritis. Assoziation oder Induktion? Pathologe 10: 21–26
28. Topfmeier P, Mateblowski M (1989) Therapie von Ulcus ventriculi et duodeni mit Wismutsubnitrat: Multizentrische Untersuchung mit Ulkowis. Gastroenterohepatologie 7: 2–12
29. Vestweber A, Kusche J, Günther HW, Rohde H (1990) Basisches Wismutgallat – bewährt, vergessen, wiederentdeckt. Med Welt 41: 745–748
30. Warren JR, Marshall BJ (1983) Undefined curved bacilli on gastric epithelium in active chronic gastritis. Lancet 1: 1273–1275
31. Weber J, Riemann JF (1988) Campylobacter pylori ein pathogenetischer Faktor für die Gastritis und die peptische Ulkuskrankheit? Med Welt 39: 763–766
32. Wegener M, Börsch G (1988) Nicht-ulzeröse Dyspepsie. Dtsch. Med. Wochenschr 113: 1767–1773
33. Williams B, Luckas M, Ellingham JHM (1988) Do young patients with dyspepsia need investigation? Lancet ii: 1349–1351

Colloidal Bismuth Subcitrate Versus Amoxicillin in the Treatment of *Helicobacter pylori*-Associated Type B Gastritis

G. Bisi, L. Accorsi, V. Rollo, G. Merighi, and S. Gullini

Introduction

Notwithstanding the many specific treatments against *Helicobacter pylori* (HP)-positive antral gastritis [1–13], HP detection on the gastric mucosa of patients with gastroduodenal peptic disease is still an open question. The purpose of this open noncontrolled study was to compare, the efficacy of two established treatments, i.e., colloidal bismuth subcitrate (CBS; De-Nol) and amoxicillin in HP-positive patients with histologically confirmed type B gastritis.

Methods

A group of 50 patients (27 men, mean age 44.3 years; 23 women, mean age 51.4) with endoscopical diagnosis of aspecific phlogosis of the gastric mucosa and histologic diagnosis of HP-associated gastritis were entered into the study. They were referred to our hospital because of different gastrointestinal symptoms (nausea, belching, retrosternal and gastric pyrosis, meteorism, borborygmus, postprandial cephalalgia). All the symptoms were graded from 0 to 3 on the basis of their severity. Details of life habit and previous antacid therapy were noted; no patient had a prior history of therapy with amoxicillin.

During endoscopy, performed with a Fujinon EVG-F instrument, four biopsy specimens of antral mucosa were taken in order to study the histology of the base pathology and to detect HP. Simethicone was not utilized during endoscopy; all the instruments were sterilized with a commercial quaternary compound (alkyl-benzyl-dimethylammonium) after each examination. Biopsy specimens were placed into a sterile test tube containing Bouin's solution and routinely processed (hematoxylin and eosin staining method). The specimens were also stained with Giemsa to detect the microorganisms on the gastric epithelium. This was the only method used during trial screening and subsequent assessments since, in our opinion, the enzymatic method then available was unable to ensure the required reliability [14].

Subjects with gastric and duodenal ulcers, with signs of serious renal and/or hepatic failure, or with marked alterations of the cardiovascular apparatus, and

Gastroenterology and Digestive Endoscopy Department, S. Anna Hospital, Corso Giovecca 203, Ferrara, Italy

J.M. Pajares et al. (Eds.)
H. pylori and Gastroduodenal Pathology

subjects with a previous history of allergy to penicillin were excluded. Patients were informed on the nature and the aims of the study, conducted in accordance with the principles of the Declaration of Helsinki, and gave their oral consent.

Before and after the study, patients underwent laboratory tests for hepatic and renal functions. All the patients were randomized to receive either CBS (two tablets twice a day for 4 weeks = 25 patients, group A) or amoxicillin (one capsule 500 mg twice a day for 15 days = 25 patients, group B). Administration of any H2 inhibitor or mucosa protectors was stopped for the whole study period (3 months). Only antacids were allowed.

The protocol provided for a basal examination (endoscopy, histologic study of the gastric mucosa, HP detection, and clinical laboratory tests) to select the patients. At the first assessment (after 30 days), clinical and endoscopic assessments with HP detection and histologic study of the gastric mucosa were performed. After 90 days, the same examinations were performed (second assessment).

Results

In group A the basal endoscopy showed 25 patients with edema and hyperemia of the gastric mucosa, in five cases associated with antral erosions and in two without macroscopic lesions. After 4 weeks the endoscopic picture proved unchanged in 23 patients and improved in two, while at the second assessment 23 patients showed edema and hyperemia of the gastric mucosa and two had antral erosions.

In group B the basal endoscopy revealed 20 patients with edema and hyperemia of the gastric mucosa, in four cases associated with antral erosions and one case without macroscopic lesions. At the first assessment, the endoscopic picture proved unchanged in 14 cases (56%) and improved in 11 cases (44%). After 12 weeks the endoscopic results were substantially analogous to those previously revealed: 16 patients showed signs of edema and hyperemia of the gastric mucosa, six had antral erosions, and three were normal.

In group A the histologic examination of the antral biopsy specimens showed 15 cases of inactive chronic superficial gastritis, eight active chronic superficial gastritis, and two cases of active chronic superficial gastritis, associated with intestinal metaplasia. At the first assessment nine cases of inactive chronic superficial gastritis, four of active chronic superficial gastritis, and two cases of active chronic superficial gastritis associated with intestinal metaplasia were found. At the final assessment nine cases of inactive chronic superficial gastritis, 12 of active chronic superficial gastritis, and four cases of active chronic superficial gastritis associated with intestinal metaplasia were present. In group B the basal histologic examination of the antral mucosa showed 14 cases of inactive chronic superficial gastritis, seven of active chronic superficial gastritis, two of chronic atrophic gastritis, and two of chronic superficial gastritis associated with intestinal metaplasia. The histologic average score was unchanged both at the first and the second assessment.

All the patients presented various dyspeptic symptoms at the beginning (see "Methods"). After 4 weeks all the group A patients showed a marked improvement in dyspeptic symptoms, with a significant reduction of the symptoms score ($p < 0.05$). At the second assessment the symptom score was further improved. In group B all the patients showed a significant improvement in dyspeptic symptoms both after 4 and 12 weeks.

Helicobacter pylori was detected on the gastric mucosa of all subjects before the start of treatment. After 4 weeks the HP clearance rate was 60% (15/25) in group A and 40% (10/25) in the group B. At the second assessment HP was present in 22 out of 25 patients of group A (88%) and in 24 out of 25 patients (96%), with a HP eradication of 12% and 4%, respectively.

Discussion

From the results of this open uncontrolled study, the following considerations can be made:

1. In contrast to the conclusions drawn by Glupczynski et al. [2] who found that HP vanished in 72% of the cases after treatment (although with a higher dosage of amoxicillin), the present study showed a very low HP clearance (40%). After 3 months all the patients except one were HP positive, as already shown by Borsch et al. [3]. Thus the treatment with a low dosage (1 g/day) of amoxicillin appears to be, in our experience, of low efficacy. HP detection at the second assessment is more likely due to recurrence rather than to reinfection, as already demonstrated by Langemberg et al. [4] by the DNA analysis of HP strains.
2. The low efficacy of amoxicillin monotherapy in HP eradication could be explained by its topical activity which, as it gets rid of all the surface microorganisms, could be ineffective against the small reserves situated deeper in the more protected areas of the mucosa (gastric crypts). HP migration could take place to recolonize the whole mucosa. In fact, the antibiotic could not act at the proper concentration because the contact with the bacteria is difficult due to the protection afforded by the mucus layer and the location far from the blood supply. In vitro amoxicillin inhibits the growth of HP with a minimal inhibitory concentration ranging between 0.015 and 0.125 μg/ml [5–7], which is largely below that achieved in gastric biopsy specimens after oral administration of a 500 mg amoxicillin capsule [8]. These facts can also explain the differences found between the in vitro and in vivo action [9] and, in part, our disappointing results.
3. The difficulty of obtaining a topographically homogeneous sample of the antral mucosa may explain the quantitative variations found in inflammatory infiltrate whether HP are present or not. In our experience, the poor HP clearance rate did not lead to morphohistologic modifications, at least in the short term. Our results, which are in contrast with these already published by Glupczynski and McNulty [2, 10], can be read as a confirmation of the multifactorial involvement in the pathogenesis of gastritis.

4. The improvement in the symptom score observed at the first assessment and confirmed at the second assessment, in discordance with endoscopy, histology, and HP status, leads to the important question of a possible placebo effect. In fact, it cannot be neglected that patients voluntarily enrolled in a trial are psychologically involved to a higher degree and, because of the strong dependence on the physician, they feel better.

Conclusions

In our opinion, until the etiopathogenetic role of HP has been established for sure and the source of infection and the environmental stores of the bacteria are known, antibiotic monotherapy to treat HP-positive antral gastritis appears not to be advisable. The use of a multiple drug treatment seems to be even less auspicious [3, 11] as it ensures short-term efficacy, but, in the majority of cases, it is associated with an early return to the acute state. Furthermore, we cannot exclude that such drastic treatment can give rise to antibiotic resistance as already described [12, 15, 16]. Until the routes by which HP infection takes place have been established and new and more specific drugs are available which are able to eliminate HP and prevent its consequences, CBS appears to be, in our opinion, the monotherapy of choice in HP-associated gastritis because of the already established property of coating, swelling, and lysing HP [13–17].

References

1. Axon ATR (1989) Campylobacter pylori-therapy review. Scand J Gastroenterol 24 [Suppl 160]: 35–38
2. Glupczynski Y, Burette A, Labbe M, Deprez C, Deruck M, Deltenre M (1988) Campylobacter pylori-associated gastritis: a double blind placebo controlled trial with amoxycillin. Am J Gastroenterol 83: 365–372
3. Borsch G, Mai U, Muller M (1988) Monotherapy or polichemotherapy in the treatment of Campylobacter pylori related gastroduodenal disease. Scand J Gastroenterol 23 [suppl 142]: 101–106
4. Langenberg W, Rauws EAJ, Widjojokusumo A et al. (1986) Identification of Campylobacter pyloridis isolates by restriction endonuclease DNA analysis. J Clin Microbiol 24: 414–417
5. Goodwin CS, Blake P, Blincow E (1986) The minimum inhibitory and bactericidal concentrations of antibiotics and anti-ulcer agents against Campylobacter pyloridis. J Antimicrob Chemother 17: 309–314
6. McNlty CAM, Dent J, Wise R (1985) Susceptibility of clinical isolates of Campylobacter pyloridis to 11 antimicrobial agents against Campylobacter pyloridis. Antimicrob Agents Chemother 28 :019837–838
7. Lambert T, Mgraud F, Gerbaud G et al. (1986) Susceptibility of Campylobacter pyloridis to 20 antimicrobial agents. Antimicrob Agents Chemother 30 :019510–511
8. McNulty CAM, Dent JC, Ford GA et al. (1987) Antimicrobial concentrations in gastric mucosa. In: 4th international workshop on campylobacter infections, Göteborg, June 16–18

9. Bayerdorffer E, Ottenjann R (1988) The role of antibiotics in Campylobacter pylori associated peptic ulcer disease. Scand J Gastroenterol. 23 [Suppl 142]: 93–100
10. McNulty CAM, Gearty JC, Crump B et al. (1986) Campylobacter pylori and associated gastritis: investigator blind, placebo controlled trial of bismuth salycilate and eritromycin ethylsuccinate. Br Med J 293:019645–649
11. Borody TJ, Cole P, Noonan S, Morgan A, Ossip G, Raysey J, Brandl S (1988) Long-term Campylobacter pylori recurrence post eradication. Gastroenterology 94:019A43
12. Glupczynski Y, Bruck C, Burette A, Labbe M, Deltenre M, Avesani V (1987) Comparative in vitro activity of 21 antimicrobial and anti-ulcer agents against clinical isolates of Campylobacter pyloridis. 4th international workshop on campylobacter infections, Goteborg, June 16–18
13. Lambert JR, Hansky J, Davidson A, Pinkerd K, Stockman K (1985) Campylobacter-like organisms (CLO) in vitro and in vivo susceptibility to antimicrobial and antiulcer therapy. Gastroenterology 92:0191518
14. Basso O, Gullini S, Macario F, Cantarini D, Boccia S, Ghinelli F (1987) Evaluation of sensitivity and specificity of CLO test for detection of Campylobacter pyloridis. Ital J Gastroenterol 19 [Suppl 3] :
15. Bayerdorffer E, Simon TH, Bastlein CH, Ottenjan R, Kasper G (1987) Bismuth/Ofloxacin combination for duodenal ulcer. Lancet ii:0191467–1468
16. Goodwin CS, Armstrong JA (1986) Will antibacterial chemotherapy be efficacious for gastritis and peptic ulcers? J Antimicrob Chemother 17:0191–4
17. McNulty AM (1987) The treatment of Campylobacter associated gastritis. Am J Gastroenterol 82:019245–247

Response of *Helicobacter pylori*-Associated Gastritis to Colloidal Bismuth Subcitrate Versus Ranitidine in Patients with Duodenal Ulcer: A Detailed Histological Assessment

J.I. Wyatt

Introduction

Only a minority of patients treated with monotherapy using colloidal bismuth subcitrate (CBS) will be eradicated of *Helicobacter pylori*. However, temporary clearance of the bacteria is achieved in a higher proportion of patients and may be related to ulcer healing by CBS. It is not known what factors contribute to this variation in response rate, although previous work by McNulty et al. [1] suggested that elimination of *H. pylori* might be more successful in patients with initially more severe gastritis.

The current study forms part of a large multicentre study performed in Switzerland, comparing CBS with ranitidine in patients with *Helicobacter* gastritis and duodenal ulceration, which will be reported in full elsewhere. The aims of the investigation reported here were:

1. To determine whether any histological features of the pre-treatment gastric biopsies could predict clearance of *H. pylori*.
2. Whether *H. pylori* persists in gastric pits after treatment, before relapse is recognised.

Methods

This was a histological study of antral biopsies from patients with duodenal ulcer participating in a multicentre, double-blind treatment trial of CBS (De-Nol, Gist-Brocades, The Netherlands coated tablets 240 mg b.d.) versus ranitidine (300 mg b.d.). Only patients with adequate antral biopsies before and after treatment, who were initially *H. pylori* positive by CLO test and histology, and who had complete data sets were included (53/120 patients).

Histological sections of gastric antral biopsies stained with haematoxylin and eosin were graded for degree of mononuclear cell infiltrate (0–3), neutrophil infiltration (0–4) and for lymphoid follicles. The distribution and density of *H. pylori* was graded on sections stained with Giemsa, as has been previously described [2]. Briefly, the surface epithelium, upper pits and deep pits were each scored from 0 (no bacteria) to 4 (dense coating of bacteria) and the scores added

Pathology Department, St. James's University Hospital, Leeds, LS9 7TF, UK

J.M. Pajares et al. (Eds.)
H. pylori and Gastroduodenal Pathology

to give a possible total range of 0–12. This density score gives a non-linear grading of the density of *H. pylori*. Previous evaluation has shown that the score was reproducible to ± 2 points, and that 80% of antral biopsies taken from the same patient at the same time had scores within ± 2 of each other.

Results

Of the 53 patients included, ulcer healing was achieved after 4–8 weeks of treatment by 20/22 patients treated with CBS and 29/31 with ranitidine. *H. pylori* was cleared after 4 weeks (biopsies negative by both CLO test and histology) in ten patients on CBS and one on ranitidine ($p < 0.0005$). Ulcer healing was not related to clearance of *H. pylori*.

Inflammatory Scores

Both mononuclear cells (Fig. 1) and neutrophil scores (Fig. 2) had fallen significantly in the CBS group compared with the ranitidine group at 4 weeks: this was largely accounted for by decreasing inflammation in the patients cleared of *H. pylori*. The presence of lymphoid follicles in biopsies showed no relation to the effect of treatment.

The severity of inflammation before treatment did not significantly affect clearance, although there was a trend for the patients with less active gastritis to be more often cleared of *H. pylori* after CBS.

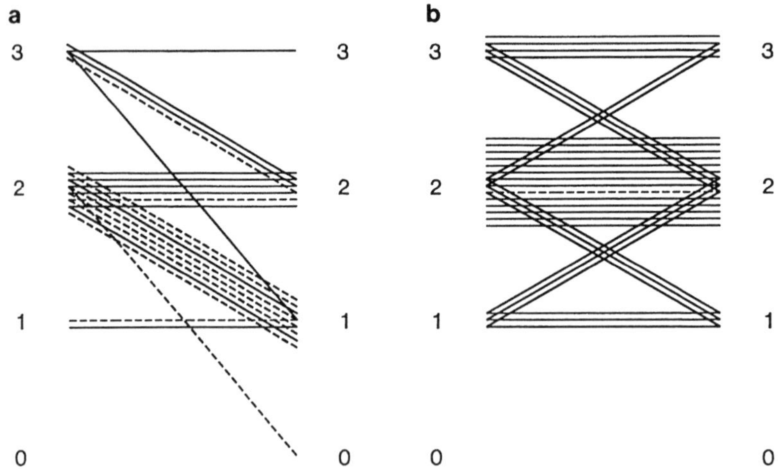

Fig. 1a, b. Mononuclear cell scores before and after treatment with CBS (**a**; $n = 22$) and ranitidine (**b**; $n = 31$). *Dashed line*, patients cleared of *H. pylori*; *solid line*, patients not cleared of *H. pylori*

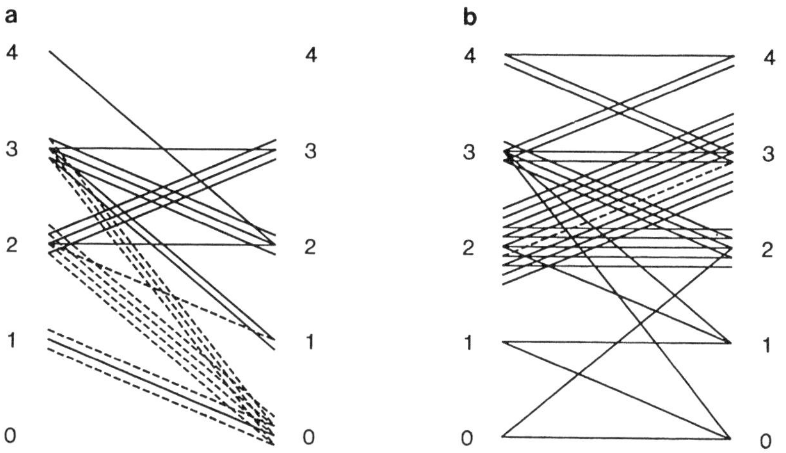

Fig. 2a, b. Neutrophil scores before and after treatment with CBS (**a**; $n = 22$) and ranitidine (**b**; $n = 31$). *Dashed line*, patients cleared of *H. pylori*; *solid line*, to patients not cleared of *H. pylori*

H. pylori Density Scores

Clearance of *H. pylori* was not dependent on either the density or the mucosal distribution of bacteria prior to treatment (Fig. 3). Also, there was no morphological evidence that *H. pylori* could persist preferentially in the gastric pits after CBS treatment.

Interestingly, of the patients not cleared of *H. pylori* by CBS, there was no clear evidence of suppression of the bacteria. However, there was more variation in *H. pylori* density scores before and after treatment in both treatment groups

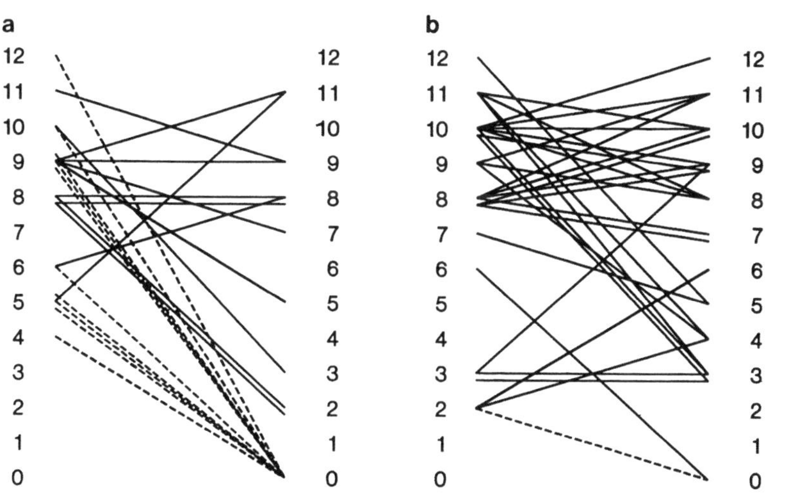

Fig. 3a, b. *H. pylori* density scores before and after treatment with CBS (**a**; $n = 22$) and ranitidine (**b**; $n = 31$). *Dashed line*, patients cleared of *H. pylori*; *solid line*, patients not cleared of *H. pylori*

Table 1. *H. pylori* scores before and after treatment in patients who remained *H. pylori* positive

	CBS ($n = 12$)	Ranitidine ($n = 30$)
No change (scores within ± 2)	7	21
Number expected if bacterial density constant	10	24
H. pylori score decreased by > 2	4	7
H. pylori score increased by > 2	1	3

in patients who remained *H. pylori* positive than would have been expected from the 80% that was found when concurrent pairs of antral biopsies from the individual patients had been studied (Table 1). This suggests some fluctuation of *H. pylori* density with time.

CLO Test after Treatment

Of the 42 patients who remained *H. pylori* positive after treatment, 41 (98%) were positive by histology and 37 (88%) by CLO test. This reduced sensitivity of the CLO test after treatment occurred predominantly in patients with low *H. pylori* scores on post-treatment histology (Fig. 4).

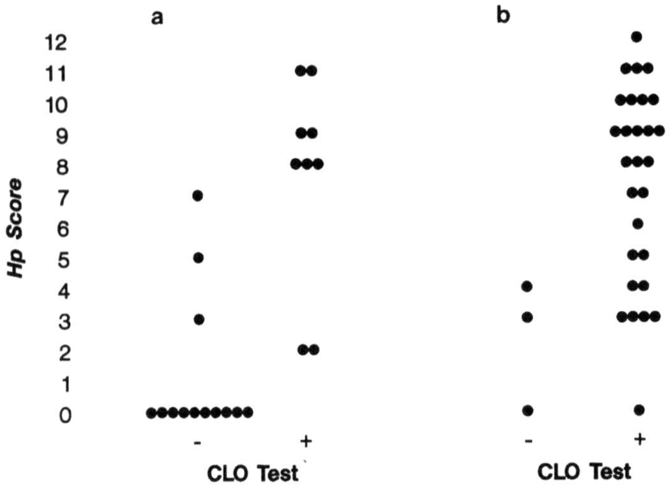

Fig. 4a, b. *H. pylori* density score versus CLO test result in post-treatment biopsies in patients treated with CBS (**a**; $n = 22$) and ranitidine (**b**; $n = 31$)

Conclusions and Discussion

1. Clearance of *H. pylori* was observed after 4 weeks of treatment in 10/22 patients taking CBS and 1/31 taking ranitidine; ulcer healing was independent of *H. pylori* status and was achieved in 20/22 and 29/31 patients, respectively.
2. There was a non-significant trend for patients with less active gastritis to be more often cleared of *H. pylori* by CBS. This was opposite to the effect observed by McNulty et al. [1] and suggests that the pre-treatment severity of gastritis is not important for the effectiveness of CBS (although it may be more relevant for antibiotics having a systemic mode of action).
3. Clearance of *H. pylori* by CBS was not affected by the pre-treatment bacterial density, nor did *H. pylori* appear to persist in gastric pits after treatment. There was a tendency for bacterial density to fall in both groups during anti-ulcer treatment – an effect also described by Hui et al. [3] in patients taking sucralfate using a 0–3 grading system for *H. pylori*. These authors also found a reduction in the activity of gastritis during duodenal ulcer healing, although this was not observed in our present study.
4. The CLO test was unreliable in determining clearance if used within 1 week of treatment. This confirms previous reports [4] and is related to insensitivity of the urease test in patients with low bacterial loads.

References

1. McNulty CAM, Eyre-Brook IA, Uff JS et al. (1989) Triple therapy is not always 95% effective. In: Ruiz Palacios GM (ed) Campylobacter V: proceedings of the 5th international workshop on Campylobacter infections
2. Wyatt JI, Rathbone BJ, Heatley RV (1986) Local immune response to gastric Campylobacter in non-ulcer dyspepsia. J Clin Pathol 39:863–870
3. Hui WM, Lam SK, Ho J et al. (1989) Effect of sucralfate and cimetidine on duodenal ulcer associated antral gastritis and C. pylori. Am J Med 86 (Suppl 64):60–65
4. Deltenre M, Glupczynski Y, De Prez C et al. (1989) The reliability of urease tests, histology and culture in the diagnosis of Campylobacter pylori infection. Scand J Gastroenterol 24 (Suppl 60): 19–24

Eradication of *Helicobacter pylori*: Summary of Workshop

C. O'Morain[1] and G. Börsch[2]

Helicobacter pylori has been implicated in a wide spectrum of gastroduodenal diseases. These range from asymptomatic gastritis, non-ulcerative dyspepsia, gastric and duodenal ulcers to a tentative association with gastric cancer. It is important in analysing results of treatment studies to eradicate *H. pylori* that patients may have a variety of diagnoses. There have been some data to suggest that the strain of *H. pylori* associated with duodenal ulcer may be more virulent than the strain associated with asymptomatic gastritis. This has been demonstrated at a genomic level. In addition, cytotoxic enzymes that the bacteria produce, in particular phospholipase, are greater in patients with duodenal ulcer than in patients with gastritis.

Bismuth Salts

Colloidal bismuth subcitrate (CBS) has many properties which enable it to heal ulcers. These include an ability to coat the surface of an ulcer to form a protective layer allowing the underlying mucosa to heal, and it also enhances prostaglandin production, increases alkaline secretion and improves the quality of the mucous layer. More recently it has been shown to have anti-bacterial properties in vitro. There are many forms of bismuth salts. CBS is the most effective formulation in healing ulcers. Many studies show it to be as effective as H2 antagonists. The advantage of bismuth as treatment of duodenal ulcers compared to H2 antagonists is that the relapse rate is less in patients who have been followed up over a long period of time. The reason for this is probably its effect on eradicating *H. pylori*.

CBS has advantages over other bismuth preparations in eradicating *H. pylori*. This includes its solubility which allows it to gain access to the bacteria buried deep in the gastric pits. It is not dependent on the level of pH for its activity. One of the drawbacks related to CBS is its systemic absorption. However, there is a transient rise in serum bismuth shortly after oral administration but this does not reach toxic levels.

Different formulations of CBS have been developed in an effort to maximise its effect in eradicating *H. pylori*. It was originally available in liquid form, but this was found unpalatable by most patients. Chew tablets and, more recently,

[1] Department of Gastroenterology, Adelaide/Meath Hospital, Trinity College, Dublin, Ireland
[2] Elisabeth-Krankenhaus, Akademisches Lehrkrankenhaus, Moltkestr. 61, 4300 Essen, FRG

J.M. Pajares et al. (Eds.)
H. pylori and Gastroduodenal Pathology

swallow tablets have been introduced. There are preliminary studies using it in capsule form. There has been no significant difference between these different formulations in eradicating *H. pylori*.

It is important when analysing results of eradication of *H. pylori* that 4 weeks elapse before taking a biopsy for culture. This applies particularly to bismuth as it has a bacterostatic rather than a bactericidal effect. Culture of an antral biopsy is the gold standard for assessing eradication. However, non-invasive methods using the ^{13}C-urea breath test show promise and may be more useful in long-term studies as these would avoid endoscopy. ^{13}C is a stable isotope which is an advantage over the ^{14}C-urea breath tests. Serology is useful in epidemiology studies but is of little value in assessing eradication, as it takes too long for antibody titres to fall.

Combination Treatments

Most studies show that bismuth salts can only eradicate *H. pylori* in 10%–30% of patients. Administering bismuth four times a day yields better eradication rates than giving it twice a day. The logical step to improve eradication would be to add antibiotics to the treatment. *H. pylori* is sensitive to a wide range of antibiotics in vitro. The pH level should be considered in choosing an appropriate antibiotic as its activity may be pH dependent. The bacteria is ideally suited to its environment as it is buried deep in the gastric pits and protected from the acid environment of the stomach by the mucous layer. Different formulations of bismuth, such as dispersal capsules and liquid formulations, come into contact with the bacteria, increasing the efficiency in eradicating the organism.

Treatment may be unsuccessful as *H. pylori* may be resistant to the antibiotics chosen. Metronidazole appears to be the most favoured drug and is used in combination therapy for eradication of *H. pylori*. Metronidazole is not dependent on the level of pH for its activity and is secreted into the gastric juices. Resistance to metronidazole may occur because the patient may have taken metronidazole previously and resistance can also develop during treatment. The resistance rates vary from country to country but, in general, appear to be higher in underdeveloped countries than in developed countries. Resistance to antibiotics is also encountered in tuberculosis treatment. *H. pylori* is also slow-growing and has some similarities with tuberculosis in that it needs more than one drug to eradicate it. There are some side effects associated with metronidazole. These include a metallic taste and a sensory neuropathy if used for a long period of time, and it induces a feeling of nausea if taken with alcohol.

Dual therapy with CBS and metronidazole or other antibiotics such as amoxycillin increases the eradication rates up to 70%. The optimal dosage and duration of treatment have not been established, but the longer antibiotics are taken and the greater the dosage, the more side effects are likely to occur. It may be best that patients take medication at the start of treatment as this improves compliance. Some regimens suggest taking the antibiotic 2 weeks after commencing CBS but by then patients may be relatively asymptomatic and may neglect taking their medication.

Triple therapy using another antibiotic with metronidazole is the most effective treatment in eradicating *H. pylori*. This may prevent the development of resistance. However, using three drugs adds to the complexity and reduces compliance. There have been other regimens suggested, including quadruple regimens, but side effects are common on these. The most effective treatment has not been determined and it is suggested that sensitivity tests should be performed before prescribing antibiotics.

Omeprazole

Omeprazole is currently the most effective therapy in the acute treatment of peptic ulcer. It inhibits gastric acid secretion by blocking the hydrogen potassium adenosine triphosphatase enzyme (the proton pump). Omeprazole has been shown in vitro to have an inhibitory effect on the growth of *H. pylori*. It also has an effect in vivo on *H. pylori* as it temporarily suppresses *H. pylori*. However, if biopsies are taken 4 weeks after the patient discontinues omeprazole, *H. pylori* is still present. Omeprazole induces achlorhydria and in this environment *H. pylori* may be less biologically active as it does not need to secrete urease to keep its micro-environment alkaline. It has been postulated that urease may be cytotoxic by causing back diffusion of the hydrogen into the cell. The level of acid may allow other organisms to grow, thus suppressing *H. pylori*. Patients with duodenal ulcers treated with omeprazole also show improvement with the associated gastritis while on omeprazole treatment. This is not observed when patients are treated with H2 antagonists. H2 antagonists have no effect on *H. pylori* and do not suppress *H. pylori*. Omeprazole in its present formulation has an acid-resistant drug coating so it is not released into the stomach. It has to be absorbed into the small intestine and then acts on the parietal cells at a cellular level. It is possible that omeprazole in a different formulation could be active against *H. pylori* in vivo.

At a higher dosage, omeprazole may be released into the stomach and may help to eradicate *H. pylori*. By inducing achlorhydria omeprazole may allow antibiotics that are dependent on pH for their activity to become more active. Achlorhydria may also allow other bacteria to proliferate and suppress the growth of *H. pylori*.

Effect of Eradication

The principal interest in the eradication of *H. pylori* is in its role in peptic ulcer disease. It has been confirmed by several studies that successful eradication of *H. pylori* results in a cure of the disease. It exerts its effect locally by increasing the pH as urease is secreted beneath the mucous layer in the antral mucosa. This gives a false message which stimulates gastrin secretion and acts on the parietal cell causing acid hypersecretion in the duodenum. This changes the environment which might result in gastric metaplasia. Gastric metaplasia in the duodenum occurs in response to an injurious toxic environment. This gastric metaplasia in turn may be come by *H. pylori* from the antrum. *H. pylori* can release cytotoxic enzymes which damage the mucosa and give rise to ulceration in the duodenum. Patients who are infected with *H. pylori* have higher levels of gastrin and pepsinogen. Successful eradication of *H. pylori* results in normalising of gastrin levels and a fall in acid secretion.

Conclusion

Eradication of *H. pylori* is a major therapeutic challenge. The prize is enormous as eradication will result in a cure for peptic ulcer. *H. pylori* infection can stimulate gastrin and pepsin release which leads to acid hypersecretion. Gastrin is secreted into the circulation, binds the receptor site of the parietal cell and stimulates acid secretion. Acid hypersecretion leads to conditions in the duodenum in which gastric metaplasia may occur. *H. pylori* can colonise this ectopic tissue which can result in tissue damage and ulceration.

References

1. Bayer-dorffer E, Mannes GA, Hochter W, Weingart J, Heldwein W, Muller-Lissner S, Oertel H, Blendinger Ch, Kuntzen O, Bornschein W, Malfertheiner P, Wilkening J, Ruckdeschel G, Pfaller P, Von Wulffen H, Kopchke W, Stolte M (1990) Relapse rate of H. pylori (HP) positive duodenal ulcers (DUf) following antibacterial therapy. Rev Esp Enferm Dig 78 [Suppl]: 238
2. Bode G, Malferheiner P, Lehnhardt G, Ditschuneit H (1990) Influence of 6 different bismuth salts on growth morphology and viability of Helicobacter pylori. Rev Esp Enferm Dig 78 [Suppl]: 276
3. Chittajallu RS, Dorian CA, McColl KEL (1990) Serum pepsinogen 1 in duodenal ulcer: effect of eradication of H. pylori and correlation with serum gastrin and antral gastritis. Rev Esp Enferm Dig 78 [Suppl]: 226
4. Daskalopoulos G, Carrick J, Lee A, Noar M (1990) Comparison of three triple therapies in the eradication of H. pylori. Rev Esp Enferm Dig 78 [Suppl]: 226
5. Daskalopoulos G, Carrick J, Lee A, Noar M (1990) Comparison of three triple therapies in the eradication of H. pylori. Rev Esp Enferm Dig 78 [Suppl]: 292
6. DeKoster E, Nyst JF, Glupczynski Y, Deprez C, DeReuck M, Deltenre M (1990) H. Pylori treatment: one week CBS + Omeprazole + amoxycillin + minocyclin. Rev Esp Enferm Dig 78 [Suppl] : 257
7. De Koster E, Burette A, Nyst JF, Glupczynski Y, Deprez C, Deltenre (1990) H. pylori treatment: the macrolide trail: one week erythromycin + CBS + omeprazole. Rev Esp Enferm Dig 78 [Suppl]: 258
8. Eberhardt R, Topfmeier P, Mateblowski M (1990) Effect of bismuth subnitrate on Helicobacter pylori and on healing and relapse rate of peptic ulcer. Rev Esp Enferm Dig 78 [Suppl]: 274
9. Glupczynski Y, Bourdeaux L, Verhas M, Balegamire B, De Vos D, Hennart Ph, Butzler JP (1990) Evaluation of double or triple antibiotic/bismuth therapy of Helicobacter pylori in central Africa. Rev Esp Enferm Dig 78 [Suppl]: 242
10. Graham DY, Lew GM, Klein PD, Evans DG, Evans DJ Jr, Alpert LC, Opekun AR, Malaty HM (1990) Factors affecting the eradication of Helicobacter pylori infection with triple therapy. Rev Esp Enferm Dig 78 [Suppl]: 254

11. Kosunen TU, Rautelin H, Seppala K, Valtonen V, Vainio U (1990) Metronidazole resistant Helicobacter pylori. Rev Esp Enferm Dig 78 [Suppl]: 261
12. Labenz J, Gyenes E, Peitz U, Borsch G (1990) Ciprofloxacin-omeprazole treatment for eradication of Helicobacter pylori. Rev Esp Enferm Dig 78 [Suppl]: 224
13. Lambert JR, Lin SK, Schembri M, Nicholson L, Korman MG (1990) Helicobacter pylori therapy randomized study of denol/antibiotic combinations. Rev Esp Enferm Dig 78 [Suppl]: 251
14. Logan RPH, Gummett PA, Misiewicz JJ, Polson RJ, Johnson P, Baron JH (1990) 1, 2, or 4 weeks colloidal bismuth subcitrate of H. pylori. Rev Esp Enferm Dig 78 [Suppl]: 270
15. Logan RPH, Gummett PA, Misiewicz JJ, Polson RJ, Johnson P, Baron JH (1990) 1, 2, or 4 weeks colloidal bismuth subcitrate for H. pylori. Rev Esp Enferm Dig 78 [Suppl]: 270
16. Lopez Lavid C, Sanz JC, Martin E, Castanos R, Jimenez I, Lopez-Brea M (1990) Helicobacter pylori: biotypes and correlation with antimicrobial susceptibility. Rev Esp Enferm Dig 78 [Suppl]: 200
17. Oderda G, Lerro P, Dell'Olio D, Poli E, Ansaldi N (1990) Amoxycillin + Tinidazole in H. pylori gastritis in children: how long to treat? Rev Esp Enferm Dig 78 [Suppl]: 226
18. Paradis A, Goldie J, Veldhuyzen van Zanten SJO, Richardson H, Hunt RH (1990) The in vitro inhibitory effect of omeprazole on Helicobacter pylori: a bimodal distribution? Rev Esp Enferm Dig 78 [Suppl]: 207
19. Paradis A, Goldie J, Veldhuyzen van Zanten SJO, Richardson H, Hunt RH (1990) The in vitro inhibitory effect of omeprazole on Helicobacter pylori. Rev Esp Enferm Dig 78 [Suppl]: 207
20. Patchett S, Beattie S, Keane C, O'Morain C (1990) Treatment of H. pylori associated PUD. A safe and effective regime Rev Esp Enferm Dig 78 [Suppl]: 268
21. Pretolani S, Bonvicini F, Careddu N, Cilla D, Acampora P, Gasbarrini A, Gasbarrini G Resistant ulcers and Helicobacter pylori gastritis: effect of treatment with omeprazole. Rev Esp Enferm Dig 78 [Suppl]: 226
22. Sobala GM, Wyatt JI, Dixon MF, Rathbone BJ, Axon ATR (1990) Active duodenitis associated with subsequent eradication failure of H. pylori. Rev Esp Enferm Dig 78 [Suppl]: 240
23. Suerbaum S, Leying H, Hemmerle B, Klemm K, Operkuch W (1990) Antibacterial activity of pantoprazole, omeprazole, and other (H^+/K^+) ATPase inhibitors against Helicobacter pylori. Rev Esp Enferm Dig 78 [Suppl]: 256
24. vander Voort LHM, van der Bos AP, Kamsteeg H (1990) In vitro bactericidal effects of CBS on Helicobacter pylori. Rev Esp Enferm Dig 78 [Suppl]: 221
25. Vogt K, Hahn H (1990) Urease production by helicobacter pylori. Rev Esp Enferm Dig 78 [Suppl]: 44
26. Westblom TU, Duriex DE (1990) H_2-Blockers increase antibiotic concentrations in gastric mucosa. Rev Esp Enferm Dig 78 [Suppl]: 262

Treatment of *Helicobacter pylori* Infection: Challenges and Future Prospects

J.M. Pajares

From the practical point of view of the gastroenterologists and general practitioners there are still some questions awaiting sastifactory and complete answers. Which patients to treat and what therapy to use are the most important for any physician. As far as the first question is concerned patients with duodenal ulcer are most likely to benefit from anti-*Helicobacter pylori* therapy, mainly in order to prevent the ulcer relapse. Nowdays, many data support the fact that the eradication of *H. pylori* prevents the recurrence of duodenal ulcer in almost 75% of the patients.

In this regard, therapy to eradicate *H. pylori* will be a part of the therapy of *H. pylori*-positive patients with duodenal ulcer. However, it is unclear whether one should treat all patients or well-defined groups. At the same time, many patients become reinfected early and many develop resistance to the antibiotics. In the future, new data should be produced concerning the avoidance of resistance, the prevention of reinfection, the development of new antibiotics with better specifity for killing *H. pylori* bacteria, and the identification of subgroups of peptic ulcer patients in whom anti-*H. pylori* therapy should be applied. It should also be mentioned that pathophysiology of the peptic ulcer in terms of gastric motility, gastric release, mucous secretion, prostaglandins and other protective factors of the gastric mucosa must be investigated in both patients and in model animals infected by *H. pylori.*

Chronic gastritis is another *H. pylori*-associated pathology to be treated by eradication therapy. From the clinician's point of view, there are three con siderations: (a) the treatment of all patients with chronic gastritis; (b) the treatment of only the symptomatic patients; (c) considering the fact that histological lesions (i.e., intestinal metaplasia and dysplasia) of chronic gastritis are considered to be preneoplastic lesions, their treatment should prevent further progression to carcinoma of the stomach. Present data confirm that approximately one third of the symptomatic *H. pylori*-positive patients with chronic gastritis get relief from their symptoms with anti-*H. pylori* therapy. A correlation between *H. pylori* eradication and an improvement of gastric mucosa inflammation leading to a relief of subjective symptoms has been found in some while in others there is no correlation.

Investigation to assess the pathophysiology of chronic gastritis has demonstrated possible different mechanisms which produce symptoms. Therefore, not every patient with chronic gastritis should be treated in clinical practice. Future

Hospital de la Princesa, Universidad Autonoma, C/Diego de León 62, 28006 Madrid, Spain

J.M. Pajares et al. (Eds.)
H. pylori and Gastroduodenal Pathology

clinical research must define quite clearly the clinical syndrome related to *H. pylori* infection of the gastric mucosa and the group of patients to be treated.

The third aim is the treatment of *H. pylori*-positive patients with chronic gastritis to prevent the development of gastric carcinoma. In this regard many papers, most of them on epidemiology, seem to prove a positive correlation between the prevalence of infection and gastric carcinoma rates. If one accepts this fact, then, if the infection, is cured or prevented the cancer rates will decrease. Thus, future scientific work should be carried out to treat *H. pylori*-positive patients with chronic gastritis to prevent the development of gastric carcinoma. However, owing to the long duration of the neoplastic process, the efficacy of the treatment should be evaluated on either the appearance or the amelioration of the intestinal metaplasia and dysplasia of the gastric mucosa, as regards both intermediate end-points and preneoplastic histological lesions.

Another problem to solve is which therapy to choose. Many drugs and drug combinations have been tried; they are generally classified as mono-dual, and triple therapy regimens. Among them, triple therapy seems to be the most effective and reliable and the least expensive. Decisions regarding how often to re-evaluate the patients should also be made. Also, should we treat the relatives of *H. pylori*-positive patients to prevent re-infection? Many other questions of important clinical interest need appropriate answers and make this field very attractive not only for the gastroenterologist, but also for the pediatrician, general practitioner, and for the clinician in general.

Subject Index

MIX
Papier aus verantwortungsvollen Quellen
Paper from responsible sources
FSC® C105338

If you have any concerns about our products,
you can contact us on
ProductSafety@springernature.com

In case Publisher is established outside the EU,
the EU authorized representative is:
Springer Nature Customer Service Center GmbH
Europaplatz 3, 69115 Heidelberg, Germany

Printed by Libri Plureos GmbH
in Hamburg, Germany